Analog Electronics

Analog Electronics

Second edition

Ian Hickman, BSc. (Hons), CEng., MIEE, MIEEE

Newnes

OXFORD AUCKLAND BOSTON JOHANNESBURG MELBOURNE NEW DELHI

Newnes
An imprint of Butterworth-Heinemann
Linacre House, Jordan Hill, Oxford OX2 8DP
225 Wildwood Avenue, Woburn, MA01801-2041
A division of Reed Educational and Professional Publishing Ltd

A member of the Reed Elsevier plc group

First published 1990
2nd edition 1999
Transferred to digital printing 2004
© Ian Hickman 1990, 1999

British Library Cataloguing in Publication Data
Hickman, Ian
  Analog electronics
  1. Analog function circuits
  1. Title
  621.3815'3

ISBN 0 7506 4416 8

Typeset by David Gregson Associates, Beccles, Suffolk

'Digital is easy. Analog ... that's professional.'

Roland Moreno, inventor of the smart card,
in an interview with *Electronics Times*

A note for North American readers

There are only a few differences in the terminology of electronics between Europe and North America, for example, plate for anode. The term anode or plate in any case goes back almost to the beginning of the century, when transatlantic communication was slow and travel less common. With more modern terms, such as base, emitter and collector, the usage is uniform throughout the English speaking world. Indeed the English terms often show up in thin disguise in other languages: for example, before being replaced by 'sweeper', wobbulator turned up in French as wobbulateur and in German as wobbelgenerator.

Component types are another matter. Whereas in North America JEDEC (2N- - - -) numbers predominate, in Europe Pro Electron number are common. However, the big semiconductor houses on both sides of the Atlantic, for example Motorola and Philips, nowadays produce devices under both numbering systems. Chapter 3 uses both types as examples. The data books of all the major device manufacturers give cross-reference tables of exact or near equivalent devices between the two systems, and between those and their in-house numbers – examples of the latter will also be found throughout the book, for example in Chapter 8.

There are doubtless many minor differences of usage of which I am not aware – and of course almost everyone has his of her own pet way of transliterating names like Chebyshev – but whilst these may sound odd to the American ear, I hope they will not in any way obscure the meaning.

Contents

Preface

Electronics has been my profession for well over a quarter of a century and my hobby for even longer. Over that whole period, I have been an avid collector of knowledge of the subject, so that by now my card index system contains references to hundreds of articles published during that time. Now references are all very well, but one often needs information in a hurry, so it has been my practice, more often than not, also to save the article itself. Thus I now have, stored in many bulging files, an invaluable hoard of articles, photocopies and originals, from dozens of magazines, books and learned journals. For some years the feeling has been growing that I should not sit on all this information, but should share it around. Of course, it is all freely available already, in the various publications in which it originally appeared, but that makes it a very diffuse body of knowledge and consequently very elusive. In this book I have tried to bring some of it together, concentrating on what I have found over the years to be the most useful, and seeking to explain it as simply as possible. Whether or not I have succeeded, the reader must judge for himself.

This book is not a textbook, but I hope nevertheless that you will learn a good deal from it. Textbooks have traditionally presented a great deal of information compressed within a relatively confined space – a format which is appropriate in conjunction with a course including lessons or lectures, at a school, polytechnic or university. However, it makes life very difficult for the student, however keen, who is working on his own with no one to consult when something is not clear. It must be said also that some textbooks seem to delight in the most abstract treatment of the subject, dragging in degree-level maths at every turn, even when a more concrete approach – using simple vector diagrams, for example – would be perfectly satisfactory and much more readily comprehensible to normal mortals. On occasions even, one might be excused for thinking this or that particular textbook to be mainly an ego trip for the author.

Now make no mistake, maths is an essential tool in electrical engineering in general and in electronics in particular. Indeed, the research laboratories of all the large electronics companies employ at least one 'tame mathematician' to help out whenever an engineer finds himself grappling with the mathematical aspects of a problem where his own maths is too rusty. For the practising electronic engineer (unless also a born mathematician) can no more expect to be fluent in all the mathematical techniques he may ever need, or indeed may have learnt in the past, than the mathematician can expect to be abreast of all the latest developments in electronics (it takes the engineer all his time to do that!). It seems particularly appropriate therefore to attempt to explain analog electronic circuits as simply as possible, appealing as far as possible to nothing more complicated than basic algebra and trigonometry, with which I assume the reader of this book to be familiar. This has been done successfully in the past. Older readers may recall the articles by 'Cathode Ray', the pen-name of a well known writer of yesteryear on electronics, which appeared over many years in the magazine *Wireless World*. The approach adopted in this book is not essentially different. The pace is more leisurely and discursive than in a typical textbook, the aim being to take the reader 'inside' electronic circuits so that he can see what makes them tick – how and why exactly they do what they do. To this end, vector diagrams are particularly useful; they illustrate very graphically what is going on, enabling one to grasp exactly how the circuit works rather than simply accepting that if one slogs through the maths, the circuit does indeed behave as the textbooks say. There will of course be those whose minds work in a more academic, mathematical way, and these may well find their needs served better by conventional textbooks.

With this brief apology for a style which some will undoubtedly find leisurely to the point of boredom, but which will I hope materially assist others, it only remains to mention two minor points before passing on to the main body of the book. First, I must apologize to British and many other non-US readers for spelling 'analog' throughout in the North American manner: they will in any case be used to seeing it spelt thus, whereas 'analogue' looks very quaint to North American eyes. Second, the following pages can be read at different levels. The technically minded adolescent, already interested in electronics in the early years of secondary or high school, will find much of practical interest, even if the theory is not appreciated until later. Technicians and students at technical colleges and polytechnics will all find the book useful, as also will electronics undergraduates. Indeed, many graduates and even post-graduates will find the book very handy, especially those who come into electronics from a different background, such as a physics degree.

Writing the following pages has turned out to be a not inconsiderable task. My sincere thanks are due first to my ever-loving (and long-suffering) wife, who shared the typing load, and also to those who have kindly vetted the work. In particular, for checking the manuscript for howlers and for many helpful suggestions, I must thank my colleagues Pete C., Dave F., Tim S. and especially my colleague and friend of more than a quarter of a century's standing, Mick G. Thanks also to Dave Watson who produced the 'three-dimensional wire grid' illustrations of poles and zeros in Appendix 4 and elsewhere. For permission to reproduce circuit diagrams or other material, supplied or originally published by them, my thanks are also due to all the following:

C. Barmaper Ltd
EDN
Electronic Design
Electronic Engineering
Electronic Product Design
Electronics World (formerly *Wireless World*)
ETI
Ever Ready Company (Great Britain) Ltd
Hewlett-Packard Journal
Maplin Electronic Supplies Ltd
Maxim Integrated Products UK Ltd
Microwave Journal
Microwaves & RF
Motorola Inc.
New Electronics
Philips Components Ltd (formerly Mullard Ltd)
Practical Electronics
Practical Wireless

Ian Hickman
Eur. Ing

1   Passive components

The passive components used in electronic circuits all make use of one of the three fundamental phenomena of resistance, capacitance and inductance. Just occasionally, two may be involved, for example delay cable depends for its operation on both capacitance and inductance. Some components depend on the interaction between an electrical property and, say, a mechanical property; thus a piezoelectric sounder operates by virtue of the small change in dimension of certain types of ceramic dielectric when a voltage is applied. But most passive components are simply resistors, capacitors or inductors. In some ways inductance is the most subtle effect of the three, since with its aid one can make transformers, which will be described later in this chapter.

Resistors

Some substances, for example metals (particularly copper and aluminium – also gold, but that's a bit expensive for everyday use), conduct electricity well; these substances are called *conductors*. They are distinct from many others called *insulators*, such as glass, polystyrene, wax, PTFE etc., which in practical terms do not conduct electricity at all. In fact, their resistivity is about 10^{18} or a million million million times that of metals. Even though copper, say, conducts electricity well, it exhibits some resistance to the flow of electricity and consequently it does not conduct perfectly; energy is lost in the process, appearing in the form of heat. In the case of a wire of length l metres and cross-sectional area A square metres, the current I in amperes which flows when an electrical supply with an electromotive force (EMF) of E volts is connected across it is given by

$$I = E \left/ \left(\frac{l}{A} \rho \right) \right. \tag{1.1}$$

where ρ (lower-case Greek letter rho) is a property of the material of the wire, called *resistivity*. In the case of copper the value of ρ is 1.55×10^{-8} Ωm in other words, the resistance between opposite faces of a solid cube of copper of 1 m side is 0.0155 $\mu\Omega$. The term $(l/A)\rho$ is called the *resistance* of the wire, denoted by R. So one may write

$$R = \frac{l}{A} \rho \tag{1.2}$$

Combining (1.1) and (1.2) gives $I = E/R$, the form in which most people are familiar with *Ohm's law* (see Figure 1.1). As mentioned earlier, when current flows through a resistance, energy is dissipated as heat. The rate at which energy is

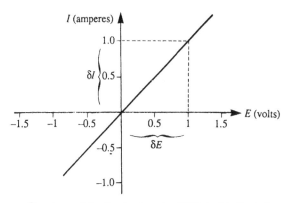

The slope of the line is given by $\delta I/\delta E$. In this illustration $\delta I = 1$ A and $\delta E = 1$ V, so the conductance $G = 1$ S. The S stands for siemens, the unit of conductance, formerly called the mho. $G = 1/R$.

Figure 1.1   Current through a resistor of R ohms as a function of the applied voltage. The relation is linear, as shown, for a perfect resistor. At DC and low frequencies, most resistors are perfect for practical purposes.

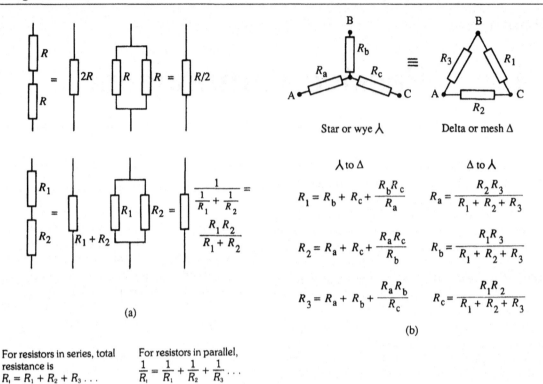

(a)

(b)

For resistors in series, total resistance is
$R_t = R_1 + R_2 + R_3 \ldots$

For resistors in parallel,
$$\frac{1}{R_t} = \frac{1}{R_1} + \frac{1}{R_2} + \frac{1}{R_3} \ldots$$

Figure 1.2   Resistors in combination.
   (a) Series parallel (also works for impedances).
   (b) The star–delta transformation (also works for impedances, enabling negative values of resistance effctively to be produced).

dissipated is measured in watts, where one watt equals one joule per second. If a current of I amperes flows through a resistance of R ohms, the power dissipated is given by $W = I^2 R$. Using Ohm's law it also follows that $W = EI = E^2/R$, where E is the EMF necessary to cause the current I to flow through the resistance R. Clearly from (1.2), if a second identical wire is connected in series with the first (doubling l) the resistance is doubled, whilst if it is connected in parallel (doubling A) the resistance is halved (Figure 1.2 also shows the useful 'star–delta' equivalence).

Electronic engineers use resistors from a fraction of an ohm up to millions of ohms. Low-value resistors up to a few thousand ohms are often *wirewound*, although pure copper wire is seldom used owing to its high temperature coefficient of resistance, namely +0.4% per degree centigrade. At one time, wirewound resistors with values up to

1 MΩ (one million ohms) were available, but were expensive owing to the vast number of turns of very fine wire needed to achieve this resistance. Nichrome (an alloy of chromium and nickel) is used for high-power resistors designed to dissipate several or many watts, whilst precision wirewound resistors may use constantan or manganin (alloys of copper with nickel or manganese respectively). Such resistors have an extremely low temperature coefficient of resistance; they are available manufactured to a tolerance of better than 0.05% and are stable to within one part per million (1 PPM) per year. Such resistors are used as reference standard resistors in measurements and standards laboratories. In many electronic circuits, resistors with a tolerance of 1, 2 or 5% are entirely satisfactory; indeed, in the era of thermionic valves 20% was the norm.

In the interests of economy, most low-power

resistors up to 1 W rating are not wirewound, and indeed the resistive element is frequently non-metallic. *Carbon composition* resistors have a cylindrical resistance element made of an insulating compound loaded with carbon, usually protected by a moulded phenolic covering. Such resistors were universally used at one time and are still widely employed in the USA. The resistance tends to rise as the resistor ages, owing to the absorption of moisture: the effect is less pronounced where the resistor is run at or near its rated dissipation and operates for long periods. Carbon composition resistors not only are inexpensive but also behave very well at radio frequencies, unlike wirewound resistors and to a lesser extent spiralled film resistors.

The next big improvement in resistor technology was the *carbon film* resistor, popularly known in the early days as a Histab resistor owing to its improved ageing characteristic. It was available in 5, 2 and 1% tolerances, and the 5% variety is still widely used in the UK and Europe as a general purpose low-wattage resistor. Manufacture is highly automated, resulting in a low-cost resistor that is very reliable when used within its rated voltage and power limits. (Note that for resistance values much above 100 kΩ, it is not possible in the case of a carbon film resistor to dissipate its rated power without exceeding its rated working voltage.) The carbon film is deposited pyrolytically on a ceramic rod, to a thickness giving an end-to-end resistance of a few per cent of the required final value. End caps and leads are then fitted and a spiral groove is automatically machined in the carbon film. The machine terminates the cut when the required resistance is reached, and a protective insulating lacquer is applied over the film and end caps. Finally the resistance and tolerance are marked on the body, usually by means of the standard code of coloured bands shown in Appendix 1.

Metal oxide resistors are manufactured in much the same way as carbon film, except that the resistive film is tin oxide. They exhibit a higher power rating, size for size, than carbon film, and when derated to 50 or 25% of maximum they exhibit a degree of stability comparable to Histab or semiprecision types respectively.

Resistors are mass produced in certain preferred values, though specialist manufacturers will supply resistors of any nominal value, at a premium. Appendix 1 shows the various E series, from E6 which is appropriate to 20% tolerance resistors, to E96 for 1%.

Resistors of 1% tolerance are readily available in metal film and metal glaze construction. Metal glaze resistors use a film of glass frit and metal powder, fused onto a ceramic core, resulting in a resistor with good surge and short-term overload capability and good stability even in very low and very high resistance values. *Metal film* resistors have a conducting film made entirely of metal throughout and consequently offer a very low noise level and a low voltage coefficient.

The latter can be a very important consideration in critical measurement or very low-distortion applications. Ohm's law indicates that the current through a resistor is directly proportional to the voltage across it; in other words, if the current is plotted against the voltage as in Figure 1.1, the result should be a perfectly straight line, at least if the rated dissipation is not exceeded. Hence a resistor is described as a 'linear' component. It can more accurately be described by a power series for current as follows:

$$I = \frac{(E + \alpha E^2 + \beta E^3 + \gamma E^4 + \cdots)}{R} \qquad (1.3)$$

If α, β, γ and the coefficients of higher powers of E are all zero, the item is a perfectly linear resistor. In practice, α is usually immeasurably small. Coefficient β will also be very small, but not necessarily zero. For instance, the contact resistance between individual grains of carbon in a carbon composition resistor can vary slightly with the current flowing, i.e. with the applied voltage, whilst with film resistors the very small contact resistance between the film and the end caps can vary likewise. A quality control check used in resistor manufacture is to apply a pure sinusoidal voltage of large amplitude across sample resistors and check the size of any third-harmonic component generated – indicating a measurable value of β. Contact resistance variation can also be responsible for the generation of an excess level of random noise in a resistor, as can ragged edges of the spiral adjustment cut in a film resistor.

It is sometimes convenient to connect two or more resistors in *series* or *parallel,* particularly when a very low or very high resistance is required. It has already been noted that when two equal resistors are connected in series, the resultant resistance is twice that of either resistor alone, and if they are connected in parallel it is half. In the general case of several resistors of different values, the results of series and parallel combinations are summarized in Figure 1.2a. So, for example, to obtain a resistance of $0.33\ \Omega$ (often written as 0R33) three $1\ \Omega$ (1R0) resistors in parallel may be used. Not only does this arrangement provide three times the power rating of a single resistor, it also offers a closer initial tolerance. In values down to 1R0, resistors are available with a 1% selection tolerance; whereas for values below 1R0, 5 or 10% is standard. This would be an inconveniently large tolerance in many applications, for example the current sensing shunt in a linear laboratory power supply. The parallel resistor solution may, however, involve a cost penalty, for although three IR0 resistors will usually be cheaper than a higher-power 0R33 resistor, the assembly cost in production is higher.

Series resistors may be used likewise either to obtain a value not otherwise readily available (e.g. 200M); or to obtain a closer tolerance (e.g. two 1% 750K resistors where a 1M5 resistor is only available in 5% tolerance); or to gain twice the working voltage obtainable with a single resistor. Unequal value resistors may be combined to give a value not otherwise readily obtainable. For example, E96 values are usually restricted to resistors above 100R. Thus a 40R resistance may be produced by a 39R resistance in series with 1R0, a cheaper solution than three 120R resistors in parallel. Likewise, a 39R 1% resistor in parallel with 1K0 is a cheaper solution for 37R5 at 2% than two 75R 1% resistors in parallel, as the 1K0 resistor may be 5 or 10% tolerance. If you don't believe it, do the sums! In addition to its initial selection tolerance, a resistor's value changes with ageing, especially if used at its maximum dissipation rating. This must be borne in mind when deciding whether it is worth achieving a particular nominal value by the above means.

Variable resistors are available in various technologies: wirewound, carbon film, conductive plastic, cermet etc. Both ends of the resistive track are brought out to contacts, in addition to the 'slider' or 'wiper'. When the component is used purely as a *variable resistor*, connections are made to one end of the track and the wiper. It may be useful to connect the other end of the track to the wiper since then, in the event of the wiper going open-circuit for any reason, the in-circuit resistance will only rise to that of the track rather than go completely open-circuit. When the component is used as a *potentiometer,* the wiper provides a signal which varies between the voltage at one end of the track and that at the other – usually maximum and zero respectively (Figure 1.3). Thus the voltage at the output depends upon the position of the wiper. But what about the effect of the resistance of any circuit we may wish to connect to the wiper? Well, this is as convenient a point as any for a digression to look at some of the corollaries of Ohm's law when connecting sources of electricity to loads of one sort or another, e.g. batteries to bulbs or whatever.

Figure 1.4a shows an ideal battery or voltage source, and Figure 1.4b a more realistic one with a finite 'internal resistance'. It would clearly be imprudent to short-circuit the ideal battery, since Ohm's law indicates that with a resistance of zero ohms between its terminals the resultant current would be infinite – smoke and sparks the order of the day. To be more precise, the foregoing scenario must be fictional: for if the voltage source really has zero internal resistance there must always be E volts between its terminals, however much current it supplies; whereas if the short-circuit really has zero resistance there can be no voltage between the source's terminals, however much current flows. Shades of the irresistible force and the immovable object! In practice a source, be it battery or power supply, will always have some *internal* or *source resistance,* say R_s. In principle one can measure R_s by noting the open-circuit voltage E and measuring the short-circuit current I_{sc} through an ammeter. Then $R_s = E/I_{sc}$. In practice this only works approximately, for the ammeter itself will have a

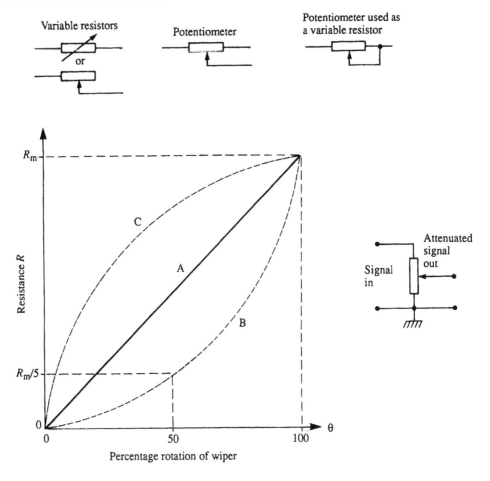

(A) Linear law.
(B) Log law (20% log shown; some potentiometers have a 10% log law). Used for volume controls.
(C) Reverse log law.

Figure 1.3 Variable resistors and potentiometers.

small but finite resistance: nevertheless you can, in the case of a dry (Leclanché primary type) battery, get a reasonable estimate of its source resistance. (It is best not to try this with batteries having a low internal resistance, such as lead-acid or Ni-Cd types.) Naturally it pays to short-circuit the battery through the ammeter for no longer than is absolutely necessary to note the reading, as the procedure will rapidly discharge the battery. Furthermore, the current will in all probability be gradually falling, since with most types of battery the internal resistance rises as the battery is discharged. In fact, the end of the useful life of a common or garden primary (i.e. non-rechargeable) battery such as the zinc-carbon (Leclanché) variety is set by the rise in internal resistance rather than by any fall in the battery's EMF as measured off load. (Measuring the open-circuit voltage and the short-circuit current to determine the internal resistance is even less successful in the case of a laboratory stabilized power supply, where R_s may be zero or even negative, but only up to a certain rated output current.)

The observant reader will not fail to notice that the current flowing in the load resistance in Figure 1.4c must also be responsible for dissipating

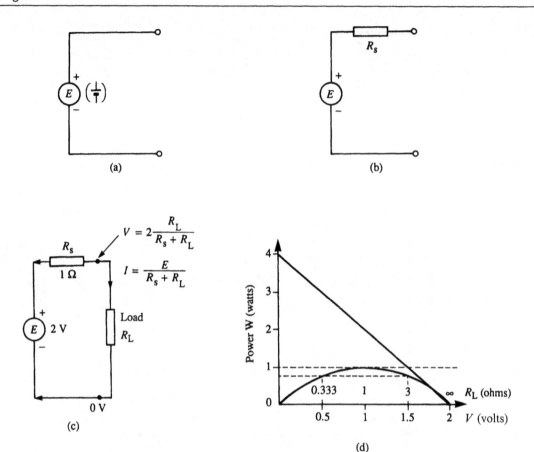

Figure 1.4   The maximum power theorem.
(a) Ideal voltage source.
(b) Generator or source with internal resistance R_s.
(c) Connected to a load R_L.
(d) $E = 2\,V$, $R_s = 1\,\Omega$. Maximum power in the load occurs when $R_L = R_s$ and $V = E/2$ (the matched condition), but only half the power is supplied to the load. On short-circuit, four times the matched load power is supplied, all dissipated in the battery's internal resistance R_s.

energy in the internal resistance of the source itself. Figure 1.4d shows the power (rate of energy) dissipation in the source resistance and the load for values of load resistance from zero to infinity.

It can be seen that the maximum power in the load occurs when its resistance is equal to the internal resistance of the source, that the terminal voltage V is then equal to half the source EMF E, and that the same power is then dissipated in the source's internal resistance as in the load. This is called the *matched* condition, wherein the efficiency, defined as the power in the load divided by the total power supplied by the source, is just 50%. This result is usually dignified with the title of the *maximum power theorem*. The matched condition gives the greatest possible power in the load, but only at the expense of wasting as much again in the internal resistance of the source. In many cases, therefore, the source is restricted to load resistances much higher than its own internal resistance, thus ensuring that nearly all of the power finishes up where it is really wanted – in the load. Good examples of this are a radio transmitter and a hand flashlamp; an even more telling example is a 660 MW three-phase turbo-alternator!

Now Ohm's law relates the current through a resistor to the applied EMF at any instant and consequently, like the maximum power theorem, applies to both AC and DC. The AC waveform shown in Figure 1.5 is called a sinusoidal waveform, or more simply a *sine wave*.

It is the waveform generated across the ends of a loop of wire rotating in a uniform magnetic field, such as the earth's field may be considered to be, at least over a localized area. Its frequency is measured in cycles per second or hertz (Hz), which is the modern term. As a necessary result of Ohm's law, not only is the current waveform in a resistive circuit the same shape as the voltage waveform, but also its peaks and troughs line up with the voltage waveform as shown in Figure 1.5. The sine wave shown contains alternating energy at one frequency only, and is the only waveshape with this important property. An audio-frequency sine wave reproduced through a loudspeaker has a characteristically round dull sound, like the flue pipes of a flute stop on an organ. In contrast, a sawtooth waveform or an organ-reed stop contains many overtones or *harmonics*.

Returning to the potentiometer, which might be the *volume control* in a hi-fi reproducing organ music or whatever, to any circuitry connected to the wiper of the potentiometer it will appear as a source of an alternating EMF, having some internal resistance. When the wiper is at the zero potential (ground or earth) end of the track, this source resistance is zero. At the other end of the track, the source resistance seen 'looking back' into the wiper circuit is equal to the resistance of the track itself in parallel with the source resistance of whatever circuit is supplying the signal to the volume control. If this source resistance is very much higher than the resistance of the track, then the resistance looking back into the wiper simply increases from zero up to very nearly the track resistance of the potentiometer as the volume is turned up to maximum. In the more likely case where the source resistance is much lower than the track resistance – let's assume it is zero – then the highest resistance seen at the wiper occurs at midtrack and is equal to one-quarter of the end-to-end track resistance. If the potentiometer is indeed a volume control, then midtrack position

won't in fact correspond to midtravel, as a volume control is designed with a non-linear (approximately logarithmic) variation of track resistance. This gives better control at low volumes, as the ear does not perceive changes of loudness linearly.

Preset potentiometers for circuit adjustment on test, on the other hand, almost invariably have linear tracks, often with multiturn leadscrew operation to enable very fine adjustments to be made easily. Potentiometers for user operation, e.g. tone and volume controls, are designed for continued use and are rated at greater than 100 000 operations, whereas preset controls are only rated for a few hundred operations.

Capacitors

Capacitors are the next item on any shopping list of passive components. The conduction of electricity, at least in metals, is due to the movement of electrons. A current of one ampere means that approximately 6242×10^{14} electrons are flowing past any given point in the conductor each second. This number of electrons constitutes one coulomb of electrical *charge,* so a current of one ampere is alternatively expressed as a rate of charge movement of one coulomb per second.

In a piece of metal an outer electron of each atom is free to move about in the atomic lattice. Under the action of an applied EMF, e.g. from a battery, electrons flow through the conductors forming the circuit towards the positive pole of the battery (i.e. in the opposite direction to the conventional flow of current), to be replaced by other electrons flowing from the battery's negative pole. If a capacitor forms part of the circuit, a continuous current cannot flow, since a capacitor consists of two plates of metal separated by a non-conducting medium – even a vacuum, for example (Figure 1.6a). If a battery is connected across the plates, its EMF causes some electrons to leave the plate connected to its positive pole or terminal and an equal number to flow onto the negative plate, as indicated in Figure 1.6c. A capacitor is said to have a *capacitance C* of one farad (1 F) if an applied EMF of one volt stores one coulomb (1 C) of charge. The capacitance is proportional to *A*, the area of the plates in Figure 1.6a, and

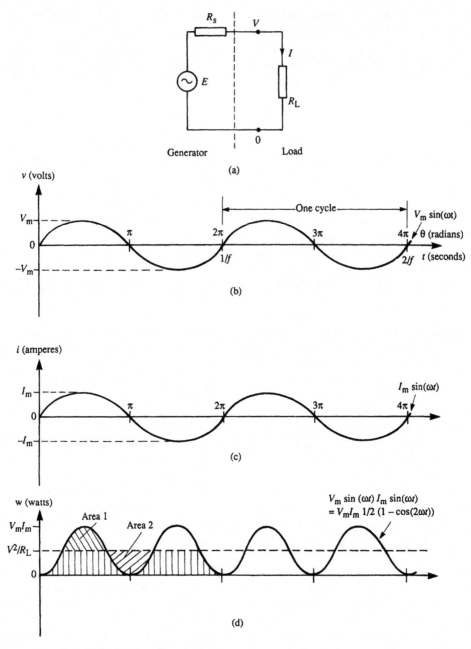

(a)

(b)

(c)

(d)

One cycle corresponds to 360° (or 2π radians), e.g. 1 revolution of a loop of wire in a magnetic field. If the waveform has a frequency of f Hz then each cycle lasts $1/f$ seconds. Thus there are $\omega = 2\pi f$ radians per second. Note that there are two power peaks per cycle of the applied voltage, so the angular frequency of the power waveform is $2\omega t$ radians per second.

Peak power load $= V_m I_m = V_m{}^2/R_L = I_m{}^2 R_L$, occurs at $\theta = \pi/2, 3\pi/2$ radians etc. Power in load at $\theta = \omega t = 0, \pi, 2\pi$ etc. is zero. Since area 1 equals 2, average power in load is $(1/2)(v_m{}^2/R_L) = V^2/R_L$, where $V = V_m/\sqrt{2}$. V is called the effective or root mean square (RMS) voltage.

Figure 1.5   Alternating voltage and power in a resistive circuit.

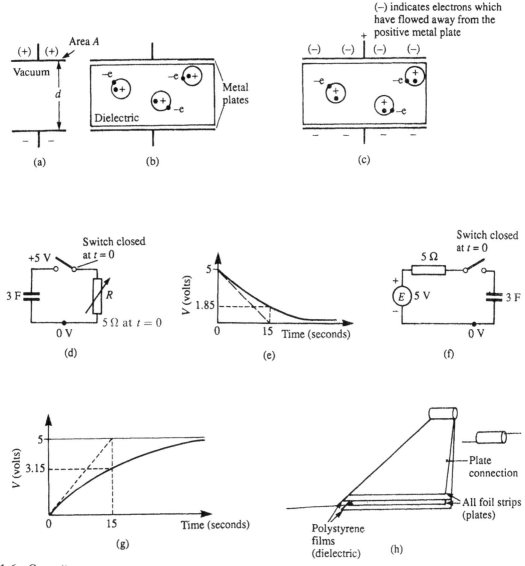

Figure 1.6 Capacitors.

inversely proportional to their separation d, so that $C = k(A/d)$ (provided that d^2 is much smaller than A). *In vacuo* the value of the constant k is 8.85×10^{-12} F/m, and it is known as the *permittivity of free space* ε_0. Thus *in vacuo* $C = \varepsilon_0(A/d)$. More commonly, the plates of a capacitor are separated by material of some kind – air or a solid substance – rather than the vacuum of free space. The permittivity of air is for practical purposes the same as that of free space.

As mentioned earlier, an insulator or *dielectric* is a substance such as air, polystyrene, ceramic etc. which does not conduct electricity. This is because, in an insulator, all of the electrons are closely bound to the respective atoms of which they form part. But although they cannot be completely detached from their parent atoms (except by an electrical force so great as to rupture and damage the dielectric), they can and do 'give' a little (as in Figure 1.6c), the amount being directly proportional to the applied voltage. This net displacement of charge in the dielectric enables a larger charge to be stored by the capacitor at a given voltage than if the plates were *in vacuo*. The ratio by which the stored charge is increased is known as the *relative permittivity* ε_r. Thus $C = \varepsilon_0 \varepsilon_r (A/d)$. So when a battery is connected to a capacitor there is a transient electrical current round the circuit, as electrons flow from the positive plate of the capacitor to the positive terminal of the battery, and to the negative plate from the negative terminal.

It was stated earlier that if the total transient flow of current needed to charge a capacitor to one volt amounts to a total charge of one coulomb, the capacitor is said to have a capacitance of one farad. More generally, the charge stored on a capacitor is proportional to both the size of the capacitor and the applied EMF; so $Q = CV$, where Q coulombs is the charge stored when a voltage V exists between the terminals of a capacitor of C farads. In electronics capacitors as small as 10^{-12} farad (called one picofarad and written 1 pF) up to a few thousand microfarads or more are used. You will also encounter nanofarads ($1 \text{ nF} = 10^{-9} \text{ F}$) and microfarads ($1 \text{ μF} = 10^{-6} \text{ F}$). Capacitors as large as 500 000 μF are found in computer power supplies, where it is necessary to store considerable energy, whilst small capacitors up to several farads are now readily available for memory back-up purposes.

So just how much energy can a capacitor store? This can be answered by connecting a resistor across a charged capacitor and finding out how much heat the electrical energy has been converted into by the time the capacitor is completely discharged. Imagine a 3 F capacitor charged up to 5 V (Figure 1.6d). The stored charge Q is given by CV, in this case 15 coulombs or 15 C. (It is just unfortunate that we use C F to mean a capacitor of value C measured in units of farads, and Q C for a charge of value Q measured in units of coulombs!) Well then, imagine a 5 ohm resistor connected across the capacitor and see what happens. Initially, the current I will of course be just 1 A, so the capacitor is being discharged at a rate of 1 C per second. At that rate, after 1 s there would be 14 C left, so the voltage would be $V = Q/C =$

4.67 V. After 15 s the charge would be all gone, there would be zero voltage across the capacitor, as indicated by the dashed line in Figure 1.6e. But of course as the voltage across the capacitor falls, so too must the current through the 5 Ω resistance as shown by the full line. In fact, after a time $T = CR$ seconds (15 s in this case) the current will only have fallen to 37% of its original value. But to come back to that point in a minute, though; meanwhile concentrate for the moment on working out the stored energy. At any moment the power being dissipated in the resistor is $I^2 R$, so initially it is 5 W or 5 joules per second. Suppose a 5 Ω *variable* resistor is used, and its value linearly reduced to zero over 15 s. Then the initial 1 A will be maintained constant for 15 s, by the end of which time the capacitor will be discharged. The initial heat dissipation in the resistor will be 5 J per second, falling linearly to zero, just like the resistance, since I^2 is constant. So the average power is 2.5 W maintained for 15 s, or a total of 37.5 J.

Starting with twice the capacitance, the initial rate of voltage drop would only have been 0.167 V per second; and, reducing the resistance to zero over 30 s to maintain the current constant at 1 A as before, the average power of 2.5 W would have been maintained for twice as long. So the energy stored by a capacitor at a given voltage is *directly* proportional to its capacitance. Suppose, however, that the 3 F capacitor had initially been charged to 10 V; then the initial charge would be 30 C and the initial current through the 5 Ω resistor would be 2 A. The initial power dissipation $I^2 R$ would be 20 W and the discharge time 15 s (reducing the resistance steadily to zero over that period, as before). So with an average power of 10 W, the stored energy appearing as heat in the resistor is now 150 J or four times as much. Thus the energy stored in a capacitor is proportional to the *square* of the voltage. In fact, quite simply the stored energy is given by

$$J = \tfrac{1}{2} CV^2$$

You may wonder about that $\tfrac{1}{2}$: shouldn't there be another $\tfrac{1}{2} CV^2$ lurking about somewhere? Well, certainly the sums agree with the formula. Going back to the 3 F capacitor charged to 5 V, the formula gives 37.5 J – and that is indeed what it was.

Suppose that, instead of discharging the capacitor, it is charged up to 5 V from an initially discharged state (Figure 1.6f). If it is charged via a 5 Ω resistor the initial current will be 1 A, and if the resistance is linearly reduced to zero over 15 s, a total charge of 15 C will be stored in the capacitor (Figure 1.6g). With a constant current of 1 A and an average resistance of 2.5 Ω, the heat dissipation in the resistor will be 37.5 J. Furthermore at the end of 15 s there will be 37.5 J of energy stored in the capacitor, so the 5 V battery must have supplied 75 J, as indeed it has: 5 V at 1 A for 15 s equals 75 J. So one must expend CV joules of energy to store just half that amount in a capacitor.

If a fixed 5 Ω resistor is used, the voltage across the capacitor will have reached only 63% of its final value in a time CR seconds, as shown by the solid line in Figure 1.6g. In theory it will take an infinite time to reach 5 V, since the nearer it gets, the less the potential difference across the resistor and hence the lower the current available to supply the remaining charge. However, after a period of 5 CR seconds the voltage will be within 1% of its final value, and after 12 CR within one part in a million. But this doesn't alter the fact that of a total energy CV^2 joules provided by the battery, only half is stored in the capacitor and the other half is dissipated as heat in the resistor. Of course one could charge the capacitor directly from the battery without putting a resistor in series. However, the only result is to charge the capacitor and store the $\frac{1}{2}CV^2$ joules more quickly, whilst dissipating $\frac{1}{2}CV^2$ joules as before but this time in the internal resistance of the battery. This makes a capacitor rather inconvenient as an energy storage device. Not only is charging it from a fixed voltage source such as a battery only 50% efficient, but it is only possible to recover the stored energy completely if one is not fussy about the voltage at which it is accepted – for example, when turning it into heat in a resistor. Contrast this with a secondary (rechargeable) battery such as a lead-acid accumulator or a Ni-Cd (nickel-cadmium) battery, which can accept energy at a very nearly constant voltage and return up to 90% of it at the same voltage. Another disadvantage of the capacitor as an energy store is leakage. The dielectric of a capacitor is ideally a perfect insulator. In practice its resistivity, though exceedingly high, will not be infinite. The result is that a discharge resistor is effectively built into the capacitor, so that the stored charge slowly dies away as the positive charge on one plate is neutralized by the leakage of electrons from the other. This tendency to self-discharge is called the *shunt loss* of a capacitor.

Figure 1.6e shows how the voltage falls when a capacitor is discharged: rapidly at first, but ever more slowly as time advances. The charge on the capacitor at any instant is proportional to the voltage, and the rate of discharge (the current through the resistor) is likewise proportional to the voltage. So at each instant, the rate of discharge is proportional to the charge remaining at that instant. This is an example of the *exponential function,* a fundamental concept in many branches of engineering, which may be briefly explained as follows.

Suppose you invest £1 at 100% compound interest per annum. At that very favourable rate, you would have £2 at the end of the first year, £4 at the end of the second and so on, since the yearly rate of increase is equal to the present value. Suppose, however, that the 100% annual interest were added as 50% at the end of each six months; then after one year you would have £2.25. If (100/12)% were added each month, the year-end total would be £2.61. If the interest were added not monthly, daily or even by the minute, but *continuously,* then at the end of a year you would have approximately £2.718 28. The *rate* of interest would always be 100% per annum, and at the start of the year this would correspond to £1.00 per annum. But as the sum increased, so would the rate of increase in terms of pounds per annum, reaching £2.718 28 ... per annum at the year end. The number 2.718 28 ... is called exponential e. The value of an investment of £P at n% compound *continuous* interest after t years is £$P\,e^{(n/100)t}$, often written £$P\exp((n/100)t)$.

Now, going back to the resistor and capacitor circuit: the rate of discharge is proportional to the charge (and hence voltage) remaining; so this is simply compound interest of −100% per annum!

So $V = V_0 e^{-1t}$, where t is measured in units of time of CR seconds, not years, and V_0 is the capacitor voltage at time t_0, the arbitrary start of time at the left of the graph in Figure 1.6c. To get from units of CR seconds to seconds simply write $V = V_0 e^{-t/CR}$. Thus if $V_0 = 1$ V, then after a time $t = CR$ seconds, it will have discharged to $e^{-1} = 1/2.718\,28 = 0.37$ V approximately.

Remember that in a circuit with a direct current (DC) source such as a battery, and containing a capacitor, only a transient charging current flows; this ceases entirely when the capacitor is fully charged. So current cannot flow continuously in one direction through a capacitor. But if first a positive supply is connected to the capacitor, then a negative one, and so on alternately, charging current will always be flowing one way or the other. Thus an alternating voltage will cause an alternating current apparently to flow through a capacitor (see Figure 1.7). At each and every instant, the stored charge Q in the capacitor must equal CV. So the waveform of charge versus time is identical to that of the applied voltage, whatever shape the voltage waveform may be (provided that C is constant, which is usually the case). A current is simply the rate of movement of charge; so the current must be zero at the voltage peaks, where the amount of charge is momentarily not changing, and a maximum when the voltage is zero, where the charge is changing most rapidly. In fact for the sinusoidal applied voltage shown, the current waveform has exactly the same shape as the voltage and charge waveforms; however, unlike the resistive circuit (see earlier, Figure 1.5), it is moved to the left – advanced in time – by one-quarter of a cycle or 90° or $\pi/2$ radians. The current waveform is in fact a cosine wave, this being the waveform which charts the rate of change of a sine wave. The sine wave is a waveform whose present rate of change is equal to the value of the waveform at some point in the future, namely one-quarter of a cycle later (see Figure 1.7). This sounds reminiscent of the exponential function, whose present rate of change is equal to its present value: you might therefore expect there to be a mathematical relation between the two, and indeed there is.

One complete cycle of a sinusoidal voltage corresponds to 360° rotation of the loop of wire in a magnetic field, as mentioned earlier. The next rotation is 360° to 720°, and so on; the frequency is simply the number of rotations or cycles per second (Hz). The voltage v at any instant t is given by $v = V_m \sin(\omega t)$ (assuming v equalled zero at t_0, the point from which t is measured), where V_m is the voltage at the positive peak of the sine wave and ω is the angular frequency, expressed in radians per second (ω is the lower-case Greek letter omega). There are 2π radians in one complete cycle, so $\omega = 2\pi f$ radians per second.

One can represent the *phase* relationship between the voltage and current in a capacitor without needing to show the repetitive sinusoidal waveforms of Figure 1.7, by using a *vector diagram* (Figure 1.8a). The instantaneous value of the voltage or current is given by the projection of the appropriate vector onto the horizontal axis. Thus at the instant shown, corresponding to just before $\theta = \pi/2$ or $t = 1/4f$, the voltage is almost at its maximum positive value, whilst the current is almost zero. Imagine the voltage and current vectors rotating anticlockwise at $\omega/2\pi$ revolutions per second; then the projection of the voltage and current vectors onto the horizontal axis will change in sympathy with the waveforms shown in Figure 1.7b and d. Constantly rotating vectors are a little difficult to follow, but if you imagine the paper rotating in the opposite direction at the same speed, the diagram will appear frozen in the position shown. Of course this only works if all the vectors on the diagram are rotating at the same speed, i.e. all represent things happening at the same frequency. So one can't conveniently show the power and energy waveforms of Figure 1.7e on the vector diagram.

Now the peak value of the capacitor current I_m is proportional to the peak value of the voltage V_m, even though they do not occur at the same instant. So one may write $I_m = V_m/X_c$, which looks like Ohm's law but with R replaced by X_c. X_c is called the *reactance* of the capacitor, and it differs from the resistance of a resistor in two important respects. First, the sinusoidal current through a capacitor resulting from an applied sinusoidal voltage is advanced by a quarter of a cycle. Second, the reactance of a capacitor varies

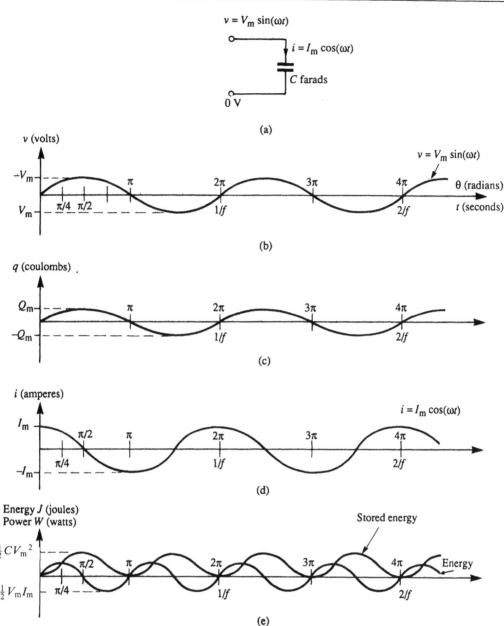

(a)

(b)

(c)

(d)

(e)

Stored energy $J = \frac{1}{2}Cv^2 = \frac{1}{2}C[V_m \sin(\omega t)]^2 = \frac{1}{2}CV_m^2 \frac{1}{2}[1 - \cos(2\omega t)]$

Energy flow $= vi = V_m \sin(\omega t)I_m \cos(\omega t) = \frac{1}{2}V_mI_m \sin(2\omega t)$ watts.

The maximum energy stored is $\frac{1}{2}CV_m^2 = \frac{1}{2}C(-V_m)^2$ joules; the energy stored is zero at $\theta = \omega t = 0, \pi, 2\pi$ etc.

The rate of energy flows is iv watts. This is positive (into the capacitor) over the first quarter-cycle, 0 to $\pi/2$, as v and i are both positive. It peaks at $\pi/4$, where $W = \frac{1}{2}V_mI_m$, the point at which the stored energy is increasing most rapidly. By $\pi/2$, where energy flow W is zero again, $\frac{1}{2}CV^2$ joules of energy have been stored. The energy flow $W = vi$ now becomes negative as the capacitor gives up its stored energy over the remainder of the first half-cycle of voltage. The zero net energy means that no power is dissipated, unlike the resistive case of Figure 1.5.

Figure 1.7 Alternating voltage and power in a capacitive circuit.

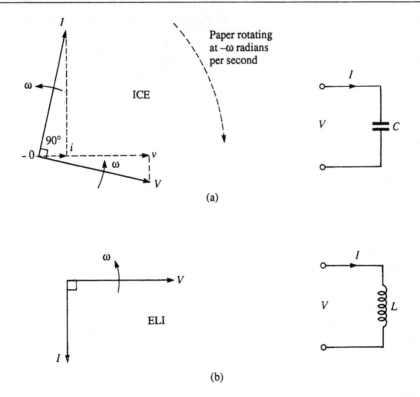

Figure 1.8 Phase of voltage and current in reactive components.
(a) ICE: the current *I* leads the applied EMF *E* (here *V*) in a capacitor. The origin 0 represents zero volts, often referred to as ground.
(b) ELI: the applied EMF *E* (here *V*) across an inductor *L* leads the current *I*.

with frequency. For if the voltage and charge waveforms in Figure 1.7 were of twice the frequency, the rate of change of charge (the current) at $t = 0$ would be twice as great. Thus the reactance is inversely proportional to the frequency, and in fact $X_c = 1/\omega C$. Recalling that the instantaneous voltage $v = V_m \sin(\omega t)$, note that $i = V_m \sin(\omega t)/(1/\omega C) = V_m \sin(\omega t)\omega C$. However, that can't be right, since *i* is not in fact in phase with *v*, see Figure 1.7, but in advance or 'leading' by 90°. A way round this difficulty is to write $i = V_m \sin(\omega t) j\omega C$, where j is an 'operator' and indicates a 90° anticlockwise rotation of a vector. This makes the formula for *i* agree with the vector diagram of Figure 1.8a. So from now on, just tack the j onto the reactance to give $X_c = 1/j\omega C$, and you will find that the 90° displacement between current and voltage looks after itself. Also, for convenience, from now on, drop the subscript m

and simply write *V* for the maximum or peak voltage V_m.

Capacitors are used for a number of different purposes, one of which has already been mentioned: energy storage. They are also used to pass on alternating voltage signals to a following circuit whilst blocking any associated constant voltage level, or to bypass unwanted AC signals to ground. It has always been a problem to obtain a very high value of capacitance in a reasonably small package, and a number of different constructions are used to meet different requirements. If the two plates of the capacitor in Figure 1.6(a) each have an area of *A* square metres and are separated by *d* metres *in vacuo,* the capacitance, it was noted earlier, is given by $C = \varepsilon_0(A/d)$ farads (where ε is the lower-case Greek letter epsilon) approximately, if d^2 is small compared with *A*. So if $A = 1\,m^2$ and $d = 1\,mm$, the capacitance is

8.89×10^{-9} farads or 889 pF in a vacuum and just 0.06% higher in air. Capacitance values much larger than this are frequently required; so how can they be achieved? Simply by using a solid dielectric with a relative permittivity of ε_r as shown earlier. The dielectric can be a thin film of plastic, and ε_r is then typically in the range 2 to 5. The resultant increase in capacitance is thus not large but, with a solid film to hold the plates apart, a much smaller value of d is practicable.

The plates may be long narrow strips of *aluminium foil* separated by the dielectric – say polystyrene film 0.02 mm thick – and the sandwich is rolled up into a cylinder as in Figure 1.6h. Note that unlike in Figure 1.6a, both sides of each plate now contribute to the capacitance, with the exception of the outer plate's outermost turn. The foil is often replaced by a layer of metal evaporated onto the film (this is done in a vacuum), resulting in an even more compact capacitor; using this technique, 10 μF capacitors fitting four to a matchbox can be produced. The dielectric in a foil capacitor being so thin, the electric stress in the dielectric – measured in volts per metre – can be very high. If the voltage applied across the capacitor exceeds the rated working voltage, the dielectric may be punctured and the capacitor becomes a short-circuit. Some foil capacitors are 'self-healing'. If the dielectric fails, say at a pinhole, the resultant current will burn out the metallization in the immediate vicinity and thus clear the short-circuit. The external circuitry must have a certain minimum resistance to limit the energy input during the clearing process to a safe value. Conversely, if the circuit resistance is too high the current may be too low to clear the short-circuit.

For values greater than 10 μF, electrolytic capacitors are usually employed. Aluminium foil electrolytic capacitors are constructed much like Figure 1.6h; the film separating the plates is porous (e.g. paper) and the completed capacitor is enclosed within a waterproof casing. The paper is impregnated with an electrolyte during manufacture, the last stage of which is 'forming'. A DC forming voltage is applied to the capacitor and a current flows through the electrolyte. The resultant chemical action oxidizes the surface of the positive plate, a process called *anodization*. Aluminium

oxide is a non-conductor, so when anodization is complete no more current flows, or at least only a very small current called the *leakage current* or shunt loss. For low-voltage electrolytics, when new, the leakage current at 20°C is usually less than 0.01 μA per μF, times the applied voltage, usually quoted as $0.01CV\,\mu A$. The forming voltage is typically some 20% higher than the capacitor's rated maximum working voltage. The higher the working voltage required, the thicker the oxide film must be to withstand it. Hence the lower the capacitance obtainable in a given size of capacitor; the electrolyte is a good conductor, so it is the thickness of the oxide which determines the capacitance. Working voltages up to 450 V or so are the highest practically obtainable. In use, the working voltage should never be exceeded; nor should the capacitor's polarity be reversed, i.e. the red or + terminal must always be more positive than the black or − terminal. (This restriction does not apply to the 'reversible electrolytic', which consists of two electrolytic capacitors connected in series back to back and mounted in a single case.)

Before winding the aluminium foils and paper separators together, it is usual to etch the surface of the foils with a chemical; this process can increase the surface area by a factor of ten or even thirty times. The resultant large surface area and the extreme thinness of the oxide layer enable a very high capacitance to be produced in a small volume: a 7 V working 500 000 μF capacitor as used in mains power supplies for large computers may be only 5 to 8 cm diameter by 10 to 15 cm long. The increase in surface area of the electrodes produced by etching is not without its disadvantages. With heavy etching, the aluminium strip develops a surface like a lace curtain, so that the current flows through many narrow necks of metal. This represents a resistive component internal to the capacitor, called the *series loss,* resulting in energy dissipation when an *alternating* or *ripple* current flows. So while heavy etching increases the capacitance obtainable in a given size of capacitor, it does not increase *pro rata* the rated ripple current that the capacitor can support. The plates of an aluminium electrolytic capacitor are manufactured from extremely pure aluminium, better than 'four nines pure', i.e. 99.99%, as any impurity can be a centre for erosion leading to increased leakage current and eventual failure. The higher the purity of the aluminium foil the higher its cost, and this is reflected in the price of the capacitor. Whilst a cheaper electrolytic may save the manufacturer a penny or two, a resultant early equipment failure may cost the user dear.

In radio-frequency circuits working at frequencies up to hundreds of megahertz and beyond, capacitors in the range 1 to 1000 pF are widely employed. These commonly use a thin *ceramic* disc, plate, tube or multilayer structure as the dielectric, with metallized electrodes. The lower values, up to 100 pF or so, may have an NPO dielectric, i.e. one with a low, nominally zero, temperature coefficient of capacitance. In the larger values, N750, N4700 or N15000 dielectrics may be used (N750 indicates a temperature coefficient of −750 parts per million per degree C), and disc ceramics for decoupling purposes with capacitances up to 470 nF (0.47 μF) are commonly available. As with resistors, many circuits nowadays use capacitors in surface mount packaging.

Variable capacitors are available in a variety of styles. Preset variable capacitors, often called *trimmers,* are used in production for the setting-up adjustments to tuned circuits. They may use solid or air dielectric, according to the application, as also do the variable capacitors used in radio sets for tuning.

Inductors and transformers

The third type of passive component mentioned at the beginning of the chapter is the inductor. This is designed to exploit the magnetic field which surrounds any flow of current, such as in a wire – or indeed in a stroke of lightning. This is illustrated in Figure 1.9a, which shows lines of magnetic force surrounding the current in a wire. The lines form closed loops and are shown more closely packed together near the wire than further out, to indicate that the magnetic field is strongest near the wire. However, the lines are merely a crude representation of the magnetic field, which is actually continuous throughout space: some writers talk of tubes rather than lines, to indicate this. If the wire is bent into a circular loop, the magnetic field

becomes doughnut shaped as in Figure 1.9c. A series of closely spaced turns form a *solenoid* – the resultant field is then concentrated as represented in Figure 1.9d. The form and strength of the magnetic field is determined by the strength of the current causing it and the way the current flows, as is clear from Figure 1.9a to d. In the case of a solenoid of length l, if one ampere flows in the wire and there are N turns, the magneto-motive force (MMF), denoted by F, is just N amperes.

This is often written as N ampere turns, but the number of turns is really irrelevant: the solenoid could equally well consist of a single turn of copper tape of width l carrying N amperes. The effect would be the same if it were possible to ensure that the current flow of N amperes was equally distributed across the width of the tape. Dividing the tape up into N turns in series, each carrying one Nth of the total current, is a convenient subterfuge for ensuring this.

If a long thin solenoid is bent round into a *toroid* (Figure 1.9e), then instead of returning round the outside of the solenoid the magnetic lines of force are closed upon themselves entirely within the solenoid and there is no external field. The strength of the magnetic field H A/m within the toroid depends upon the strength of the magneto-motive force per unit length, in fact $H = I/l$ amperes per metre, where l is the length of the toroid's mean circumference and I is the effective current – the current per turn times the number of turns. The magnetic field causes a uniform magnetic flux density, of B webers per square metre, within the toroidal winding. The ratio B/H is called the *permeability of free space* μ_0, and its value is $4\pi \times 10^{-7}$. So $B = \mu_0 H$ in a vacuum or air, or even if the toroid is wound on a solid core, provided the core material is non-magnetic. If the cross-sectional area of the core is A m^2, the total magnetic flux Φ Wb (webers) is simply given by $\Phi = BA$.

If the toroid is provided with a ferromagnetic core – Faraday experimented with a toroid wound on an anchor ring – it is found that the flux density and hence the total flux is greatly increased. The ratio is called the *relative permeability* of the material μ_r. Thus in general $B = \mu_0 \mu_r H,$ where

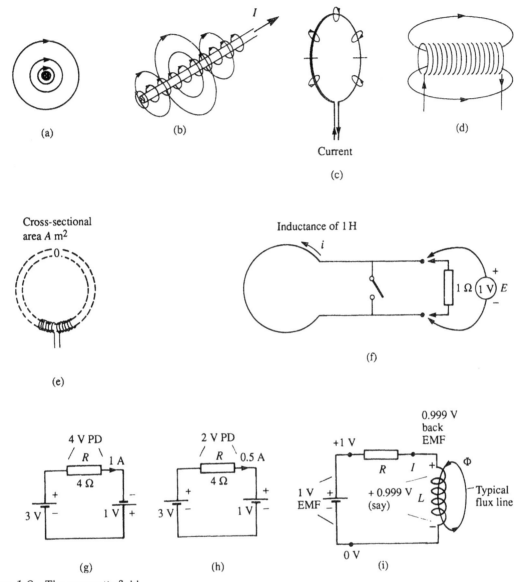

Figure 1.9 The magnetic field.

(a) End view of a conductor. The cross indicates current flowing into the paper (a point indicates flow out). By convection, the lines of flux surrounding the conductor are as shown, namely clockwise viewed in the direction of current flow (the corkscrew rule).

(b) The flux density is greatest near the conductor; note that the lines form complete loops, the path length of a loop being greater the further from the wire.

(c) Doughnut-shaped (toroidal) field around a single-turn coil.

(d) A long thin solenoid produces a 'tubular doughnut', of constant flux density within the central part of the coil.

(e) A toroidal winding has no external field. The flux density *B* within the tube is uniform over area *A* at all point around the toroid, if the diameter of the solenoid is much smaller than that of the toroid.

(f) Changing current in single-turn coil.

(g) EMF and PD: sources combining.

(h) EMF and PD: sources opposing, and energy storage.

(i) Energy storage in inductor field.

$\mu_r = 1$ for a vacuum, air and non-magnetic materials. In the cases shown in Figure 1.9a to d the flux density (indicated by the closeness of spacing of the flux lines) varies from place to place, but at each point $B = \mu_0 H$. $B = \mu_0 \mu_r H$ can be expressed in terms of total flux Φ and MMF F as follows:

$$\frac{\Phi}{A} = u_0 u_r \frac{F}{l}$$

so that

$$\Phi = \frac{F}{l/(\mu_0 \mu_r A)}$$

The term $l/\mu_0\mu_r A$ is called the *reluctance S,* of the magnetic path, and it has the units of amperes per weber. Thus, in a magnetic circuit, flux equals MMF divided by reluctance; this is a similar sort of result to current equals EMF divided by resistance in an electric circuit. Just as the individual resistances around an electric circuit can be added up when working out the total EMF needed to cause a current I to flow, so in a magnetic circuit (e.g. a core of magnetic material with a permeability μ_r, having an airgap) the reluctances can be added up to find the total MMF needed to cause a given total flux.

So far the field produced by a constant current of I amperes has been considered; but what happens when the current changes? Indeed, how does the current come to flow in the first place? (Figure 1.9c rather begs the question by assuming that the current is already flowing.) Consider what happens when an EMF of one volt is connected to a large single-turn coil, as in Figure 1.9f. Assume for the moment that the coil has negligible resistance. Nothing in this universe (except a politician's promises) changes instantaneously, so the moment after connecting the supply the current must be the same as the moment before, i.e. zero. Clearly one can expect the current to increase thereafter, but how fast? Assume that the current increases at one ampere per second, so that after one second the MMF F is just one ampere turn, and that the reluctance $S = 1$, so that the resulting flux Φ is one weber. (In fact, for this to be so, the coil would have to be very large indeed or immersed in a magnetic medium with a huge relative permeability, but that is a minor practical point

which does not affect the principle of the thing.) Having assumed the coil to have negligible resistance, the current will ultimately become very large; so why isn't it already huge after just one second? The reason is that the steadily increasing flux induces an EMF in the coil, in opposition to the applied EMF: this is known as *Lenz's law*. If the flux Φ increases by a small amount $d\Phi$ in a fraction of a second dt, so that the rate of increase is $d\Phi/dt$, then the *back EMF* induced in the single-turn coil is

$$E_B = -\frac{d\Phi}{dt} \tag{1.4}$$

or $-N\,d\Phi/dt$ in the case of an N-turn coil. But

$$\Phi = \frac{\text{MMF}}{\text{reluctance}} = \frac{F}{S}$$

So for a one-turn coil, $\Phi = I/S$. As the current and flux are both increasing, for this to remain true their rates of change must also be equal, i.e.

$$\frac{d\Phi}{dt} = \frac{1}{S}\frac{dI}{dt} \tag{1.5}$$

Substituting (1.4) in (1.5) gives $E_B = -(1/S)(dI/dt)$ for a one-turn coil. In the more general case where the MMF $F = NI$, then $\Phi = NI/S$ and so $d\Phi/dt = (N/S)(dI/dt)$. This rate of flux increase will induce a voltage $E_B = -d\Phi/dt$ in series in each of the N turns of the coil, so

$$E_B = -N\frac{d\Phi}{dt} = -N\frac{N}{S}\frac{dI}{dt} = \frac{-N^2}{S}\frac{dI}{dt} \tag{1.6}$$

The term N^2/S, which determines the induced voltage resulting from a unit rate of change of current, is called the *inductance L* and is measured in henrys: that is,

$$L = \frac{N^2}{S} \text{ henrys}$$

You must keep the difference between an electromotive force (EMF) and a potential drop or difference (PD) very clearly in mind, to understand the minus sign in equation (1.4). To illustrate this, consider two secondary batteries and a resistor connected as in Figure 1.9g. The total EMF round the circuit, counting clockwise, is $3+1$ volts, and this is balanced by the PD of IR volts across the resistor. The batteries supply a total of

4 W of power, all of which is dissipated in the resistor. If now the polarity of the 1 V battery is reversed, as in Figure 1.9h, the total EMF acting is 3 − 1 V, so the current is 0.5 A. The 3 V battery is now supplying $3 \times 0.5 = 1.5$ W, but the dissipation in the resistor I/R is only 1 W. The other 0.5 watts is disappearing into the 1 V battery; but it is not being dissipated, it is being stored as chemical energy. The situation in Figure 1.9i is just the same; the applied EMF of the battery is opposed by the back EMF of the inductor (which in turn is determined by the inductance and the rate of increase of the current), whilst energy from the battery is being stored in the steadily increasing magnetic field. If the internal resistance of the battery and the resistance of the inductor are vanishingly small, the current will continue to increase indefinitely; if not, the current will reach a limit set by the applied EMF and the total resistance in the circuit.

Returning now to Figure 1.9f, if the switch is closed one second after connecting the battery, at which time the current has risen to 1 A, then there is no voltage across the ends of the coil. No back EMF means that $d\Phi/dt$ must be zero, so dI/dt is also zero. Hence the current now circulates indefinitely, its value frozen at 1 A – provided our coil really has zero resistance. (In the meantime, disconnect the battery and replace it with a 1 Ω resistor; you will see why in a moment.) Thus energy stored in the magnetic field is preserved by a short-circuit, just as the energy stored in a capacitor is preserved by an open-circuit. Now consider what happens on opening the switch in Figure 1.9f, thus substituting the 1 Ω resistor in place of the short-circuit. At the moment the switch opens the current will still be 1 A; it cannot change its value instantaneously. This will establish a 1 V PD across the resistor, of the opposite polarity to the (now disconnected) battery; that is, the top end of the resistor will be negative with respect to the lower end. The coil is now acting as a generator, feeding its stored energy into the resistor – initially at a rate of 1 joule per second, i.e. 1 W. How much energy is there stored in the field, and how long before it is all dissipated as heat in the resistor? Initially the current must be falling at 1 A per second, since there is 1 V across

the resistor, and $E = -L\,dI/dt$ (where the inductance is unity in this case). Of course dI/dt is itself now negative (current *decreasing*), as the polarity reversal witnesses. After a fraction of a second, the current being now less than one ampere, the voltage across the resistor will have fallen likewise; so the rate of decrease of current will also be lower. The fall of current in the coil will look just like the fall of voltage across a discharging capacitor (Figure 1.6e, solid line). Suppose, however, that the resistor is a variable resistor and its resistance increases, keeping the value *inversely* proportional to the current. Then IR will be constant at 1 V and the current will fall linearly to zero in 1 second, just like the dashed line in Figure 1.6e.

Since the induced voltage across the resistor has, by this dodge, been maintained constant at 1 V, the energy dissipated in it is easily calculated. On opening the switch the dissipation is $1\,V \times 1\,A$, and this falls linearly to zero over one second. So the average power is 0.5 W maintained for one second, giving a stored energy of 0.5 J. If the inductance had been 2 H and the current 1 A when the switch was opened, the initial rate of fall would have been 0.5 A per second and the discharge would have lasted 2 s, dissipating 1 J in the resistor (assuming its value was adjusted to maintain 1 V across it as before). Thus the stored energy is proportional to the inductance L. On the other hand, if the current was 2 A when the switch was opened, the voltage across the 1 Ω resistor would have been 2 V, so the rate of fall would need to be 2 A/s (assuming 1 H inductance). Thus the initial dissipation would have been 4 W, falling to zero over 1 s, giving a stored energy of 2 J. So the stored energy is proportional to the square of the current. In fact, the stored energy is given by

$$J = \tfrac{1}{2}LI^2$$

This result is reminiscent of $J = CV^2/2$ for a capacitor.

An inductor can be and often is used as an energy store, as will appear in a later chapter in the context of power supplies, where certain limitations to the inductor's power storing ability will become apparent. In particular, the energy stored in the magnetic field of a short-circuited inductor is rapidly lost due to dissipation in the resistance of its windings. The ratio of inductance to series loss L/r, where r is the resistance of the inductor's winding, is much lower than the ratio C/R, where R is the shunt loss, for a high-quality capacitor. At very low temperatures, however, the electrical resistivity of certain alloys and compounds vanishes entirely – a phenomenon known as *superconductivity*. Under these conditions an inductor can store energy indefinitely in its magnetic field, as none is dissipated in the conductor.

In addition to use as energy storage devices, inductors have several other applications. For example, inductors with cores of magnetic material are used to pass the direct current output of a rectifier to later circuitry whilst attenuating the alternating (hum) components. Air or ferrite cored inductors (RF chokes) are used to supply power to radio-frequency amplifier stages whilst preventing RF power leaking from one stage to another via the power supply leads. This application and others make use of the AC behaviour of an inductor. Just as the reactance of a capacitor depends upon frequency, so too does that of an inductor. Since the back EMF $E_B = N \, d\Phi/dt = -L \, di/dt$, it follows that the higher the frequency, the smaller the alternating current required to give a back EMF balancing the applied alternating EMF. In fact, the *reactance* X_L of an inductor is given by $X_L = 2\pi f L = \omega L$ where f is the frequency in hertz, ω is the angular velocity in radians per second, and L is the inductance in henrys. This may be represented vectorially as in Figure 1.8b, from which it can be seen that when the voltage is at its positive peak, the current is zero but increasing. If you draw the waveforms for an inductor corresponding to those of Figure 1.7 for a capacitor, you will find that the current is increasing (or becoming less negative) all the time that the applied voltage is positive and vice versa, and that the net energy flow is zero. Again, you can look after the 90° phase shift between the voltage and *lagging* current by using the j operator and writing $X_L = j\omega L$, thus keeping the sums right. With a capacitor, the voltage produces a *leading* current; the exploitation of this difference is the basis of a particularly important application, namely tuned circuits, which will be considered later. Note that if multiplying by j signifies a 90°

anticlockwise displacement of a vector, multiplying by j again will result in a further 90° anticlockwise rotation. This is equivalent to changing the sign of the original vector. Thus $j \times j = -1$, a result which will be used extensively later.

In the meantime, imagine two identical lengths of insulated wire, glued together and bent into a loop as in Figure 1.9f. Virtually all of the flux surrounding one wire, due to the current it is carrying, will also surround the other wire. Now connect a battery to one loop – called the *primary* – and see what happens to the other loop – called the *secondary*. Suppose the *self-inductance* of the primary is 1 H. Then an applied EMF of 1 V will cause the current to increase at the rate of 1 A per second, or conversely the rate of change of 1 A per second will induce a back EMF of 1 V in the primary; it comes to the same thing. But all the flux produced by the primary also links with the secondary, so an EMF identical to the back EMF of the primary will be induced in the secondary. Since a dI/dt of 1 A/s (a rate of increase of 1 A per second) in the primary induces an EMF of 1 V in the secondary, the two windings are said to have a *mutual inductance M* of 1 H. If the two coils were placed slightly apart so that only a fraction c (a half, say) of the flux caused by the primary current linked with the secondary, then only 0.5 V would be induced in the secondary and the mutual inductance would be only 0.5 H.

In the above example the two coils were identical, so that the self-inductance of the secondary was also 1 H. In the general case, the maximum value of mutual inductance M between two unequal coupled inductors L_1 and L_2 is given by $M = \sqrt{(L_1 L_2)}$, whilst if only some of the flux of one winding links with the other winding then $M = k\sqrt{(L_1 L_2)}$, where k is less than unity. As a matter of interest $k = c\sqrt{(S_1 S_2)}$, where c is as before the fraction of the primary flux linking the secondary, and S_1, S_2 are the reluctances of the primary and secondary magnetic circuits respectively.

Coupling between coils by means of mutual inductance is used in band-pass tuned circuits, which are briefly mentioned in the chapter covering r.f. In this application, quite small values of coupling coefficient k are used. Right now it is time to look at coupled circuits where k is as large as possible, i. e. where c is unity, so that all the flux of the primary links the secondary and vice versa.

Figure 1.10a shows a *transformer* – two coils wound on a ferromagnetic core, which results in much more flux per ampere turn, owing to the lower reluctance of the magnetic path. The resultant high value of inductance means that only a small 'magnetizing current' I_m flows. This is $90°$ out of phase with the alternating voltage, E_a at 50 or 60 Hz say, applied to the primary winding. (There is also a small in-phase current I_c due to the core loss. This together with I_m makes up the primary off-load current I_{pol}.) Since $E_B = -L\,dI/dt = -N\,d\Phi/dt$, and all of the flux Φ links both windings, it follows that the ratio of secondary voltage E_s to the primary back EMF E_{pB} is equal to the *turns ratio*:

$$\frac{E_s}{E_{pB}} = \frac{N_s}{N_p} \qquad (1.7)$$

If a resistive load R is now connected to the secondary, a current E_s/R will flow, since E_s appears to the load like a source of EMF. By itself, this current would produce a large flux in the core. However, the flux cannot change, since the resultant primary back EMF E_{pB} must balance the fixed applied EMF E_a (see Figure 1.10b and c). Consequently an additional current I_p flows in the primary to provide an MMF which cancels out the MMF due to the secondary current. Hence $I_p N_p = I_s N_s$ so

$$\frac{I_s}{I_p} = \frac{N_p}{N_s} \qquad (1.8)$$

If, for example, $N_s/N_p = 0.1$, i.e. there is a ten-to-one step-down turns ratio, then from (1.7) the secondary voltage will only be one-tenth of the primary voltage. Further, from (1.8) I_p will only be one-tenth of I_s. The power delivered to the load is $I_s^2 R$ and the power input to the transformer primary is $I_p^2 R'$ where R' is that resistor which, connected directly to E_a, would draw the same power as R draws via the transformer (assuming for the moment that the transformer is perfectly efficient). Since in this example I_p is only a tenth of I_s and E_s is only a tenth of E_a then R' must equal $(R \times 100)$ ohms. Hence a resistance (or indeed a

reactance) connected to one winding of a transformer appears at the other transformed by the square of the turns ratio.

So far an almost perfect transformer has been considered, where $E_a I_p = E_s I_s$, ignoring the small magnetizing current I_o, which flows in the primary when the transformer is off load. The term 'magnetizing current' is often used loosely to mean I_{pol}. The difference is not large since I_c is usually much smaller than I_m. In a perfect transformer, the magnetizing inductance would be infinity, so that no primary current at all would flow when the transformer was off load. In practice, increasing the primary inductance beyond what is necessary makes it more difficult to ensure that virtually all the primary flux links the secondary, resulting in undesirable *leakage reactance*. Further, the extra primary and secondary turns increase the winding resistance, reducing efficiency. Nevertheless the 'ideal transformer', with its infinite primary inductance, zero leakage inductance and zero winding resistance, if an unachievable goal, is useful as a touchstone.

Figure 1.10d shows the equivalent circuit of a practical power transformer, warts and all. R_c represents the *core loss,* which is caused by hysteresis and eddy currents in the magnetic core. *Eddy currents* are minimized by building the core up from thin stampings insulated from each other, whilst *hysteresis* is minimized by stamping the laminations from special transformer-grade steel. The core loss R_c and the magnetizing inductance L_m are responsible for the current I_{pol} in Figure 1.10c. They are shown connected downstream of the primary winding resistance R_{wp} and leakage inductance L_{lp} since the magnetizing current and core loss actually reduce slightly on full load. This is because of the extra voltage drop across R_{wp} and L_{lp} due to I_p. A useful simplification, usually valid, is to refer the secondary leakage inductance and winding resistance across to the primary, *pro rata* to the square of the turns ratio (see Figure 1.10e, L_l and R_w). Using this simplification, Figure 1.10f shows the vector diagram for a transformer with full-rated resistive load: for simplicity a unity turns is depicted so that $E_s = E_a$ approximately. Strictly, the simplification is only correct if the ratio of leakage inductance to total inductance and the 'per

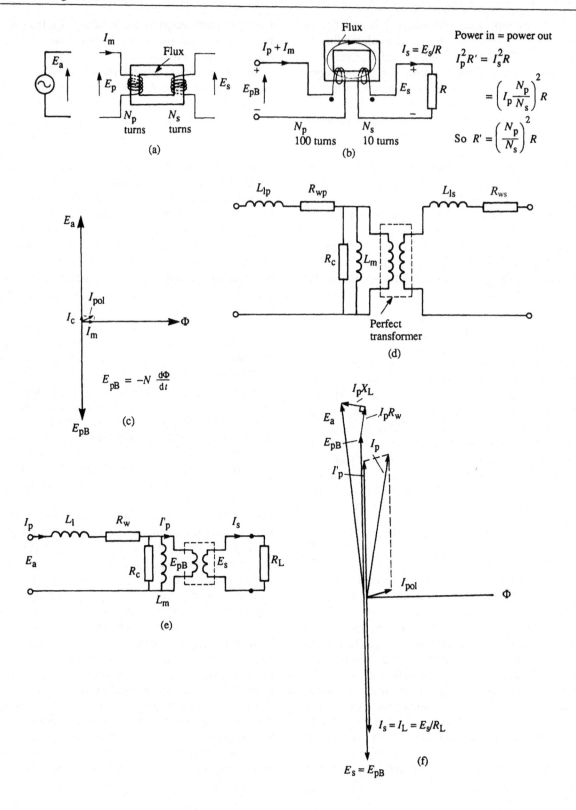

$E_{pB} = -N \dfrac{d\Phi}{dt}$

(c)

Power in = power out

$I_p^2 R' = I_s^2 R$

$= \left(I_p \dfrac{N_p}{N_s} \right)^2 R$

So $R' = \left(\dfrac{N_p}{N_s} \right)^2 R$

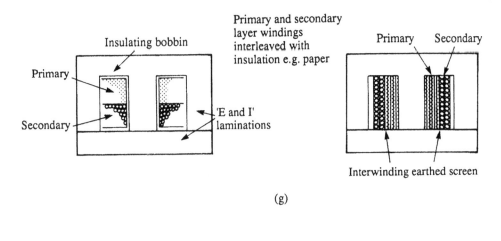

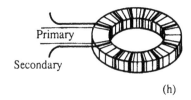

(h)

Figure 1.10 Transformers.

unit' resistance (the ratio of winding resistance to $E_{\text{rated}}/I_{\text{rated}}$) is the same for both windings. The transformer designer will of course know the approximate value of magnetizing inductance, since he will have chosen a suitable core and number of turns for the application, making allowance for tolerances on the core permeability, hysteresis and eddy current losses. The precise value of magnetizing inductance is then unimportant, but it can be measured if required on an inductance bridge or meter, with the secondary open-circuit. The total leakage inductance referred to the primary can be found by repeating the measurement with the secondary short-circuited.

In the case of a power transformer the answers will only be approximate, since the magnetizing inductance, core loss and leakage reactance vary somewhat with the peak flux level and hence with the applied voltage.

Small transformers with laminated three-limb cores as in Figure 1.10g, designed to run most if not all of the time at maximum rated load, were traditionally designed so that the core (hysteresis and eddy current) loss roughly equalled the full-load *copper loss* (winding resistance loss). The increasingly popular toroidal transformer (Figure 1.10h) exhibits a very low core loss, so that at full load the copper loss markedly predominates. In addition, the stray magnetic field is much lower than with three-limb cores and there is less tendency to emit annoying audible hum. Originally commanding a premium over conventional transformers on account of these desirable properties, toroidal transformers are now produced at such a volume and level of automation that there is little price differential. Both types are built down to a price, which means minimizing the core size and number of turns per volt, leading to a high peak flux density.

In small transformers of, say, 50 to 100 W rating, R_{w} (referred to the secondary) is often nearly one-tenth of the rated load resistance. So the full-load output voltage is only 90% of the off-load value – described as *10% regulation*. Taking account of core loss as well, the full-load efficiency of such transformers barely reaches 90%. For very small transformers with a rating of just a few watts, the regulation may be as poor as 30% and the efficiency less than 70%, whereas for a large mains distribution transformer the corresponding figures might be 2% and 98%.

Note that in Figure 1.10e, if the rated I_s flowed in a purely capacitive or inductive load instead of R_L, the losses in the transformer would be just as great. Therefore the rated secondary load for a transformer is always quoted in terms of the rated secondary volt-ampere product (VA) rather than in watts. Furthermore, the secondary current rating is strictly root mean square (RMS) or effective current. Thus with a non-linear load, e.g. a capacitor input rectifier circuit (see Chapter 10), the transformer must be derated appropriately to avoid overheating, since the RMS value of the current will be much greater than that of a sinusoidal current of the same mean value.

In some ways, power transformers are easy to design, at least in the sense that they are only required to work over a very limited range of frequencies, say 45–65 Hz or sometimes 45–440 Hz. *Signal transformers*, on the other hand, may be required to cover a 1000:1 bandwidth or more, say 20 Hz to 20 kHz or 1 to 1000 MHz. For these, special techniques and construction methods, such as sectionalized interleaved windings, may be used. By these means it is possible to produce a transformer covering 30 Hz to 2 MHz, almost a 100 000:1 bandwidth.

An interesting example of a signal transformer is the *current transformer*. This is designed with a primary which is connected in series with a heavy current circuit, and a secondary which feeds an AC milliammeter or ammeter with a replica (say 0.1%) of the primary current, for measurement purposes.

With the power transformers considered so far, designed for a specified rated primary voltage, the safe off-load condition is with the secondary open-circuit; the secondary load connected then defines the secondary current actually drawn. In the case of a current transformer, designed for a specified maximum primary current, where the current is determined by the external primary circuit and not by the load, we are in the topsy-turvy constant current world. The safe off-load condition is with the secondary short-circuited, corresponding to no voltage drop across the primary. The secondary is designed with enough inductance to support the maximum voltage drop across the measuring circuit, called the *burden*, with a magnetizing current which, referred to the primary, is less than 1% of the primary current. It is instructive to draw out the vector diagram for a current transformer on load, corresponding to the voltage transformer case of Figure 1.10e and f.

Questions

1. A 7.5 V source is connected across a 3.14 Ω resistor. How many joules are dissipated per second?

2. What type of wire is used for (a) high wattage wirewound resistors, (b) precision wirewound resistors?

3. A circuit design requires a resistance of value 509 Ω ± 2%, but only E12 value resistors, in 1%, 2% and 5% tolerance, are available. What value resistor, in parallel with a 560 Ω resistor, is needed to give the required value? Which of the three tolerance values are suitable?

4. What type of capacitor would usually be used where the required capacitance is (a) 1.5 pF, (b) 4700 μF?

5. A 1 μF capacitor is charged to 2.2 V and a 2.5 μF capacitor to 1.35 V. What is the stored energy in each? What is the stored energy after they have been connected in parallel? Explain the difference.

6. A 1 MΩ resistor is connected between the two terminals of the capacitors, each charged to the voltage in Question 5 above. How long before the voltages across each capacitor are the same to within (a) 10%, (b) 0.1%?

7. A black box with three terminals contains three star-connected capacitors, two of 15 pF, one of 30 pF. Another black box, identical to all measurements from the outside, contains three delta-connected capacitors. What are their values? (This can be done by mental arithmetic.)

8. Define the reluctance S of a magnetic circuit. Define the inductance of a coil with N turns, in terms of S.

9. The reactance of 2.5 cm of a particular piece of wire is $16\,\Omega$ at $100\,\text{MHz}$. What is its inductance?

10. An ideal transformer with ten times as many primary turns as secondary is connected to $240\,\text{V}$ AC mains. What primary current flows when (a) a $56\,\Omega$ resistor or (b) a $10\,\mu\text{F}$ capacitor is connected to the secondary?

Chapter

2   Passive circuits

Chapter 1 looked at passive components – resistors, capacitors and inductors – individually, on the theoretical side exploring their characteristics, and on the practical side noting some of their uses and limitations. It also showed how, whether we like it or not, resistance always turns up to some extent in capacitors, inductors and transformers. Now it is time to look at what goes on in circuits when resistors, capacitors and inductors are deliberately combined in various arrangements. The results will figure importantly in the following chapters.

CR and LR circuits

It was shown earlier that whilst the resistance of a resistor is (ideally, and to a large extent in practice) independent of frequency, the reactance of capacitors and inductors is not. So a resistor combined with a capacitor or an inductor can provide a network whose behaviour depends on frequency. This can be very useful when handling signals, for example music reproduction from a disc or gramophone record. On early 78 RPM records, Caruso, Dame Nellie Melba or whoever was invariably accompanied by an annoying high-frequency hiss (worse on a worn record), not to mention the clicks due to scratches. In the case of an acoustic gramophone, the hiss could be tamed somewhat by stuffing a cloth or small cushion up the horn (the origin of 'putting a sock in it', perhaps). But middle and bass response was also unfortunately muffled, as the attenuation was not very frequency selective. When electric pick-ups, amplifiers and loudspeakers replaced sound boxes and horns, an adjustable tone control or 'scratch filter' could be provided, enabling the listener selectively to reduce the high-frequency response, and with it the hiss, pops and clicks.

Figure 2.1a shows such a *top-cut* or treble-cut circuit, which, for the sake of simplicity, will be driven from a source of zero output impedance (i.e. a constant voltage source) and to fed into a load of infinitely high-input impedance, as indicated. What is v_o, the signal voltage passed to the load at any given frequency, for a given alternating input voltage v_i? Since both resistors and capacitors are linear components, i.e. the alternating current flowing through them is proportional to the applied alternating voltage, the ratio v_o/v_i is independent of v_i; it is a constant at any given frequency. This ratio is known as the *transfer function* of the network. At any given frequency it clearly has the same value as the output voltage v_o obtained for an input voltage v_i of unity. One can tell by inspection what v_o will be at zero and infinite frequency, since the magnitude of the capacitor's reactance $X_c = 1/2\pi fC$ will be infinite and zero respectively. So at these frequencies, the circuit is equivalently as shown in Figure 2.1b, offering no attenuation at 0 Hz and total attenuation at ∞ Hz (infinite frequency) respectively. Since the same current i flows through both the resistor and the capacitor, $v_o = iX_c$ and $v_i = i(R + X_c)$. A term such as $R + X_c$ containing both resistance and reactance is called an *impedance Z*. Recalling that $X_c = 1/j\omega C$, then

$$\frac{v_o}{v_i} = \frac{1/j\omega C}{R + (1/j\omega C)} = \frac{1}{1 + j\omega CR}$$

Thus v_o/v_i *is a function of* $j\omega$, i.e. the value of v_o/v_i depends upon $j\omega$. The shorthand for function of $j\omega$ is $F(j\omega)$; so, in the case of the circuit of Figure 2.1a, $F(j\omega) = 1/(1 + j\omega CR)$. Figure 2.1c shows what is going on for the particular case where $1/\omega C = R$, i.e. at the frequency $f = 1/2\pi CR$. Since the same current i flows through both R and C, it is a convenient starting point for the vector diagram.

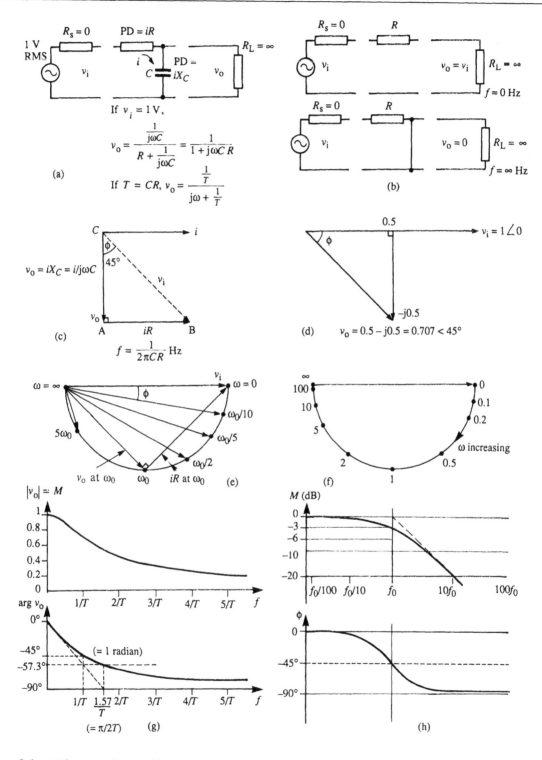

Figure 2.1   *CR* low-pass (top-cut) lag circuit.

Next you can mark in the potential drop iX_c across the capacitor. The PD iR is added to this as shown, giving v_i. Note that $iX_c = i(1/j\omega C) = -ji(1/\omega C)$. Recalling that j indicates a 90° anticlockwise rotation, $-ji(1/\omega C)$ will simply be rotated in the opposite direction to $ji(1/\omega C)$, i.e. downwards. The PD iR, on the other hand, is in phase with i, so the two PDs must be added vectorially as shown to obtain v_i. CAB is a right-angled triangle, so by Pythagoras's theorem the magnitude of v_i is $i\sqrt{(X_c^2 + R^2)}$. So for the case illustrated, where the reactance of the capacitor in ohms is numerically equal to R,

$$\frac{v_o}{v_i} = \frac{iX_c}{i\sqrt{(X_c^2 + R^2)}} = \frac{1}{\sqrt{[(X_c/X_c)^2 + (R/X_c)^2]}}$$

$$= \frac{1}{\sqrt{(1+1)}} = 0.707$$

with the phase angle shown.

You can get the same result by algebra from the transfer function rather than by geometry from the vector diagram. Starting with $F(j\omega) = 1/(1 + j\omega CR)$, the trick is to get rid of the awkward term in the denominator by multiplying top and bottom by the 'complex conjugate' of $1 + j\omega CR$. The complex conjugate of $P + jQ$ is simply $P - jQ$, or in this case $1 - j\omega CR$. So

$$F(j\omega) = \frac{1(1 - j\omega CR)}{(1 + j\omega CR)(1 - j\omega CR)} = \frac{1 - j\omega CR}{1 + \omega^2 C^2 R^2}$$

remembering that $j^2 = -1$. When $1/\omega C = R$

$$F(j\omega) = \frac{1 - j1}{1 + 1} = \tfrac{1}{2} - j\tfrac{1}{2}$$

This expresses the output voltage for unity reference input voltage, in *cartesian* or $x + jy$ form (Figure 2.1d). The terms x and y are called the in-phase and quadrature – or sometimes (misleadingly) the real and imaginary – parts of the answer, which can alternatively be expressed in *polar* coordinates. These express the same thing but in terms of the magnitude M and the phase angle Φ (or modulus and argument) of v_o relative to v_i, written $M\angle\Phi$. You can see from Figure 2.1d that the magnitude of $v_o = \sqrt{(0.5^2 + 0.5^2)} = 0.707$ (Pythagoras again) and that the phase angle of v_o relative to v_i is $-45°$ or -0.785 radians. So at

$f = 1/2\pi CR$, $F(j\omega)$ evaluates to $0.707\angle-45°$, the same answer obtained by the vector diagram. But note that, unlike the voltage vectors CA and AB in Figure 2.1c, you will not find voltages of 0.5 and $-j0.5$ at any point in the circuit (Figure 2.1a). As is often the case, the geometric (vector) solution ties up more directly with reality than the algebraic. In general, a quantity $x + jy$ in cartesian form can be converted to the magnitude and phase angle polar form $M\angle\Phi$ as follows:

$$M = \sqrt{(x^2 + y^2)} \quad \phi = \tan^{-1}(y/x), \text{ i.e. } \tan\phi = y/x$$

To convert back again,

$$x = M\cos\phi \qquad y = M\sin\phi$$

Thus in Figure 2.1d, $v_o = 0.707\cos(-45°) + 0.707\sin(-45°)$.

If the top-cut or *low-pass* circuit of Figure 2.1a were connected to another similar circuit via a *buffer amplifier* – one with infinite input impedance, zero output impedance and a gain of unity at all frequencies – the input to the second circuit would simply be the output of the first. The transfer function of the second circuit being identical to that of the first, its output would also be $0.707\angle-45°$ relative to its input at $f = 1/2\pi CR$, which is itself $0.707\angle-45°$ for an input of $1\angle0°$ to the first circuit. So the output of the second circuit would lag the input of the first by 90° and its magnitude would be $0.707 \times 0.707 = 0.5$ V. In general, when two circuits with transfer functions $F_1(j\omega)$ and $F_2(j\omega)$ are connected in *cascade* – the output of the first driving the input of the second (assuming no interaction) – the combined transfer function $F_c(j\omega)$ is given by $F_c(j\omega) = F_1(j\omega)F_2(j\omega)$.

At any frequency where $F_1(j\omega)$ and $F_2(j\omega)$ have the values $M_1\angle\Phi_1$ and $M_2\angle\Phi_2$ say, $F_c(j\omega)$ simply has the value $M_1M_2\angle(\Phi_1 + \Phi_2)$. This result is much more convenient when multiplying two complex numbers than the corresponding cartesian form, where $(a + jb)(c + jd) = (ac - bd) + j(ad + bc)$. On the other hand, the cartesian form is much more convenient than the polar when adding complex numbers. Returning to Figure 2.1c, it should be remembered that this is shown for the particular case of that frequency at which the reactance of the capacitor equals R ohms; call this frequency f_o and let $2\pi f_o = \omega_o$. In Figure 2.1e,

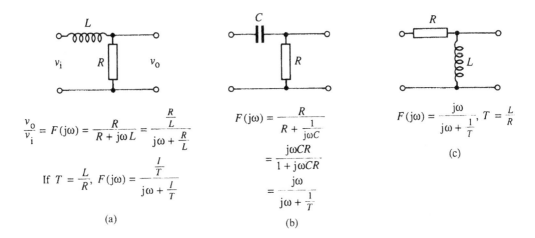

$$\frac{v_o}{v_i} = F(j\omega) = \frac{R}{R + j\omega L} = \frac{\dfrac{R}{L}}{j\omega + \dfrac{R}{L}}$$

$$\text{If } T = \frac{L}{R}, \ F(j\omega) = \frac{\dfrac{I}{T}}{j\omega + \dfrac{I}{T}}$$

(a)

$$F(j\omega) = \frac{R}{R + \dfrac{1}{j\omega C}}$$

$$= \frac{j\omega CR}{1 + j\omega CR}$$

$$= \frac{j\omega}{j\omega + \dfrac{1}{T}}$$

(b)

$$F(j\omega) = \frac{j\omega}{j\omega + \dfrac{1}{T}}, \ T = \frac{L}{R}$$

(c)

Figure 2.2 *CR* and *LR* circuits.
(a) *LR* low-pass circuit.
(b) *CR* high-pass (bass-cut) lead circuit. We can normalize the frequency f to the break, corner or 3 dB frequency f_0 where $f_0 = 1/2\pi CR$ (i.e. where $\omega_0 = 1/CR = 1/T$), by using $\omega_n = \omega/\omega_0$ instead of ω. Then when $\omega = 1/T = \omega_0$, the normalized radian frequency $\omega_n = 1$. $F(j\omega)$ then simplifies to

$$F(j\omega_n) = \frac{j\omega/\omega_0}{(j\omega/\omega_0) + (\omega_0/\omega_0)} = \frac{j\omega_n}{j\omega_n + 1}$$

or, more generally, normalized $F(s) = s/(s + 1)$.
(c) *LR* high-pass circuit. Again, if ω is normalized as above, $F(s) = s/(s + 1)$.

v_o (which also represents the transfer function if $vi = 1\underline{/0°}$) is shown for various values of ω from zero to infinity. Since the vectors iX_c and iR in Figure 2.1c are always at right angles whatever the frequency, it follows that the locus or line joining the tips of the v_o vectors for all frequencies is a semicircle, due to a theorem worked out a long time ago by a gentleman called Euclid. Figure 2.1e is a simple example of a *circle diagram* – a very useful way of looking at a circuit, as will appear in later chapters. It is even more useful if you *normalize* ω, the angular frequency. vi has already been normalized to unity, i.e. $1\underline{/0°}$ or $1\underline{/0}$ radians, making the transfer function evaluate directly to v_o at any frequency. In particular, for the example of Figure 2.1a, $F(j\omega) = (1/\sqrt{2})\underline{/-45°} = 0.707\underline{/-0.785}$ radians when $1/2\pi fC = R$, i.e. when $\omega = 1/CR$. It is useful to give the particular value of ω where $\omega = 1/CR$ the title ω_0. Dividing the values of ω in Figure 2.1e by ω_0, they simply become 0, 0.1, 0.2, 0.5, 1, 10 etc. up to ∞, as in

Figure 2.1f. As ω increases from 0 to infinity, Φ decreases from 0° to $-90°$ whilst v_o decreases from unity to zero. You now have a universal picture which applies to any low-pass circuit like Figure 2.1a. Simply multiply the ω values in Figure 2.1f by $1/2\pi CR$ to get its actual frequency response.

One can alternatively plot these changes in $|v_o|$ (signifying the modulus of v_o, i.e. the M part of $M\underline{/\Phi}$) and arg v_o (signifying the angle Φ part of M) against a linear base of frequency, as in Figure 2.1g. However, for most purposes it is better to plot v_o against a logarithmic baseline of frequency, as this enables you to see more clearly what is happening at frequencies many octaves above and below f_0, the frequency where $2\pi f_0 = \omega_0 = 1/CR = 1/T$, the frequency where v_o/v_i turned out to be $0.707\underline{/-45°}$. (Note that f_0 depends only on the product $CR = T$, not on either C or R separately.) At the same time, it is convenient to plot the magnitude on a logarithmic scale of *decibels* (unit symbol dB*). This again compresses the extremes of the range, enabling one to see very large and very small values clearly. So rather than plotting $|v_o/v_i|$ one can plot $20 \log_{10}|v_o/v_i|$ instead. This is called a *Bode plot,* after the American author H. W. Bode[1], and is shown in Figure 2.lh. Note that multiplying two numbers is equivalent to adding their logarithms. So if the value at a particular frequency of a transfer function $M\angle\Phi$ is expressed as $M_d\angle\Phi$, where M_d is the ratio M expressed in decibels, then the product $M_{d1}\angle\Phi_1 \times M_{d2}\angle\Phi_2$ is simply expressed as $(M_{d1} + M_{d2})\angle(\Phi_1 + \Phi_2)$.

At very low frequencies, the reactance of C is very high compared with R, so i is small. Consequently there is very little PD across R and the output is virtually equal to v_i, i.e. independent of frequency or – in the jargon – 'flat' (see Figure 2.1h). At very high frequencies, the reactance of C is very low compared with R; thus the current is virtually determined solely by R. So each time the frequency is doubled, X_c halves and so does the PD across it. Now $20 \log_{10} 0.5$ comes to -6 (almost exactly), so the output is said to be falling by 6 dB per octave as the frequency increases. Exactly at f_0, the response is falling by 3 dB per octave and the phase shift is then $-45°$ as shown. The slope increases to -6 dB/octave and falls to 0 dB/octave as we move further above and below f_0 respectively, the phase shift tending to $-90°$ and $0°$ likewise.

The *LR* top-cut circuit of Figure 2.2a gives exactly the same frequency response as the *CR* top-cut circuit of Figure 2.1a, i.e. it has the same transfer function. However, its input impedance behaves quite differently, rising from a pure re- sistance R at 0 Hz to an open-circuit at infinite frequency. By contrast, the input impedance of Figure 2.1a falls from a capacitive open-circuit to R as we move from 0 Hz to ∞ Hz. Similarly, the source impedances seen by the load, looking back into the output terminals of Figures 2.1a and 2.2a, also differ.

Figure 2.2b and c show *bass-cut* or *high-pass* circuits, with their response. Here the response rises (with increasing frequency) at low frequencies, at $+6$ dB per octave, becoming flat at high frequencies. Clearly, with a circuit containing only one resistance and one reactance driven from a constant voltage source and feeding into an open-circuit load, v_o can never exceed vi so the two responses shown exhaust the possibilities. However, if we consider other cases where the source is a constant current generator or the load is a *current sink* (i.e. a short-circuit), or both, we find arrangements where the output rises indefi- nitely as the frequency rises (or falls). These cases are all shown in Figure 2.3[2].

Time domain and frequency domain analysis

There is yet another, more recent representation of circuit behaviour, which has been deservedly pop- ular for nearly half a century. By way of introduc- tion recall that, for a resistor, at any instant the current is uniquely defined by the applied voltage. However, when examining the behaviour of capa- citors and inductors, it turned out to be necessary to take account not only of the voltage and current, but also of their rate of change. Thus the analysis involves currents and voltages which vary in some particular manner with the passage of time. In many cases the variation can be described by a mathematical formula, and the voltage is then said to be a determinate function of time. The formula enables us to predict the value of the voltage at any time in the future, given its present value. Some varying voltages are indeterminate, i.e. they cannot be so described, an example is the hiss-like signal from a radio receiver with the aerial disconnected. In this section, interest focuses on determinate functions of time, some of which have appeared in the analysis already. One example was the exponential function, where v_t (the voltage at

---

* Decibels, denoted by 'dB', indicate the ratio of the power at two points, e.g. the input and output on an amplifier. For example, if an amplifier has a power gain of one hundred times (100 mW output for 1 mW input, say), then its gain is 20 dB (or 2 Bels or $\times 10^2$). In general, if the power at point B is 10^N times that at point A, the power at B is $+10N$ dB with respect to A. *If the impedances at A and B are the same,* then a power gain of $\times 100$ or $+20$ dB corresponds to a voltage gain of $\times 10$ (since $W = E^2/R$). In practice, voltage ratios are often referred to in dB ($dB = 20 \times \log_{10}(v_1/v_2)$), even when the impedance levels at two points are not the same.

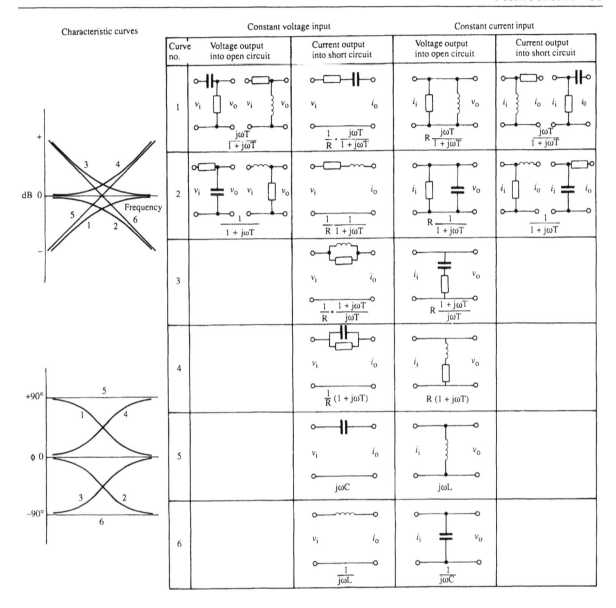

Figure 2.3 All combinations of one resistance and one reactance, and of one reactance only, and their frequency characteristics (magnitude and phase) and transfer functions (reproduced by courtesy of *Electronics and Wireless World*.

any instant *t*) is given by $v_t = v_0 \, e^{\alpha t}$, describing a voltage which increases indefinitely or dies away to zero (according to whether αt is positive or negative) from its initial value of v_0 at the instant $t = 0$. Another example of a determinate function of

time, already mentioned, is the sinusoidal function, e.g. the output voltage of an AC generator. Here $v_t = V \sin(\omega t)$, where the radian frequency ω equals 2π times the frequency f in hertz, and V is the value of the voltage at its positive peak.

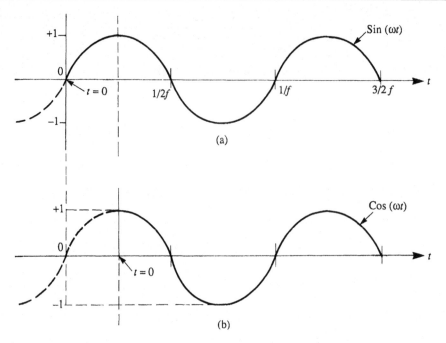

Figure 2.4   Initial transients.
(a) Sine wave connected to a circuit at t_0.
(b) Cosine wave connected to a circuit at t_0.

Considering a voltage or a circuit response specifically as a function of time is described as *time domain analysis*.

The sinusoidal waveform is particularly important owing to its unique properties. Already mentioned is the fact that its rate of change is described by the cosine wave, which has exactly the same shape but is advanced in time by a quarter of a cycle or 90° (see Figure 1.7). Now it turns out that all repetitive waveforms can be analysed into the sum of a number of sine and cosine waves of related frequencies, so it is exceedingly useful to know how a circuit responds to sine wave inputs of different frequencies. You can find out by connecting a sine wave obtained from a signal generator, for example, to the circuit's input and seeing what comes out of the output, at various frequencies. In the real world, the waveform will be connected to the circuit under test at some specific instant, say when its value is zero and rising towards its positive peak. The applied voltage is thus a sine wave as in Figure 2.4a, whereas had the connection been made at a t_0 which was $T/4$ seconds later,

where $T = 1/f$, the applied voltage (defined by reference to its phase at $t = t_0$) would be a cosine wave (Figure 2.4b). In the first case we have an input voltage exhibiting a change of slope but no change of value at t_0, whereas in the second we have an abrupt change of value of the voltage at t_0 itself, but the slope or rate of change of voltage is zero both just before and just after t_0. It is not surprising that the response of the circuit in these two cases differs, at least in the short term. However, the fine detail of the initial conditions when the signal was first applied become less important with the passage of time, and after a sufficiently long period become completely irrelevant. The response of the circuit is then said to be in the *steady state*. The difference between this and the initial response is called the *transient,* and its form will depend upon the initial conditions at t_0. For many purposes it is sufficient to know the steady state response of a circuit over a range of frequencies, i.e. the nature of the circuit's response as a function of frequency. This is known as *frequency domain analysis,* since the independent variable used to describe the circuit function is frequency rather than time. The method is based upon the Laplace transform, an operational method that in the 1950s gradually replaced the operational calculus introduced by Oliver Heaviside many years earlier[3]. The transform method eases the solution of integral/differential equations by substituting algebraic manipulation in the frequency domain for the classical methods of solution in the time domain. It can provide the full solution for any given input signal, i.e. the transient as well as the steady state response, though for reasons of space only the latter will be dealt with here.

In the frequency domain the independent variable is $\omega = 2\pi f$, with units of radians per second. The transfer function, expressed as a function of frequency $F(j\omega)$ has already been mentioned. You will recall that j, the square root of -1, was originally introduced to indicate the 90° rotation of the vector representing the voltage drop across a reactive component, relative to the current through it. However, j also possesses another significance. You may recall (in connection with Figure 1.7) that on seeing that the differential (the rate of change) of a sine wave was another waveform of exactly the same shape, it seemed likely that the sinusoidal function was somehow connected with the exponential function. Well, it turns out that

$$e^{j\theta} = \cos \theta + j \sin \theta$$
$$e^{-j\theta} = \cos \theta - j \sin \theta \qquad (2.1)$$

This is known as *Euler's identity*. So sinusoidal voltage waveforms like $V \sin \omega t$ can be represented in exponential form since using (2.1) you can write

$$\sin \omega t = \frac{e^{j\omega t} - e^{-j\omega t}}{2j} \quad \text{and} \quad \cos \omega t = \frac{e^{j\omega t} + e^{-j\omega t}}{2}$$

One can also allow for sine waves of increasing or decreasing amplitude by multiplying by an exponential term, say $e^{\sigma t}$, where σ is the lower-case Greek letter sigma. As noted earlier, if σ is positive then $e^{\sigma t}$ increases indefinitely, whilst if σ is negative then the term dies away to zero. So $e^{\sigma t}e^{j\omega t}$ represents a sinusoidal function which is increasing, decreasing – or staying the same amplitude if

σ equals zero. The law of indices states that to multiply together two powers of the same number it is only necessary to add the indices: $4 \times 8 = 2^2 \times 2^3 = 2^5 = 32$, for example. Similarly, $e^{\sigma t} e^{j\omega t} = e^{(\sigma+j\omega)t}$, thus expressing compactly in a single term the frequency and rate of growth (or decline) of a sinusoid. It is usual to use s as shorthand for $\sigma + j\omega$; s is called the complex frequency variable. A familiarity with the value of $F(s)$, plotted graphically for all values of σ and ω, provides a very useful insight into the behaviour of circuits, especially of those embodying a number of different CR and/or LR time constants. The behaviour of such circuits is often more difficult to envisage by other methods.

Frequency analysis: pole–zero diagrams

To start with a simple example, for the circuit of Figure 2.1a it was found that $F(j\omega) = 1/(1 + j\omega t)$, giving a response of $0.707\angle -45°$ at $\omega_0 = 1/T = 1/CR$. Taking the more general case using $\sigma + j\omega$,

$$F(s) = \frac{1}{1 + sT} = \frac{1/T}{s + (1/T)}$$

Figure 2.5a shows a pair of axes, the vertical one labelled $j\omega$, the horizontal one σ. Plotting the point $\sigma = -1/T$ (marked with a cross) and drawing a line joining it to the point $\omega = 1/T$ on the vertical axis, gives you a triangle. This is labelled CAB to show that it is the same as the triangle in Figure 2.1c, where $v_o/v_i = CA/CB$. As ω increases from zero to infinity, the reactance of the capacitor will fall from infinity to zero; so in both diagrams the angle BCA will increase from zero to 90°. So in Figure 2.5a, Φ represents the angle by which v_o lags v_i, reaching $-90°$ as ω approaches ∞. Likewise, since $v_o/v_i = CA/CB$, the magnitude of the transfer function is proportional to $1/CB$. Expressed in polar $(M\angle\Phi)$ form, the transfer function has evaluated to $(CA/CB)\angle\tan^{-1}(AB/AC)$, as you move up the $j\omega$ axis above the origin, where $\sigma = $ zero. In fact, if you plot the magnitude of $F(s)$, for ω from 0 to ∞ (with $\sigma = 0$), on a third axis at right angles to the σ and $j\omega$ axes, (Figure 2.5b), you simply get back to the magnitude plot of Figure 2.1g. This may sound a

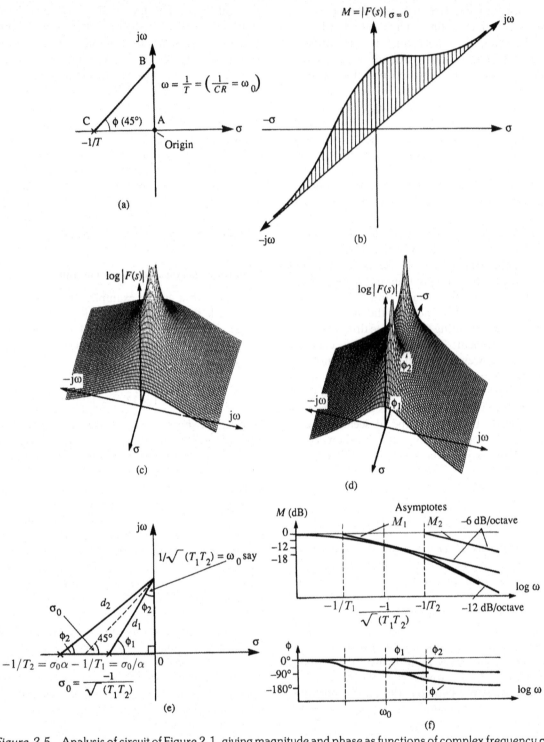

Figure 2.5   Analysis of circuit of Figure 2.1, giving magnitude and phase as functions of complex frequency $\sigma + j\omega$. ((c) and (d) reproduced by courtesy of Maxim Integrated Products UK Ltd).

complicated way to get there, but the light at the end of the tunnel is worth waiting for, so stick with it. See what happens if you plot the value of $F(s)$ as in Figure 2.5b for values of s where σ is not zero. Remember, you are plotting the value of

$$F(s) = \frac{1/T}{(\alpha + j\omega) + 1/T} = \frac{\omega_o}{(\sigma + j\omega) + \omega_o}$$

not the ratio CA/CB. (The vector diagram of Figure 2.1c only ties up with the $\sigma + j\omega$ plot for the particular case where $\sigma = 0$, i.e. for a sine wave of constant amplitude.) Consider first the case where $\sigma = 1/T$. For very large values of ω this really makes very little difference, but, at $\omega = 0$,

$$F(s) = \frac{1/T}{(1/T + j0) + 1/T} = 0.5\underline{/0^\circ}$$

Conversely, for small negative values of σ, $F(s)$ (at $\omega = 0$) is greater than unity: and, as σ reaches $-1/T$,

$$F(s) = \frac{1/T}{(-1/T + j0) + 1/T}$$

i.e. it explodes to infinity. This is shown in three-dimensional representation in Figure 2.5c. But, you may object, for values of σ more negative than $-1/T$ the picture shows $F(s)$ falling again but still positive; whereas at $\sigma = -2/T$,

$$F(s) = \frac{1/T}{(-2/T + j0) + 1/T} = -1$$

Remember, however, that $F(s)$ is a vector quantity with a magnitude M (always positive) and a phase Φ. The minus sign indicates that the phase has switched suddenly to -180°. To make this clearer, consider the value of $F(s)$ as σ becomes progressively more negative, for a value of ω constant at $0.1(1/T)$ instead of zero. The value of

$$F(s) = \frac{1/T}{(\sigma + j0.1/T) + 1/T}$$

increases, reaching a maximum value of $(1/T)/(j0.1/T) = 10\underline{/-90^\circ}$ when $\sigma = -1/T$, where you surmount the north face of the infinitely high $F(s)$ mountain. Descending the western slope, the amplitude M falls as Φ increases to -180°. Clearly the smaller the constant positive value of ω during your westward journey the higher the slope

you surmount, and the more rapidly the phase twizzles round from 0° to -180° in the vicinity of the peak. Keep this picture in mind, as you will meet it again later.

The infinitely high mountain is called a *pole* and occurs, in the low-pass case of $F(s) = (1/T)/[s + (1/T)]$ at $s = -1/T + j0$. This is the complex frequency which is the solution of: denominator of $F(s) = 0$. If there were two low-pass circuits like Figure 2.1a in cascade (but assuming a buffer amplifier to prevent any interaction between them) with different critical frequencies f_1 and f_2, then you would have

$$F(s) = \frac{1/T_1}{s + (1/T_1)} \times \frac{1/T_2}{s + (1/T_2)}$$

$$= \frac{1/T_1 T_2}{[(s + (1/T_1)][(s + (1/T_2)]}$$

$$= \frac{1/T_1 T_2}{s^2 + s[(1/T_1) + (1/T_2)] + (1/T_1 T_2)}$$

and the s-plane plot – which is known as a *pole–zero diagram* – would have two poles on the $-\sigma$ axis, at distances $1/T_1$ and $1/T_2$, to the left of the origin. At any frequency corresponding to ω on the vertical $j\omega$ axis, the response $M\underline{/\Phi} = F(j\omega)$ would be $(1/d_1 d_2)\underline{/(\Phi_1 + \Phi_2)}$, as shown in Figure 2.5d and e. This is called a *second-order* system as $F(s)$ has a term in s^2 in the denominator. As with the first-order system of Figure 2.1, the overall response falls towards zero (or $-\infty$ dB) as ω tends to infinity, but at 12 dB per octave, as shown in Figure 2.5f. This also shows the phase and amplitude responses of the two terms separately, and what is not too obvious from the Bode plot is that at a frequency $\omega = 1/\sqrt{(T_1 T_2)}$ the output lags 90° relative to the input. This can, however, be deduced from the pole–zero diagram of Figure 2.5e using secondary- or high-school geometry. Call $1/\sqrt{(T_1 T_2)}$ by the name ω_0 for short. Then consider the two right-angled triangles outlined in bold lines in Figure 2.5e. The right angle at the origin is common to both. Also $(1/T_1)/\omega_0 = \omega_0(1/T_2)$; therefore the two triangles are similar. Therefore the two angles marked Φ_2 are indeed equal. Therefore $\Phi_1 + \Phi_2 = 90^\circ$ measured anti-clockwise round the respective poles, making v_o

lag v_i by 90° at ω_0, mostly owing to the lower-frequency lag circuit represented by T_1. The point $\sigma_0 = -1/\sqrt{(T_1 T_2)}$, to which the poles are related through the parameter α by $-1/T_2 = \alpha \sigma_0$ and $-1/T_1 = \sigma_0/\alpha$, is therefore significant. If $\alpha = 1$ the two poles are coincident, as also are the two $-6\,dB$/octave asymptotes and the two 0° to $-90°$ phase curves in Figure 2.5f. As α increases, so that $-1/T_1$ moves towards zero and $-1/T_2$ moves towards infinity, the corresponding asymptotes in Figure 2.5f move further and further apart on their log scale of ω. There is then an extensive region either side of ω_0 where the phase shift dwells at around $-90°$ and the gain falls at $6\,dB$/octave. However, even if $\alpha = 1$ so that $-1/T_1 = -1/T_2$, you can never get a very sharp transition from the flat region at low frequencies to the $-12\,dB$/octave regime at high frequencies. Setting α less than 1 just does not help, it simply interchanges $-1/T_1$ and $-1/T_2$. Although one cannot achieve a sharp transition with cascaded passive CR (or LR) low-pass circuits, it is possible with circuits using also transistors or operational amplifiers (as will appear in later chapters) and also with circuits containing R, L and C.

I have been talking about pole–zero diagrams, and you have indeed seen poles (marked with crosses), though only on the $-\sigma$ axis so far; but where are the *zeros*? They are there all right but off the edge of the paper. In Figure 2.5a there is clearly a zero of $F(s)$ *at* $\omega = \infty$ since

$$F(s)_{\substack{\omega=\infty \\ \sigma=0}} = \frac{\omega_0}{(0 + j\infty) + \omega_0} = 0$$

Incidentally, $F(s)$ also becomes zero if you head infinitely far south down the $-j\omega$ axis, or for that matter east or west as σ tends to $+\infty$ or $-\infty$. Indeed the same thing results heading northeast, letting both ω and σ go to $+\infty$, or in any other direction. So the zero of $F(s)$ completely surrounds the diagram, off the edge of the map at infinity in any direction. This 'infinite zero' – if you don't find the term too confusing – can alternatively be considered as a single point, if you imagine the origin in Figure 2.5a to be on the equator, with the $j\omega$ axis at longitude 0°. Then, if you head off the $F(s)$ map in any direction, you will always arrive eventually at the same point,

round the back of the world, where the international dateline crosses the equator.

By contrast, the high-pass lead circuit of Figure 2.2b has a pole–zero diagram (not shown) with a finite zero, located at the origin, as well as a pole. The phase contribution to $F(j\omega)$ of a zero on the $-\sigma$ axis is the opposite to that of a pole, starting at zero and going to $+90°$ as you travel north from the origin along the $j\omega$ axis to infinity. However, in this case the phase is stuck at $+90°$ from the outset, as the zero is actually right at the origin. The zero is due to the s term in the *numerator* of $F(s)$, so the magnitude contribution of this term at any frequency ω is *directly* (not inversely) proportional to the distance from the point ω to the zero – again the opposite to what you find with a pole.

Note that the low-pass circuit of Figure 2.1 and the high-pass circuits of Figure 2.2b and c each have both a pole and a zero. A crucial point always to be borne in mind is that however simple or complicated $F(s)$ and the corresponding pole–zero diagram, the number of poles must always equal the number of zeros. If you can see more poles than zeros, there must be one or more zeros at infinity, and vice versa. This follows from the fact that $F(j\omega) = v_o/v_i = M\underline{/\Phi}$. Now Φ has units of radians, and a radian is simply the ratio of two lengths. Likewise, M is just a pure dimensionless ratio. So if the highest power of s in the denominator of $F(s)$ is s^3 (a *third-order* circuit with three poles), then implicitly the numerator must have terms up to s^3, even if their coefficients are zero: i.e. $F(s) = 1/(as^3 + bs^2 + cs + d)$ is really $F(s) = (As^3 + Bs^2 + Cs + D)/(as^3 + bs^2 + cs + d)$, where $A = B = C = 0$ and $D = 1$. So poetic justice and the theory of dimensions are satisfied and, just as Adam had Eve, so every pole has its zero. As a further example, now look at another first-order circuit, often called a *transitional lag* circuit. This has a finite pole and a finite zero both on the $-\sigma$ axis, with the pole nearer the origin. The transitional lag circuit enables us to get rid of (say) 20 dB of loop gain, in a feedback amplifier (see Chapter 3) without the phase shift ever reaching 90° and with negligible phase shift at very high frequencies. Figure 2.6a shows the circuit while Figure 2.6b to f illustrate the response from several different points of view. Note that K is the value of the transfer function at infinitely high frequencies, where the reactance of C is zero, as can be seen by replacing C by a short-circuit. When $s = 0, F(s)$ gives the steady state transfer function at zero frequency, where the reactance of C is infinite, so

$$F(s) = K\frac{1/T_2}{1/T_1} = \frac{T_2}{T_1}\frac{1/T_2}{1/T_1} = 1$$

as can be seen by replacing C by an open-circuit. You may recall meeting $-1\sqrt{(T_1T_2)}$ in Figure 2.5, and it turns out to be significant again here. At $\omega = \sqrt{(\omega_{01}\omega_{02})}, \Phi_p + \Phi_z = 90°$ but the actual phase shift for this transitional lag circuit is only $\Phi_p - \Phi_z$.

In Figure 2.2b, by normalizing the frequency, one finished up with the delightfully simple form $F(s) = s/(s + 1)$ – not a time constant in sight. However, this is only convenient for a simple first-order high-pass circuit, or a higher-order one where all the corner frequencies are identical. With two or more different time constants, it is best not to try normalizing, though in Figure 2.6 you could normalize by setting $\sqrt{(\omega_{01}\omega_{02}} = 1$ if you felt so inclined. Then T_1 becomes α (say) and $T_2 = 1/\alpha$.

Resonant circuits

Figure 2.7 shows an important example of a two-pole (second-order) circuit. At some frequency the circuit will be *resonant*, i.e. $|j\omega L| = |1/j\omega C|$. At this frequency, the PD across the inductor will be equal in magnitude and opposite in sign to the PD across the capacitor, so that the net PD across L and C together (but not across each separately) will be zero. The current i will then exhibit a maximum of $i = vi/R$. At resonance, then, $i[j\omega L + (1/j\omega C)] = 0$, so $\omega^2 LC = 1$ and $\omega = 1/\sqrt{(LC)}$. Give this value of ω the label ω_0. At ω_0 the output voltage $v_0 = iX_c = (v_i/R)(1/j\omega C)$, so

$$\frac{v_0}{v_i} = \frac{-j}{\omega_0 CR} = -j\frac{1}{R}\sqrt{\left(\frac{L}{C}\right)} = -j\frac{X_{co}}{R}$$

where X_{co} is the reactance of the capacitor at ω_0. Clearly, for given values of L and C, as R becomes very small the output voltage at resonance will be

much larger than the input voltage. Consequently, $(1/R)\sqrt{(L/C)}$ is often called the *magnification factor* or *quality factor* Q of the circuit, and can alternatively, be written as $Q = \omega_0 L/R = X_{L0}/R = X_{co}/R$. If, on the other hand, R is much larger than $\omega_0 L$ and $1/\omega_0 C$, the output will start to fall at 6 dB per octave when $|1/\omega C| = R$ and at 12 dB per octave at some higher frequency where $|\omega L| = R$. These results can be derived more formally, referring to Figure 2.7a, as follows:

$$F(j\omega) = \frac{v_0}{v_i} = \frac{iX_c}{i(X_L + R + X_c)}$$

$$= \frac{1/j\omega C}{j\omega L + R + 1/j\omega C} = \frac{1}{(j\omega)^2 LC + j\omega CR + 1}$$

or more generally,

$$F(s) = \frac{1}{LC^2 + CRs + 1} \tag{2.2}$$

This equation can be factorized into $1/[(s + a)(s + b)]$, where a and b are the roots of the equation

$$LCs^2 + CRs + 1 = 0 \tag{2.3}$$

When $s = -a$ or $-b$, the denominator of (2.2) equals zero, so $F(s)$ will be infinite, i.e. $-a$ and $-b$ are the positions of the poles on the pole–zero diagram. As any algebra textbook will confirm, the two roots of $ax^2 + bx + c = 0$ are given by $x = \{-b \pm \sqrt{(b^2 - 4ac)}\}/2a$, so applying this formula to (2.3) we get

$$s = \frac{-CR \pm \sqrt{(C^2R^2 - 4LC)}}{2LC} \tag{2.4}$$

Now algebra is an indispensable tool in circuit analysis, but a result like (2.4) by itself doesn't give one much feel for what is actually going on. So consider a normalized case where $L = 1$ H and $C = 1$ F, so that $\omega_0 = 1/\sqrt{(1 \times 1)} = 1$ radian per second, and consider first the case where $R = 10$ ohms. From (2.4), $s = \{-10 \pm \sqrt{(100 - 4)}\}/2 = -5 \pm 4.9$. Now we can draw the poles in as in Figure 2.7b, which also shows the corresponding 45° break frequencies on the $j\omega$ axis ω_1 and ω_2, at which each pole contributes 45° phase lag and 3 dB of attenuation. As there are no terms in s in the numerator of (2.2), no zeros are

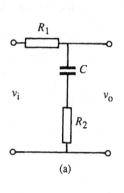

(a)

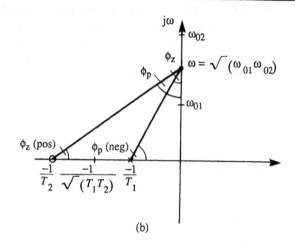

(b)

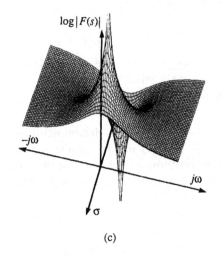

(c)

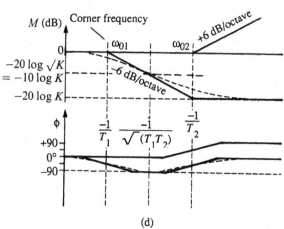

(d)

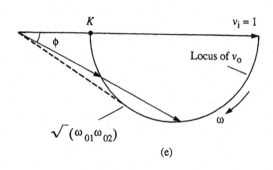

(e)

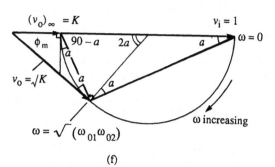

(f)

visible on the diagram: they are there, but both at infinity. The Bode diagram (Figure 2.7c) shows amplitude and phase approximated by straight-line asymptotes rather than the true curves: the pole at $\sigma = -0.1$ will cause 3 dB of attenuation and $-45°$ phase shift at $\omega = 0.1$. From ω_1 to ω_2 is just under seven octaves, i.e. $9.9/0.1 \approx 2^7$ so the phase Φ will settle down to $-90°$ over most of the middle of this range, whilst at 6 dB/octave the attenuation at ω_2, will have risen to just under $7 \times 6 = 42$ dB (actually 39.9 dB) plus another 3 dB due to the pole at $\sigma = -9.9$. At $\omega = 9.9$, Φ will have risen to $90 + 45°$, heading for $180°$ at higher frequencies still.

The picture looks in fact remarkably like the two cascaded lags of Figure 2.5d to f. I think the circle diagram goes like Figure 2.7d, almost a complete semicircle before finally diving into the origin at $-180°$, but I haven't actually tried drawing it to scale. If you substitute lower and lower values of R in (2.4), you will find that the two poles move towards each other, so it looks as though they must eventually meet.

Now look at the case where $R = 2$ ohms. Substituting this in (2.4) gives $s = \{-2\pm$

$\sqrt{(4-4)}\}/2 = 1 \pm 0$, so that the two poles are indeed coincident at $s = -1$ or at $s = 1 \pm j0$ to be precise, since s is after all the complex frequency variable. The roots a and b are both equal to $-R/2L$ (i.e. -1 in the numerical example), so

$$F(s) = \frac{1}{[s + (R/2L)][s + (R/2L)]}$$
$$= \frac{1}{[s + (R/2L)]} \frac{1}{[s + (R/2L)]}$$

that is, the same as a cascade of two identical first-order lags. Figure 2.7e shows how we can use this fact to construct points on the vector locus of v_o for three frequencies $\omega = 0.1$, $\omega = 1$ and $\omega = 10$. For example, a single lag will provide a normalized output of $0.707\underline{/-45°}$ at $\omega = 1$, and a second circle diagram erected on this base gives the final output as $0.707 \times 0.707\underline{/-45°-45°} = 0.5\underline{/-90°}$. At very high frequencies, the phase shift is clearly tending to $-180°$. This all agrees with the corresponding Bode plot, which would show a horizontal attenuation asymptote changing straight away to -12 dB/octave at $\omega = 1$, at which frequency the phase shift Φ will already have reached 90°.

Figure 2.6 Three ways of representing frequency response for transitional lag circuit.
  (a) Transitional lag circuit.

$$F(j\omega) = \frac{v_o}{v_i} = \frac{(1/j\omega C) + R_2}{(1/j\omega C) + R_2 + R_1} = \frac{1 + j\omega C R_2}{1 + j\omega C(R_1 + R_2)} = \frac{1 + j\omega T_2}{1 + j\omega T_1} = \frac{T_2[(1/T_2) + j\omega]}{T_1[(1/T_1) + j\omega]}$$

Mor generally,

$$F(s) = K\frac{s + (1/T_2)}{s + (1/T_1)} \quad \text{where} \quad K = \frac{T_2}{T_1} = \frac{1/\omega_{02}}{1/\omega_{01}} = \frac{\omega_{01}}{\omega_{02}} = \frac{CR_2}{C(R_1 + R_2)} = \frac{R_2}{R_1 + R_2}$$

(b) Pole–zero plot. Up to $\sqrt{(\omega_{01}\omega_{02})}$ the lag ϕ_p increases faster than the lead ϕ_z. Thereafter ϕ_z catches up, so the net lag falls again. The zero is shown by a circle.

(c) Pole–zero plot in three dimensions. The effect of the zero depresses the value of $F(s)$ at infinity to K. The further the zero at $-1/T_2$ is along the $-\sigma$ axis (the smaller R_2 is), the smaller is K (the response at infinity) (reproduced by courtesy of Maxim Integrated Products UK Ltd).

(d) Bode plot, showing the use of asymptotes. The -6 dB/octave asymptote is cancelled beyond ω_{02} by the $+6$ dB/octave asymptote. They crudely approximate the actual amplitude shown dashed. Likewise, phase asymptotes running from 0° below the corner frequency to $-90°$ or $+90°$ above, indicate the phase tendency if $\omega_{02} \gg \omega_{01}$.

(e) Circle diagram. For any phase angle less than ϕ_m there are two different possible values of v_o, the larger at a frequency below $\sqrt{[(1/T_1)(1/T_2)]}$, the smaller at a frequency above.

(f) Using the circle diagram. If you are better at gometry than at calculus or complex numbers, the circle diagram enables you to find the attenuation at the frquency where the phase shift ϕ is maximum. The bold triangles with bases $(v_o)_\infty = K$ and $v_i = 1$ are similar, as they have three equal angles. ϕ_m is common, each has an angle a, and each has an angle $90° + a$. Therefore when the output vector is tangential to the circle, at $\omega\sqrt{(\omega_{01}\omega_{02})}$, $K/v_o = v_o/1$, whence $v_o = \sqrt{K}$. Expressed in decibels, this amounts to half the maximum attenuation, agreeing with the Bode plot.

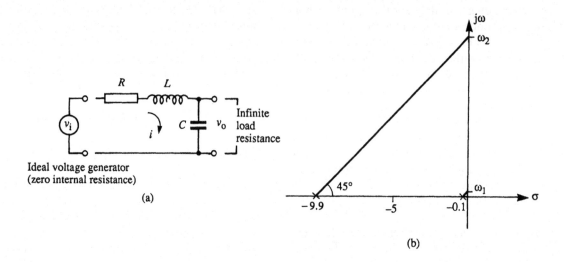

(a)

(b)

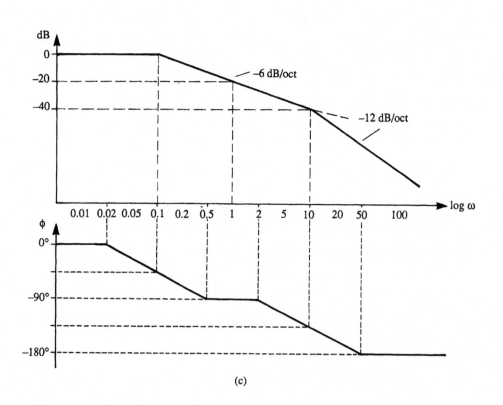

(c)

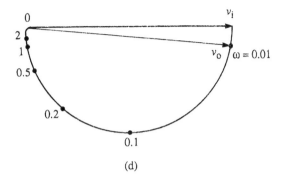

(d)

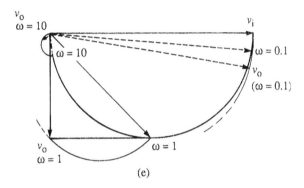

(e)

Figure 2.7 Resonant *LC* circuit analysis.

Figure 2.7e shows the circle diagram only: the pole–zero and Bode plots are easy to visualize for this case.

Now consider what happens to a normalized 1 H, 1 F tuned circuit when R is less than 2 ohms, say 1.4 ohms. From (2.4) we need to find the square root of $(1.4^2 - 4)$, i.e. $\sqrt{(-2.04)}$. When you first started learning algebra you would simply have been told that the equation didn't factorize. However, j solves the problem: $\sqrt{(-2.04)} = \sqrt{(-1)}\sqrt{(2.04)} = j1.428\,286$. So

$$s = \frac{-1.4 \pm \sqrt{(1.96 - 4)}}{2} = -0.7 \pm j0.714\,143$$

and these roots are the poles shown in Figure 2.8a. Now you will recall that the magnitude of the transfer function at any frequency ω, denoted by a

point on the $j\omega$ axis, is proportional to the product of the reciprocals of the distances from the point to each of the poles. Consider the frequency $\omega = 0.1$, i.e. one-tenth of the resonant frequency. Compared with the situation at $\omega = 0$, the distance l_1 to the upper pole is slightly less whilst the distance l_2 to the lower is slightly greater; this is shown in Figure 2.8a. So will v_o at $\omega = 0.1$ be greater or less than v_i – or is perhaps the magnitude of the transfer function still unity, as at $\omega = 0$? Imagine that each pole is a thumbtack and that a tight loop of cotton encloses these and a pencil point at the origin. As you move the pencil upwards and round, it will trace out an ellipse around the poles and entirely to the left of the $j\omega$ axis, except where it touches it at the origin. Clearly then the sum $l_1 + l_2$ increases as you move up the

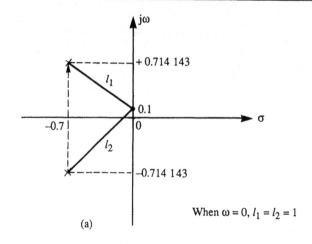

(a)

When $\omega = 0$, $l_1 = l_2 = 1$

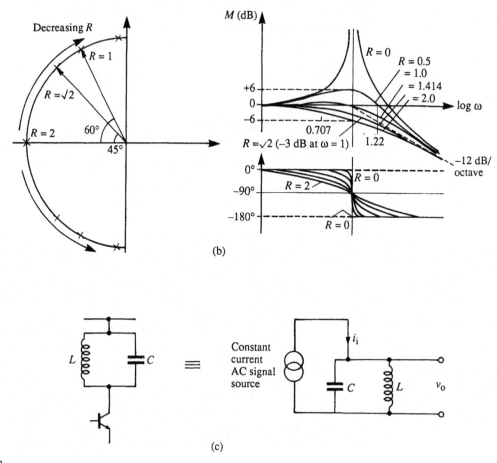

(b)

(c)

Figure 2.8
(a – b) Analysis of resonant LC circuit with complex poles.
(c) Shunt current fed parallel tuned circuit.

jω axis; so you might be tempted to think that the magnitude of v_o which is proportional to $(1/l_1)(1/l_2)$ will decrease. But not so: for in percentage terms, l_1 initially decreases faster than l_2 increases, so the magnitude of the transfer function $|F(j\omega)| = |v_o/v_i|$ actually increases. At $\omega = 0.1, (1/l_1)(1/l_2) = 1.000\,15$, i.e. there is a 'peak' of minuscule dimensions at frequencies well below the resonant frequency. At $\omega = 0.2 |F(j\omega)|$ is already back to unity.

As R decreases, the positions of the poles migrate away from each other around a semicircle of radius equal to unity centred on the origin. For normalized values of R much smaller than unity, the poles approach ever closer to the jω axis, as in Figure 2.8b. As the Bode plot shows, the output voltage at ω_0 increases indefinitely as R falls, but $|F(j\omega)|$ can never exceed $\times 2$ (or $+6$ dB) at $0.707\omega_0$ and $1.22\omega_0$ since $\sqrt{0.5}$ and $\sqrt{1.5}$ are the values of ω which make

$$\left| \frac{1}{(j\omega)^2 LC + j\omega CR + 1} \right| = 2$$

when $L = C = 1$, as R tends to zero. Likewise as you shave past the pole travelling north up the jω axis, the phase flips from zero to $-180°$ more rapidly for smaller values of R (poles nearer the axis).

You can see how easily numerical answers drop out of the expression

$$F(j\omega) = \frac{1}{(j\omega)^2 LC + j\omega CR + 1} \qquad (2.5)$$

for the particular case of a tuned circuit where $L = 1$ H and $C = 1$ F, which is of course resonant at $\omega_0 = 1/\sqrt{(LC)} = 1$ rad/s. You won't often be using a tuned circuit with those particular values, but one can simplify the maths for any old tuned circuit, using the technique – mentioned earlier – of normalization. The trick is to normalize to ω_0, i.e. to express any radian frequency under consideration not as its actual value, but by its ratio to the resonant frequency ω_0. Since $\omega_0 = 1/\sqrt{(LC)}$, just replace LC in the denominator of (2.5) with $1/\omega_0^2$: likewise $C = 1/L\omega_0^2$. With these substitutions, (2.5) becomes

$$F(j\omega) = \frac{1}{(j\omega)^2(1/\omega_0^2) + j\omega(1/L\omega_0^2)R + 1}$$

$$= \frac{1}{(j\omega/\omega_0)^2 + j(\omega/\omega_0)(R/L\omega_0) + 1}$$

So now, using ω_n to denote the ratio of the actual ω to the resonant frequency ω_0,

$$F(j\omega_n) = \frac{1}{(j\omega_n)^2 + j\omega_n(R/\omega_0 L) + 1} \qquad (2.6)$$

It was noted earlier that $\omega_0 L/R$, the ratio of the reactance of the inductor to the circuit resistance, is the voltage magnification factor Q, so (2.6) can be written as

$$F(j\omega_n) = \frac{1}{-\omega_n^2 + j\omega_n(1/Q) + 1} \qquad (2.7)$$

From this it is very clear that at resonance $(\omega_n = 1)$, $v_o/v_i = -jQ = Q\angle - 90°$.

Q has another significance that is less well known. Consider a sinusoidal current of I amperes RMS (i.e. $I\sqrt{2}$ amperes peak) and frequency f Hz flowing in an inductor of inductance L henrys and resistance R ohms. The energy stored in the inductor at the peak of the current is $(1/2)L(I\sqrt{2})^2 = LI^2$ joules. The power dissipation in the resistor is I^2R watts or I^2R joules per second, i.e. $I^2R/2\pi f$ joules per radian. So the ratio of peak energy stored to energy dissipated per radian is $LI^2/(I^2R/2\pi f) = \omega L/R$. The term $\omega L/R$ is called the quality factor Q of the inductor at the frequency ω, and is simply the ratio of reactance to resistance. It looks as though the Q of an inductor is simply proportional to frequency, but, as noted earlier, the effective resistance of an inductor tends to rise with frequency, so the Q increases with frequency rather less rapidly. In the case of a tuned circuit, of particular interest is the value of Q at ω_0, so assume for the moment that R is a constant which includes the coil's loss resistance, the series loss (if any) of the capacitor, and even the source resistance of v_i, in case it's not quite an ideal voltage source. In the case of an inductor on its own, Q was defined in terms of the energy dissipated per radian and the peak energy stored, but in the case of a loss-free tuned circuit the energy stored at ω_0 is constant. It simply circulates back and forth from $LI^2/2$ in the inductor to $CV^2/2$ in the capacitor a quarter of a cycle later, then $L(-I)^2/2$ in the inductor, and so on. With losses, energy is dissipated and so the amplitude decreases, unless – in the steady state – it is made up by energy drawn from a source.

You may see equation (2.6) expressed in yet another form. The reciprocal of Q is sometimes denoted by D, the energy dissipated per radian divided by the stored energy. Then $F(j\omega_n) = 1/\{(j\omega_n^2 + j\omega_n D + 1\}$ or, more generally, normalized $F(s) = 1/\{s^2 + Ds + 1\}$. This general form applies equally to the voltage across the capacitor of a series voltage fed tuned circuit like Figure 2.7a, to a second-order low-pass LC filter section, and to second-order active low-pass filters (which are covered in a later chapter). In the high-Q case it also applies to the shunt current fed parallel tuned circuit of Figure 2.8 (c), and it would also apply in the low-Q case if the series loss resistance was equally divided between the inductor and capacitor branches or was due solely to the finite resistance of an imperfect current source. In the case shown where the loss is entirely associated with the inductor, the frequency at which the circuit appears purely resistive is not quite the same as the frequency of maximum impedance.

A detailed study could be made of many more circuit arrangements and their representations as pole–zero diagrams, Bode plots and so on. However, to save you wading through acres of text, the results are summarized in Appendix 4. Here I have shown for each circuit the circle diagram (vector diagram as a function of frequency), the pole–zero diagrams for $F(s)$ in plan and isometric view, and the Bode plot, all together for comparison. I have not seen this done in any other book, but I think you will find it helpful. You will in any case be meeting more pole–zero plots, circle diagrams etc. in later chapters. But in the meantime, let's look at some active devices – transistors, opamps and the like, including those buffer amplifiers which have already been mentioned.

References

1. *Network Analysis and Feedback Amplifier Design*. H. W. Bode, D. Van Nostrand Company Inc., New York 1945.
2. Transfer Functions, Cathode Ray, p. 177. *Wireless World*, April 1962.
3. *State Variables for Engineers* (Chapter 3), Da Russo, Roy, Close; John Wiley and Sons Inc. 1965.

Questions

1. Define a linear component. Give examples.
2. Define impedance. How does it differ from reactance?
3. What is the complex conjugate of $x + jy$? Express the cartesian value $x + jy$ in polar $(M\angle\Phi)$ form.
4. Prove that the Q of an inductor is equal to the ratio of the energy stored, to the energy dissipated per radian.
5. An audio equipment includes a switch-selectable scratch filter (top-cut circuit), fed from a voltage source. The filter consists of a 10 K series resistor feeding a 3300 pF shunt capacitor, and connected to a load resistance of 1 MΩ. What is the frequency where the attenuation has risen to 3 dB? What is the rate of cut-off, in dB per octave, at much higher frequencies?
6. If, in question 5, the voltage source were replaced by an ideal 1 mA current source, what would be the voltage across the capacitor at (a) 5 kHz, (b) 5 Hz?
7. As a result of careful retuning, the power output of an RF amplifier was increased from 70 mW to 105 mW. What was the increase, expressed in dB?
8. A 1 V AC. generator with zero output impedance is connected through a series 159 mH inductor (with a Q of 100) to a lossless 0.159 μF shunt capacitor. What is the maximum voltage which can occur across the capacitor? At what frequency is this maximum observed?
9. In the previous question, what would have been the maximum voltage across the capacitor if the dissipation constant D of both the inductor and the capacitor had been 0.005?

How much does it differ from the previous value, and why?

10. The output of a non-inverting opamp with a gain of unity is connected through a 100 K resistor to a 4.7 nF capacitor whose other end is grounded. The input of a second, identical, circuit is connected to the junction of the resistor and the capacitor, and the input of a third such circuit connected to the resistor/capacitor of the second. What is the phase-shift between the input voltage at the first opamp and the voltage across the third capacitor, at 340 Hz? At what frequency will these two voltages be in antiphase?

Chapter

3   Active components

If passive components are the cogs and pinions of a circuit, an active component is the mainspring. The analogy is not quite exact perhaps, for the mainspring stores and releases the energy to drive the clockwork, whereas an active component drives a circuit by controlling the release of energy from a battery or power supply in a particular manner. In this sense, active devices have existed since the days of the first practical applications of electricity for communications. For although the receiving apparatus of the earliest electric telegraphs may have been passive in the sense that the indicators were operated solely by the received signal, the sending key at the originating end represented an active device of sorts. The sensitivity of the receiving end was soon improved by using a very sensitive relay to control the flow of power from a local receiving end battery – probably the earliest form of amplifier, and a clear example of an active device.

Relays are still widely used and the more sensitive varieties can provide an enormous gain, defined as the ratio of the power in the circuit controlled by the contacts to the coil power required to operate the relay – typically 30 to 60 dB. Nowadays we are more likely to think of discrete bipolar transistors and MOSFETs or linear integrated circuits like opamps as typical active devices, but generations of electronic engineers were brought up on thermionic valves.

Everybody still uses valves every day, for the cathode ray tube (CRT) in a television set is just a special type of valve, and the picture it displays was in all probability broadcast by a valved transmitter. However, this book does not cover valve circuits and applications since, CRTs apart, they are nowadays restricted to special purpose high-power uses such as broadcast transmitters, industrial RF heating, medical diathermy etc.

Turning then to semiconductors, the simplest of these is the diode.

The semiconductor diode

This, like its predecessor the thermionic diode, conducts current in one direction only. It is arguable that diodes in general are not really active devices at all, but simply non-linear passive devices. However, they are usually considered along with other active devices such as transistors and triacs, and the same plan is followed here.

The earliest semiconductor diode was a *point contact* device and was already in use before the First World War, being quite possibly contemporary with the earliest thermionic diodes. It consisted of a sharp pointed piece of springy wire (the 'cat's whisker') pressed against a lump of mineral (the 'crystal'), usually galena – an ore containing sulphide of lead. The crystal detector was widely employed as the detector in the crystal sets which were popular in the early days of broadcasting. Given a long aerial and a good earth, the crystal set produced an adequate output for use with sensitive headphones, whilst with so few stations on the air in those days the limited selectivity of the crystal set was not too serious a problem. The crystal and cat's whisker variety of point contact diode was a very hit and miss affair, with the listener probing the surface of the crystal to find a good spot. Later, new techniques and materials were developed, enabling robust preadjusted point contact diodes useful at microwave frequencies to be produced. These were used in radar sets such as AI Mk.10, an airborne interceptor radar which was in service during (and long after!) the Second World War. Germanium point contact diodes are still manufactured and are useful where a diode with a low forward voltage drop at modest current

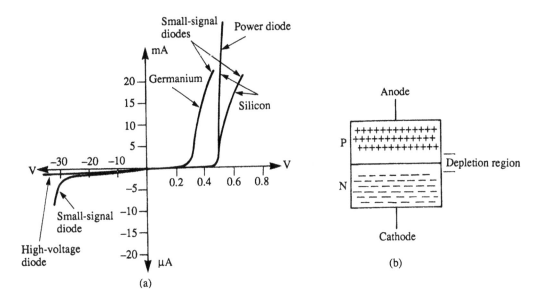

Figure 3.1 Semiconductor diodes.
(a) I/V characteristics.
(b) Diagrammatic representation of PN diode, showing majority carriers and depletion region.

(a milliampere or so) combined with very low reverse capacitance is required. However, for the last twenty years silicon has predominated as the preferred semiconductor material for both diode and transistor manufacture, whilst point contact construction gave way to junction technology even earlier.

Figure 3.1a shows the I/V characteristics of practical diodes. *Silicon* diodes are manufactured hundreds or thousands at a time, commencing with a thin wafer of single crystal silicon several inches in diameter, which is later scribed and separated to obtain the individual diodes. Silicon is one of those substances which crystallize in a cubic lattice structure; another is sodium chloride – common salt – but that is a compound, not an element like silicon. Silicon, in the form of silicon dioxide, is one of the most abundant elements in the earth's crust, occurring as quartz, in sandstone etc. When reduced to elementary silicon, purified and grown from the melt as a single crystal, it is called *intrinsic* silicon and is a poor conductor of electricity, at least at room temperature. However, if a few of the silicon atoms in the atomic lattice are replaced with atoms of the pentavalent element phosphorus (which has

five valence electrons in its outer shell, unlike quadravalent silicon which has four outer electrons), then there are electrons 'going begging', with no corresponding electron in a neighbouring atom with which to form a bond pair. These spare electrons can move around in the semiconductor lattice, rather like the electrons in a metallic conductor. The higher the *doping level,* the more free electrons and the higher the conductivity of the silicon, which is described as N type. This simply indicates that the current flow is by means of negative charge carriers, i.e. electrons. P type silicon is obtained by doping the monocrystalline lattice with a sprinkling of trivalent atoms, such as boron. Where one of these is substituted in the lattice next to a silicon atom, the latter has one of its electrons 'unpaired', a state of affairs described as a *hole*. If this is filled by an electron from a silicon atom to the right, then whilst that electron has moved to the left, the hole has effectively moved to the right. It turns out that the spare electrons in N type silicon are more mobile than the holes in P type, which explains why very high-frequency transistors are more easy to produce in NPN than PNP types – but that is jumping ahead.

To return to the silicon junction diode: the construction of this is as in Figure 3.1b, which indicates the lack of carriers (called a *depletion region*) in the immediate neighbourhood of the junction. Here, the electrons from the N region have been attracted across to the P region to fill holes. This disturbance of the charge pattern that one would expect to find throughout N type and P type material represents a potential barrier which prevents further migration of carriers across the junction. When the diode is reverse biased, the depletion region simply becomes more extensive.

The associated redistribution of charge represents a transient charging current, so that a reversed biased diode is inherently capacitive. If a forward bias voltage large enough to overcome the potential barrier is applied to the junction, about 0.6 V in silicon, then current will flow; in the case of a large-area power diode, even a current of several amperes will only result in a small further increase in the voltage drop across the diode, as indicated in Figure 3.1a. The incremental or slope resistance r_d of a forward biased diode at room temperature is given approximately by $25/I_a$ ohms. where the current through the diode I_a is in milliamperes. Hence the incremental resistance at 10 μA is 2K5, at 100 μA is 250R, and so on, but it bottoms out at a few ohms at high currents, where the bulk resistance of the semiconductor material and the resistance of the leads and bonding pads etc. come to predominate. This figure would apply to a small-signal diode: the minimum slope resistance of a high-current rectifier diode would be in the milliohm region.

Power diodes are used in power supply rectifier circuits and similar applications, whilst small *signal diodes* are widely used as detectors in radio-frequency circuits and for general purpose signal processing, as will appear in later chapters. Also worth mentioning are special purpose small-signal diodes such as the tunnel diode, backward diode, varactor diode, PIN diode, snap-off diode, Zener diode and Schottky diode; the last of these is also used as a rectifier in power circuits. However, the tunnel diode and its degenerate cousin the backward diode are only used in a few very specialized applications nowadays.

The *varactor diode* or varicap is a diode designed solely for reverse biased use. A special doping profile giving an abrupt or 'hyperabrupt' junction is used. This results in a diode whose reverse capacitance varies widely according to the magnitude of the reverse bias. The capacitance is specified at two voltages, e.g. 1 V and 15 V, and may provide a capacitance ratio of 2:1 or 3:1 for types intended for use at UHF, up to 30:1 for types designed for tuning in AM radio sets. In these applications the peak-to-peak amplitude of the RF voltage applied to the diode is small compared with the reverse bias voltage, even at minimum bias (where the capacitance is maximum). So the varactor behaves like a normal mechanical air-spaced tuning capacitor except that it is adjusted by a DC voltage rather than a rotary shaft. Tuning varactors are designed to have a low series loss r_s so that they exhibit a high quality factor Q, defined as X_c/r_s , over the range of frequencies for which they are designed.

Another use for varactors is as frequency multipliers. If an RF voltage with a peak-to-peak amplitude of several volts is applied to a reverse biased diode, its capacitance will vary in sympathy with the RF voltage. Thus the device is behaving as a non-linear (voltage dependent) capacitor, and as a result the RF current will contain harmonic components which can be extracted by suitable filtering.

The P type/intrinsic/N type diode or *PIN diode* is a PN junction diode, but fabricated so as to have a third region of intrinsic (undoped) silicon between the P and N regions. When forward biased by a direct current it can pass radio-frequency signals without distortion, down to some minimum frequency set by the lifetime of the current carriers – holes or electrons – in the intrinsic region. As the forward current is reduced, the resistance to the flow of the RF signal increases, but it does not vary over an individual cycle of the RF current. As the direct current is reduced to zero, the resistance rises towards infinity; when the diode is reverse biased, only a very small amount of RF current can flow, via the diode's reverse biased capacitance. The construction ensures that this is very small, so the PIN diode can be used as an electronically controlled RF switch or relay, on when forward biased and off when reverse biased.

It can also be used as a variable resistor or attenuator by adjusting the amount of forward current. An ordinary PN diode can also be used as an RF switch, but it is necessary to ensure that the peak RF current when 'on' is much smaller than the direct current, otherwise waveform distortion will occur. It is the long lifetime of carriers in the intrinsic region (long compared with a single cycle of the RF) which enables the PIN diode to operate as an adjustable linear resistor, even when the peaks of the RF current exceed the direct current.

When a PN junction diode which has been carrying current in the forward direction is suddenly reverse biased, the current does not cease instantaneously. The charge has first to redistribute itself to re-establish the depletion layer. Thus for a very brief period, the reverse current flow is much greater than the small steady-state reverse leakage current. The more rapidly the diode is reverse biased, the larger the transient current and the more rapidly the charge is extracted. *Snap-off diodes* are designed so that the end of the reverse current recovery pulse is very abrupt, rather than the tailing off observed in ordinary PN junction diodes. It is thus possible to produce a very short sharp current pulse by rapidly reverse biasing a snap-off diode. This can be used for a number of applications, such as high-order harmonic generation (turning a VHF or UHF drive current into a microwave signal) or operating the sampling gate in a digital sampling oscilloscope.

Small-signal *Schottky diodes* operate by a fundamentally different form of forward conduction. As a result of this, there is virtually no stored charge to be recovered when they are reverse biased, enabling them to operate efficiently as detectors or rectifiers at very high frequencies. *Zener diodes* conduct in the forward direction like any other silicon diode, but they also conduct in the reverse direction, and this is how they are normally used. At low reverse voltages, a Zener diode conducts only a very small leakage current like any other diode. But when the voltage reaches the nominal Zener voltage, the diode current increases rapidly, exhibiting a low incremental resistance. Diodes with a low breakdown voltage – up to about 4 V – operate in true Zener breakdown: this conduction mechanism exhibits a small negative temperature coefficient of

voltage. Higher-voltage diodes, rated at 6 V and up, operate by a different mechanism called *avalanche breakdown*, which exhibits a small positive temperature coefficient. In diodes rated at about 5 V both mechanisms occur, resulting in a very low or zero temperature coefficient of voltage. However, the lowest slope resistance occurs in diodes of about 7 V breakdown voltage.

Bipolar transistors

Unlike semiconductor diodes, transistors did not see active service in the Second World War; they were born several years too late. In 1948 it was discovered that if a point contact diode detector were equipped with two cat's whiskers rather than the usual one, spaced very close together, the current through one of them could be influenced by a current through the other. The crystal used was germanium, one of the rare earths, and the device had to be prepared by discharging a capacitor through each of the cat's whiskers in turn to 'form' a junction. Over the following years, the theory of conduction via junctions was elaborated as the physical processes were unravelled, and the more reproducible junction transistor replaced point contact transistors.

However, the point contact transistor survives to this day in the form of the standard graphical symbol denoting a bipolar junction transistor (Figure 3.2a). This has three separate regions, as in Figure 3.2b, which shows (purely diagrammatically and not to scale) an NPN junction transistor. With the *base* (another term dating from point contact days) short-circuited to the *emitter,* no *collector* current can flow since the collector/base junction is a reverse biased diode, complete with depletion layer as shown. The higher the reverse bias voltage, the wider the depletion layer, which is found mainly on the collector side of the junction since the collector is more lightly doped than the base. In fact, the pentavalent atoms which make the collector N type are found also in the base region. The base is a layer which has been converted to P type by substituting so many trivalent (hole donating) atoms into the silicon lattice, e.g. by diffusion or ion bombardment, as to swamp the effect of the pentavalent atoms. So holes are the

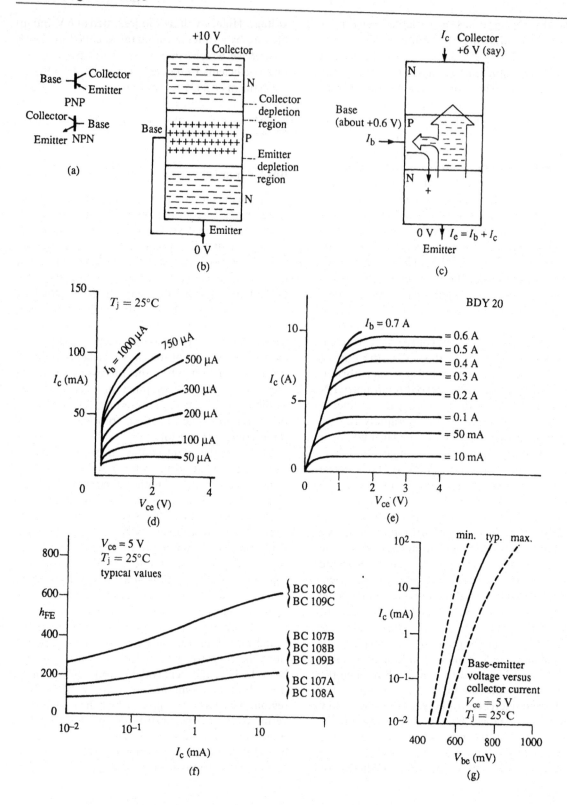

(a)

(b)

(c)

(d)

(e)

(f)

(g)

majority carriers in the base region, just as electrons are the majority carriers in the collector and emitter regions. The collector 'junction' turns out then to be largely notional; it is simply that plane for which on one side (the base) holes or P type donor atoms predominate, whilst on the other (the collector) electrons or N type donor atoms predominate, albeit at a much lower concentration.

Figure 3.2c shows what happens when the base/emitter junction is forward biased. Electrons flow from the emitter into the base region and, simultaneously, holes flow from the base into the emitter. The latter play no useful part in transistor action: they contribute to the base current but not to the collector current. Their effect is minimized by making the N type emitter doping a hundred times or more heavier than the base doping, so that the vast majority of current flow across the emitter/base junction consists of electrons flowing into the base from the emitter. Some of these electrons flow out of the base, forming the greater part of the base current. But most of them, being *minority carriers* (electrons in what should be a P type region) are swept across the collector junction by the electric field gradient existing across the depletion layer. This is illustrated (in diagrammatic form) in Figure 3.2c, while Figure 3.2d shows the collector characteristics of a small-signal NPN transistor and Figure 3.2e those of an NPN power transistor. It can be seen that, except at very low values, the collector voltage has comparatively little effect upon the collector current, for a given constant base current. For this

reason, the bipolar junction transistor is often described as having a 'pentode-like' output characteristic, by an analogy dating from the days of valves. This is a fair analogy as far as the collector characteristic is concerned, but there the similarity ends. The pentode's anode current is controlled by the g_1 (control grid) voltage, but there is, at least for negative values of control grid voltage, negligible grid current. By contrast, the base/emitter input circuit of a transistor looks very much like a diode, and the collector current is more linearly related to the base current than to the base/emitter voltage (Figure 3.2f and g). For a silicon NPN transistor, little current flows in either the base or collector circuit until the base/emitter voltage V_{be} reaches about $+0.6\,V$, the corresponding figure for a germanium NPN transistor being about $+0.3\,V$. For both types, the V_{be} corresponding to a given collector current falls by about $2\,mV$ for each degree centigrade of temperature rise, whether this is due to the ambient temperature increasing or due to the collector dissipation warming the transistor up. The reduction in V_{be} may well cause an increase in collector current and dissipation, heating the transistor further and resulting in a further fall in V_{be}. It thus behoves the circuit designer, especially when dealing with power transistors, to ensure that this process cannot lead to thermal runaway and destruction of the transistor.

Although the base/emitter junction behaves like a diode, exhibiting an incremental resistance of $25/I_e$ at the emitter, most of the emitter current

Figure 3.2 The bipolar transistor.

(a) Bipoar transistor symbols.

(b) NPN junction transistor, cut-off condition. Only majority carriers are shown. The emitter depletion region is very much narrower than the collector depletion region because of reverse bias and higher doping levels. Only a very small collector leakage current I_{cb} flows.

(c) NPN small signal silicon junction transistor, conducting. Only minority carriers are shown. The DC common emitter current gain is $h_{FE} = I_c/I_b$, roughly constant and typically around 100. The AC small signal current gain is $h_{fe} = dI_c/dI_b = i_c/i_b$.

(d) Collection current versus collector/emitter voltage, for an NPN small signal transistor (BC 107/8/9).

(e) Collector current versus collector/emitter voltage, for an NPN power transistor.

(f) h_{FE} versus collector current for an NPN small signal transistor.

(g) Collector current versus base/emitter voltage for an NPN small signal transistor.

(Parts (d) to (g) reproduced by courtesy of Philips Components Ltd.)

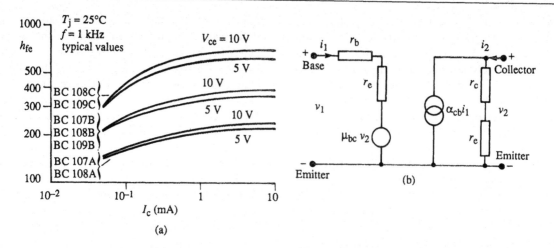

(a)

(b)

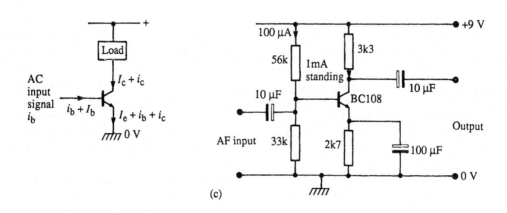

(c)

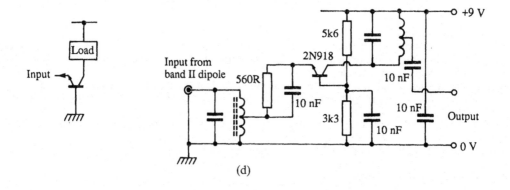

(d)

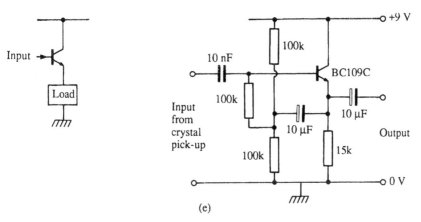

Figure 3.3 Small-signal amplifiers.
  (a) h_{fe} versus collector current for an NPN small-signal transistor of same type as in Figure 3.5f (reproduced by courtesy of Philips Components Ltd.).
  (b) Common emitter equivalent circuit.
  (c) Common emitter audio amplifier. I_b = base bias or standing current; I_c = collector standing current; i_c = useful signal current in load.
  (d) Common base RF amplifier.
  (e) Common collector high-input-impedance audio amplifier.

appears in the collector circuit, as has been described above.

The ratio I_c/I_b is denoted by the symbol h_{FE} and is colloquially called the *DC current gain* or static forward current transfer ratio. Thus if a base current of 10 µA results in a collector current of 3 mA – typically the case for a high-gain general purpose audio-frequency NPN transistor such as a BC109 – then $h_{FE} = 300$. As Figure 3.2f shows, the value found for h_{FE} will vary somewhat according to the conditions (collector current and voltage) at which it is measured. When designing a transistor amplifier stage, it is necessary to ensure that any transistor of the type to be used, regardless of its current gain, its V_{be} etc., will work reliably over a wide range of temperatures: the no-signal DC conditions must be stable and well defined. The DC current gain h_{FE} is the appropriate gain parameter to use for this purpose. When working out the stage gain or AC small-signal amplification provided by the stage, h_{fe} is the appropriate parameter, this is the *AC current gain* dI_c/dI_b. Usefully, for many modern small-signal transistors there is little difference in the value of h_{FE} and h_{fe} over a considerable range of current, as can be seen from Figures 3.5f and 3.6a (allowing for the

linear h_{FE} axis in one and the logarithmic h_{fe} axis in the other).

Once the AC performance of a transistor is considered, it is essential to allow for the effects of reactance. Just as there is capacitance between the various electrodes of a valve, so too there are unavoidable capacitances associated with the three electrodes of a transistor. The collector/base capacitance, though usually not the largest of these, is particularly important as it provides a path for AC signals from the collector circuit back to the base circuit. In this respect, the transistor is more like a triode than a pentode, and as such the Miller effect will reduce the high-frequency gain of a transistor amplifier stage, and may even cause an RF stage to oscillate due to feedback of in-phase energy from the collector to the base circuit.

One sees many different theoretical models for the bipolar transistor, and almost as many different sets of parameters to describe it: *z, g, y,* hybrid, *s,* etc. Some equivalent circuits are thought to be particularly appropriate to a particular configuration, e.g. grounded base, whilst others try to model the transistor in a way that is independent of how it is connected. Over the years numerous workers have elaborated such models, each proclaiming the advantages of his particular equivalent circuit.

Just one particular set of parameters will be mentioned here, because they have been widely used and because they have given rise to the symbol commonly used to denote a transistor's current gain. These are the *hybrid* parameters which are generally applicable to any *two-port network,* i.e. one with an input circuit and a separate output circuit. Figure 3.7a shows such a two-port, with all the detail of its internal circuitry hidden inside a box – the well-known 'black box' of electronics. The voltages and currents at the two ports are as defined in Figure 3.4a, and I have used *v* and *i* rather than *V* and *I* to indicate small-signal alternating currents, not the standing DC bias conditions. Now v_1, i_1, v_2, and i_2 are all variables and their interrelation can be described in terms of four *h* parameters as follows:

$$v_1 = h_{11}i_1 + h_{12}v_2 \qquad (3.4)$$

$$i_2 = h_{21}i_1 + h_{22}v_2 \qquad (3.5)$$

Each of the *h* parameters is defined in terms of two of the four variables by applying either of the two conditions $i_1 = 0$ or $v_2 = 0$:

$$h_{11} = \left.\frac{v_1}{i_1}\right|_{v2=0} \qquad (3.6)$$

$$h_{21} = \left.\frac{i_2}{i_1}\right|_{v2=0} \qquad (3.7)$$

$$h_{12} = \left.\frac{v_1}{v_2}\right|_{i1=0} \qquad (3.8)$$

$$h_{22} = \left.\frac{i_2}{v_2}\right|_{i1=0} \qquad (3.9)$$

Thus h_{11} is the *input impedance* with the output port short-circuited as far as AC signals are concerned. At least at low frequencies, this impedance will be resistive and its units will be ohms. Next, h_{21} is the *current transfer ratio*, again with the output circuit short-circuited so that no output voltage variations result: being a pure ratio, h_{21} has no units. Like h_{11} it will be a complex quantity at high frequencies, i.e. the output current will not be exactly in phase with the input current. Third, h_{12} is the *voltage feedback* ratio, i.e. the voltage

appearing at the input port as the result of the voltage variations at the output port (again this will be a complex number at high frequencies). Finally, h_{22} is the *output admittance*, measured – like h_{12} – with the input port open-circuit to signals. These parameters are called hybrid because of the mixture of units: impedance, admittance and pure ratios.

In equation (3.4) the input voltage v_1 is shown as being the result of the potential drop due to i_1 flowing through the input impedance plus a term representing the influence of any output voltage variation v_2 on the input circuit. When considering only small signals, to which the transistor responds in a linear manner, it is valid simply to add the two effects as shown. In fact the hybrid parameters are examples of partial differentials: these describe how a function of two variables reacts when first one variable is changed whilst the other is held constant. and then vice versa. Here, v_1 is a function of both i_1 and v_2 – so $h_{11} = \partial v_1/\partial i_1$ with v_2 held constant (short-circuited), and $h_{12} = \partial v_1/\partial v_2$ with the other parameter i_1 held constant at zero (open-circuit). Likewise, i_2 is a function of both i_1 and v_2; the relevant parameters h_{21} and h_{22} are defined by (3.7) and (3.9), and i_2 is as defined in equation (3.5). Of course the interrelation of v_1, i_1, v_2 and i_2 could be specified in other ways: the above scheme is simply the one used with *h* parameters.

The particular utility of *h* parameters for specifying transistors arises from the ease of determining h_{11} and h_{21} with the output circuit short-circuited to signal currents. Having defined *h* parameters, they can be shown connected as in Figure 3.4b. Since a transistor has only three electrodes, the dashed line has been added to show that one of them must be common to both the input and the output ports. The common electrode may be the base or the collector, but particularly important is the case where the input and output circuits have a common emitter.

Armed with the model of Figure 3.4b and knowing the source and load impedance, you can now proceed to calculate the gain of a transistor stage – provided you know the relevant values of the four *h* parameters (see Figure 3.4c). For example, for a *common emitter* stage you will need h_{ie} (the input impedance h_{11} in the common emitter

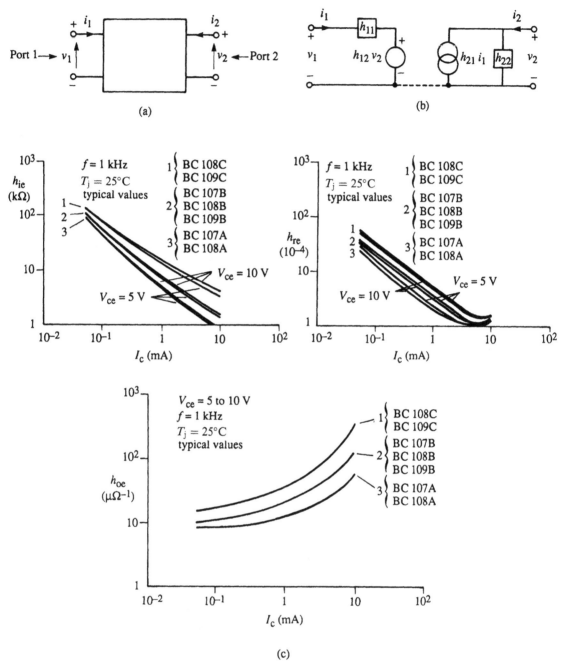

Figure 3.4 *h* parameters.

(a) Generalized two-port black box. *v* and *i* are small-signal alternating qualities. At both ports, the current is shown as in phase with the voltage (at least at low frequencies), i.e. both ports are considered as resistances (impedances).

(b) Transistor model using hybrid parameters.

(c) *h* parameters of a typical small-signal transistor family (see also Figure 3.6a) (reproduced by courtesy of Philips Components Ltd.).

configuration), h_{fe} (the common emitter forward current transfer ratio or current gain corresponding to h_{21}), h_{re} (the common emitter voltage feedback ratio corresponding to h_{12}) and h_{oe} (the common emitter output admittance corresponding to h_{22}). You will generally find that the data sheet for the transistor you are using quotes maximum and minimum values for h_{fe} at a given collector current and voltage, and may well also include a graph showing how the typical or normalized value of h_{fe} varies with the standing collector current I_c. Sometimes, particularly with power transistors, only h_{FE} is quoted: this is simply the ratio I_c/I_b, often called the DC current gain or static forward current transfer ratio. As mentioned earlier, for most transistors this can often be taken as a fair guide or approximation for h_{fe}, for example compare Figures 3.2f and 3.3a. From these it can be seen that over the range 0.1 to 10 mA collector current the typical value of h_{FE} is slightly greater than that of h_{FE}, so the latter can be taken as a guide to h_{fe}, with a little in hand for safety. Less commonly you may find h_{oe} quoted on the data sheet, whilst h_{ie} and h_{re} are often simply not quoted at all. Sometimes a mixture of parameters is quoted; for example, data for the silicon NPN transistor type 2N930 quote h_{FE} at five different values of collector current, and low-frequency (1 kHz) values for h_{ib}, h_{rb}, h_{fe} and h_{ob} – all at 5 V, 1 mA. The only data given to assist the designer in predicting the device's performance at high frequency are f_T and C_{obo}. *The transition frequency f_T* is the notional frequency at which $|h_{fe}|$ has fallen to unity, projected at -6 dB per octave from a measurement at some lower frequency. For example, f_T (min.) for a 2N918 NPN transistor is 600 MHz measured at 100 MHz, i.e. its common emitter current gain h_{fe} at 100 MHz is at least 6. C_{obo} is the *common base output capacitance* measured at $I_c = 0$, at the stated V_{cb} and test frequency (10 V and 140 kHz in the case of the 2N918).

If you were designing a *common base* or *common collector* stage, then you would need the corresponding set of h parameters, namely h_{ib}, h_{fb}, h_{rb} and h_{ob} or h_{ic}, h_{fc}, h_{rc} and h_{oc} respectively. These are seldom available – in fact, h parameters together with z, v, g and transmission parameters

are probably used more often in the examination hall than in the laboratory. The notable exception are the *scattering* parameters s, which are widely used in radio-frequency and microwave circuit design (see Appendix 5). Not only are many UHF and microwave devices (bipolar transistors, silicon and gallium arsenide field effect transistors) specified on the data sheet in s parameters, but s parameter test sets are commonplace in RF and microwave development laboratories. This means that if it is necessary to use a device at a different supply voltage and current from that at which the data sheet parameters are specified, they can be checked at those actual operating conditions.

The h parameters for a given transistor configuration, say grounded emitter, can be compared with the elements of an equivalent circuit designed to mimic the operation of the device. In Figure 3.3b r_e is the incremental slope resistance of the base/emitter diode; it was shown earlier that this is approximately equal to $25/I_e$ where I_e is the standing emitter current in milliamperes. Resistance r_c is the collector slope or incremental resistance, which is high. (For a small-signal transistor in a common emitter circuit, say a BC109 at 2 mA collector current, 15 K would be a typical value: see Figure 3.4c.) The base input resistance r_b is much higher than $25/I_e$, since most of the emitter current flows into the collector circuit, a useful approximation being $h_{fe} \times 25/I_e$. The ideal voltage generator μ_{bc} represents the voltage feedback h_{12} (h_{re} in this case), whilst the constant current generator α_{cb} represents h_{21} or h_{fe}, the ratio of collector current to base current. Comparing Figures 3.3b and 3.4b, you can see that $h_{11} = r_b + r_e$, $h_{12} = \mu_{bc}$, $h_{21} = \alpha_{cb}$ and $h_{22} = 1/(r_e + r_c)$.

Not the least confusing aspect of electronics is the range of different symbols used to represent this or that parameter, so it will be worth clearing up some of this right here. The small-signal common emitter current gain is, as has already been seen, sometimes called α_{cb} but more often h_{fe}; the symbol β is also used. The symbol α_{ce} or just α is used to denote the common base forward current transfer ratio h_{fb}: the term 'gain' is perhaps less appropriate here, as i_c is actually slightly less than i_e, the difference being the base current i_b.

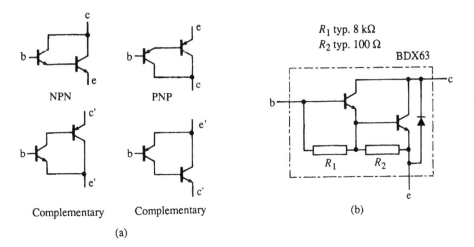

Figure 3.5
(a) Darlington connected discrete transistors.
(b) Typical monolithic NPN Darlington power transistor (reproduced by courtesy of Philips Components Ltd).

It follows from this that $\beta = \alpha/(1 - \alpha)$. The symbols α and β have largely fallen into disuse, probably because it is not immediately obvious whether they refer to small-signal or DC gain: with h_{fe} and h_{FE} – or h_{fb} and h_{FB} – you know at once exactly where you stand.

When h parameters for a given device are available, their utility is limited by two factors: first, usually typical values only are given (except in the case of h_{fe}) and second, they are measured at a frequency in the audio range, such as 1 kHz. At higher frequencies the performance is limited by two factors: the inherent capacitances associated with the transistor structure, and the reduction of current gain at high frequencies. The effect of these factors is covered in a later chapter, which deals with radio-frequency (RF) circuits.

In addition to their use as small-signal amplifiers, transistors are also used as switches. In this mode they are either reverse biased at the base, so that no collector current flows or conducting heavily so that the magnitude of the voltage drop across the collector load approaches that of the collector supply rail. The transistor is then said to be *bottomed*, its V_{ce} being equal to or even less than V_{be}. For this type of large-signal application, the small-signal parameters mentioned earlier are of little if any use. In fact, if (as is usually the case)

one is interested in switching the transistor on or off as quickly as possible, it can more usefully be considered as a charge-controlled rather than a current-controlled device. Here again, although sophisticated theoretical models of switching performance exist, they often involve parameters (such as r_{bb}', the extrinsic or ohmic part of the base resistance) for which data sheets frequently fail to provide even a typical value. Thus one is usually forced to adopt a more pragmatic approach, based upon such data sheet values as are available, plus the manufacturer's application notes if any, backed up by practical in-circuit measurements.

Returning for the moment to small-signal amplifiers, Figure 3.3c, d and e shows the three possible configurations of a single-transistor amplifier and indicates the salient performance features of each. Since the majority of applications nowadays tend to use NPN devices, this type has been illustrated. Most early transistors were PNP types; these required a radical readjustment of the thought processes of electronic engineers brought up on valve circuits, since with PNP transistors the 'supply rail' was negative with respect to ground. The confusion was greatest in switching (logic) circuits, where one was used to the anode of a cut-off valve rising to the (positive) HT rail, this being usually the logic 1 state. Almost overnight, engineers had to get used to collectors flying up to −6 V when cut off, and vice versa. Then NPN devices became more and more readily available, and eventually came to predominate. Thus the modern circuit engineer has the great advantage of being able to employ either NPN or PNP devices in a circuit, whichever proves most convenient – and not infrequently both types are used together. The modern valve circuit engineer, by contrast, still has to make do without a thermionic equivalent of the PNP transistor.

A constant grumble of the circuit designer was for many years that the current gain h_{FE} of power transistors, especially at high currents, was too low. The transistor manufacturers' answer to this complaint was the *Darlington*, which is now available in a wide variety of case styles and voltage (and current) ratings in both NPN and PNP versions. The circuit designer had already for years been using the emitter current of one transistor to supply the base current of another, as in Figure 3.5a. The Darlington compound transistor, now simply called the Darlington, integrates both transistors, two resistors to assist in rapid turn-off in switching applications, and usually (as in the case of the ubiquitous TIP120 series from Texas Instruments) an antiparallel diode between collector and emitter. Despite the great convenience of a power transistor with a value of h_{FE} in excess of 1000, the one fly in the ointment is the saturation or bottoming voltage. In a small-signal transistor (and even some power transistors) this may be as low as 200 mV, though usually one or two volts, but in a power Darlington it is often as much as 2 to 4 V. The reason is apparent from Figure 3.5b: the $V_{ce\,sat}$ of the compound transistor cannot be less than the $V_{ce\,sat}$ of the first transistor plus the V_{be} of the second.

Field effect transistors

An important milestone in the development of modern active semiconductor devices was the field effect transistor, or FET for short. These did not become generally available until the 1960s, although they were described in detail and analysed as early as 1952.

Figure 3.6a shows the symbols and Figure 3.6b and c the construction and operation of the first type introduced, the depletion mode *junction FET* or JFET. In this device, in contrast to the bipolar transistor, conduction is by means of majority carriers which flow through the *channel* between the *source* (analogous to an emitter or cathode) and the *drain* (collector or anode). The *gate* is a region of silicon of opposite polarity to the source cum channel cum drain. When the gate is at the same potential as the source and drain, its depletion region is shallow and current carriers (electrons in the case of the N channel FET shown in Figure 3.6c) can flow between the source and the drain. The FET is thus a unipolar device, minority carriers play no part in its action. As the gate is made progressively more negative, the depletion layer extends across the channel, depleting it of carriers and eventually pinching off the conducting path entirely when V_{gs} reaches $-V_p$, the pinch-off voltage. Thus for zero (or only very small) voltages of either polarity between the drain and the source, the device can be used as a passive voltage-controlled resistor. The JFET is, however, more normally employed in the active mode as an amplifier (Figure 3.6d) with a positive supply rail (for an N channel JFET), much like an NPN transistor stage. Figure 3.6e shows a typical drain characteristic. Provided that the gate is reversed biased (as it normally will be) it draws no current.

The positive excursions of gate voltage of an N channel JFET, or the negative excursions in the case of a P channel JFET, must be limited to less than about 0.5 V to avoid turn-on of the gate/source junction; otherwise the benefit of a high input impedance is lost.

In the *metal-oxide semiconductor* field effect transistor (MOSFET) the gate is isolated from the channel by a thin layer of silicon dioxide, which is a non-conductor: thus the gate circuit never conducts regardless of its polarity relative to the channel. The channel is a thin layer formed between the substrate and the oxide. In the *enhancement* (normally off) MOSFET, a channel of semiconductor of the same polarity as the source and drain is induced in the substrate by the voltage applied to the gate (Figure 3.7b). In the *depletion*

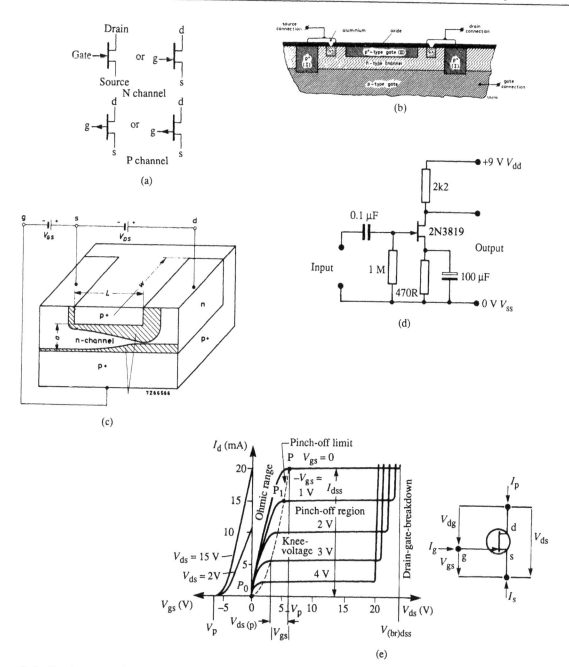

Figure 3.6 Depletion mode junction field effect transistors.
  (a) Symbols.
  (b) Structure of an N channel JFET.
  (c) Sectional view of an N channel JFET. The P⁺ upper and lower gate regions should be imagined to be connected in front of the plane of the paper, so that the N channel is surrounded by an annular gate region.
  (d) JFET audio-frequency amplifier.
  (e) Characteristics of N channel JFET: pinch-off voltage $V_p = -6\,V$.
(Parts (b), (c) and (e) reproduced by courtesy of Philips Components Ltd.)

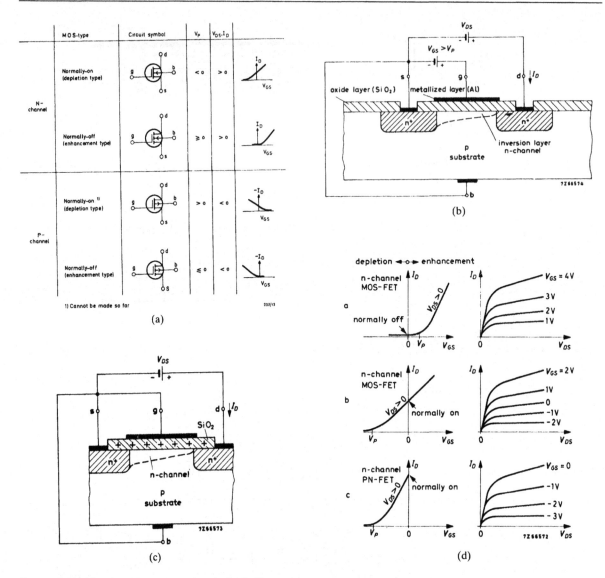

Figure 3.7   Metal-oxide semiconductor field effect transistors.
(a) MOSFET types. Substrate terminal b (bulk) is generally connected to the source, often internally.
(b) Cross-section through an N channel enhancement (normally off) MOSFET.
(c) Cross-section through an N channel depletion (normally on) MOSFET.
(d) Examples of FET characteristics: (i) normally off (enhancement); (ii) normally on (depletion and enhancement); (iii) pure depletion (JFETs only).
(Parts (a) to (d) reproduced by courtesy of Philips Components Ltd.).

(normally on) MOSFET, a gate voltage is effectively built in by ions trapped in the gate oxide (Figure 3.7c). Figure 3.7a shows symbols for the four possible types, and Figure 3.7d summarizes the characteristics of N channel types. Since it is

much easier to arrange for positive ions to be trapped in the gate oxide than negative ions or electrons, P channel depletion MOSFETs are not generally available. Indeed, for both JFETs and MOSFETs of all types, N channel devices far

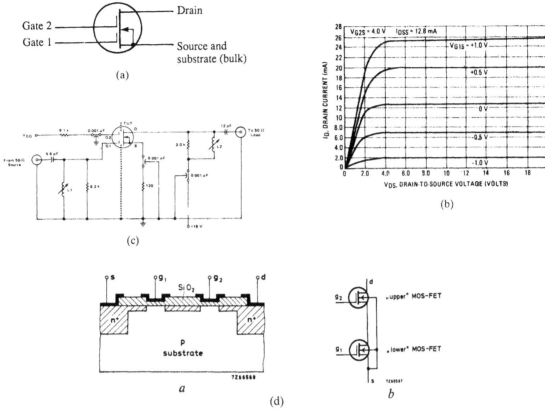

Figure 3.8   Dual-gate MOSFETS.

(a) Dual-gate N channel MOSFET symbol. Gate protection diodes, not shown, are fabricated on the chip in many device types. These limit the gate/source voltage excursion in either polarity, to protect the thin gate oxide layer from excessive voltages, e.g. static charges.

(b) Drain characteristics (3N203/MPF203).

(c) Amplifier with AGC applied to gate 2. 50 Ω source and load (3N203/MPF203).

(Parts (b) and (c) reproduced by courtesy of Motorola Inc.)

(d) Construction and discrete equivalent of a dual-gate N channel MOSFET (reproduced by courtesy of Philips Components Ltd).

outnumber P channel devices. In consequence, one only chooses a P channel device where it notably simplifies the circuitry or where it is required to operate with an N channel device as a complementary pair. (These are further described in the chapter covering audio amplifiers.)

Note that whilst the source and substrate are internally connected in many MOSFETS, in some (such as the Motorola 2N351) the substrate connection is brought out on a separate lead. In some instances it is possible to use the substrate, where brought out separately, as another input terminal. For example, in a frequency changer application,

the input signal may be applied to the gate and the local oscillator signal to the substrate. In high-power MOSFETS, whether designed for switching applications or as HF/VHF/UHF power amplitiers, the substrate is always internally connected to the source.

In the N channel *dual-gate* MOSFET (Figure 3.8) there is a second gate between gate 1 and the drain. Gate 2 is typically operated at +4 V with respect to the source and serves the same purpose as the screen grid in a tetrode or pentode. Consequently the reverse transfer capacitance C_{rss} between drain and gate 1 is only about

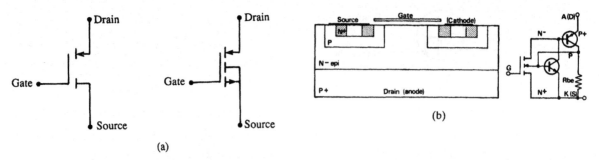

Figure 3.9   The gain enhanced MOSFET (GEMFET).
  (a) Symbols for GEMFET, COMFET (conductivity modulated FET) and other similar devices.
  (b) Structure and equivalent circuit of the GEMFET (reproduced by courtesy of Motorola Inc.).

0.01 pF, against 1 pF or thereabouts for small-signal JFETS, single-gate MOSFETs and most bipolar transistors designed for RF applications.

N channel *power* MOSFETs for switching applications are available with drain voltage ratings up to 500 V or more and are capable of passing 20 A with a drain/source voltage drop of only a few volts, corresponding to a drain/source resistance in the fully on condition of $r_{ds\,on}$ on of just a few hundred milliohms. Other devices with lower drain voltage ratings exhibit $r_{ds\,on}$ resistances as low as 0.010 ohms, and improved devices are constantly being developed and introduced. Consequently these figures will already doubtless be out of date by the time you read this. A very high drain voltage rating in a power MOSFET requires the use of a high-resistivity drain region, so that very low levels of $r_{ds\,on}$ cannot be achieved in high-voltage MOSFETS.

A development which provides a lower drain/source voltage drop in the fully on condition utilizes an additional P type layer at the drain connection. This is indicated by the arrowhead on the symbol for this type of device (Figure 3.9a). The device is variously known as a conductivity modulated power MOSFET or COMFET (trademark of GE/RCA), as a gain enhanced MOSFET or GEMFET (trademark of Motorola), and so on. Like the basic MOSFET these are all varieties of *insulated gate* field effect transistors (IGFETs). The additional heavily doped P type drain region results in the injection of minority carriers (holes) into the main N type drain region when the device switches on, supplementing the majority carrier electrons and reducing the drain region on voltage drop. However, nothing comes for free in this world, and the price paid here is a slower switch-off than a pure MOSFET; this is a characteristic of devices like bipolar transistors which use minority carrier conduction. An interesting result of the additional drain P layer is that the antiparallel diode inherent in a normal power MOSFET – and in Darlingtons – is no longer connected to the drain. Consequently COMFETS, GEMFETs and similar devices will actually block reverse drain voltages, i.e. N channel types will not conduct when the drain voltage is negative with respect to source. Indeed, the structure has much in common with an insulated gate silicon controlled rectifier (SCR) – and SCRs together with other members of the thyristor family are the next subject in this necessarily brief review of active devices.

Thyristors

The name thyristor, given because the action is analogous to that of the thyratron tube (a gas-filled triggered discharge valve), applies to a whole family of semiconductors having three or more junctions (four layers or more). These devices, which can have two, three or four external terminals, are bistable (they are either off or on) in operation and may be unidirectional or bidirectional. This extensive family of devices is summarized in Figure 3.10, which also indicates the maximum voltage and current ratings typically available in each type of device.

The best-known member of this family is the

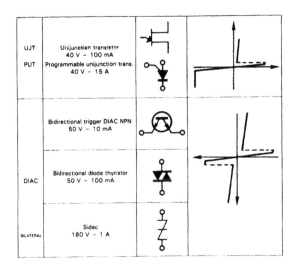

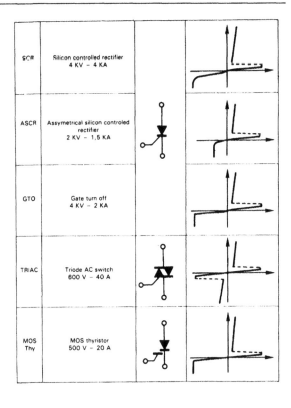

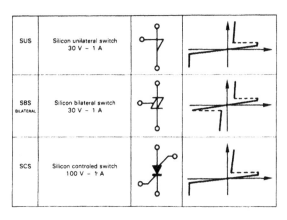

Figure 3.10 Four-layer (three-junction) device family, with typical characteristics.
   The UJT is a single-junction device but is shown here as its characteristic resembles that of the three-junction PUT. Likewise, the characteristic of the two-junction diac resembels that of the three-junction bidirectional diode thyristor. (Reproduced by courtesy of Motorola Inc.)

silicon-controlled rectifier, also known as a reverse blocking triode thyristor: it is a unidirectional conducting device having three terminals (anode, cathode and gate). This is a four-layer device (PNPN), the operation of which can be represented by two complementary transistors so interconnected that the collector current of each supplies the base current of the other (Figure 3.11a). If a current is supplied to the base of the NPN section of the SCR from an external source, it will start to conduct, its collector supplying base current to the PNP section. If the resultant col-

lector current of the PNP section exceeds the externally supplied trigger current, both transistors conduct heavily and continue to do so even if the external base current is removed. This type of SCR will only turn off again when the current through the device falls below some minimum *holding current*. This is a current which is so small that the loop gain of the two devices no longer exceeds unity, so that one or other transistor no longer receives enough base current to keep it switched on. Thus the SCR makes good use of the fact that as the current through a transistor is reduced, the

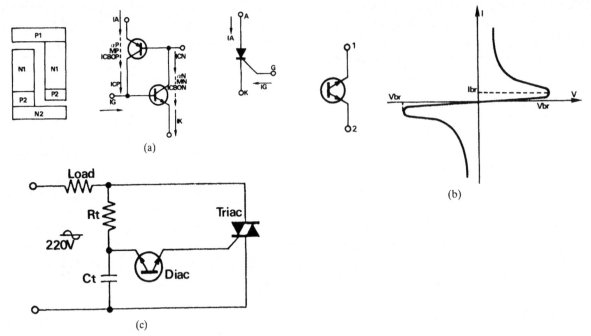

Figure 3.11
(a) The SCR modelled as a latching complementary DC coupled bipolar transistor pair with positive feedback.
(b) The diac.
(c) Typical application. (Reproduced by courtesy of Motorola Inc.)

current gain falls. The gate of an SCR should not be forward biased whilst the SCR is blocking a negative voltage at its anode, since this results in increased leakage current, causing higher dissipation. SCRs are the heavyweights of the active semiconductor world, being capable of controlling tens of kilowatts of power. A more recent development is the *gate turn-off* (GTO) thyristor, which is capable of being switched off again by injecting at the gate a pulse of current of the opposite polarity from the trigger current. The structure of the *MOS thyristor* is similar to a standard MOS power FET with the addition of a P type layer in series with the drain. This results in a four-layer NPNP structure which can be triggered on by a positive voltage at the gate of the MOSFET structure. The latter is brought out as the gate terminal, in place of the P base region of the NPN section. This results in a high-voltage, high-current device capable of controlling kilowatts of power. Whilst the steady-state gate current is zero, a low-impedance drive is required to turn the device on

rapidly, owing to the need to charge the effective gate input capacitance quickly.

The triode AC semiconductor switch or *triac* is a bilateral-controlled switch and can thus be used to control power in AC circuits. It can be triggered on by either a positive or negative input current at the gate relative to main terminal 1 (MT1), whatever the polarity of the voltage at MT2. MT1 corresponds to the cathode of SCR and MT2 to the anode; clearly the terms anode and cathode are inappropriate to a device which can conduct in either direction. As with all members of the thyristor family, a substantial pulse of trigger current is desirable in order rapidly to switch the device on fully. This ensures that the full supply voltage appears rapidly across the load, minimizing the turn-on period during which the device is passing current whilst the voltage across it is still large. This is particularly important in high-frequency applications where switching losses represent a large proportion of the device dissipation. Rapid injection of trigger current can be ensured by using a two-terminal trigger device such as a diode AC switch or *diac* (Figure 3.11b). This symmetrical device remains non-conducting, until the voltage across it – of either polarity – exceeds some breakover value (typically 20 to 30 V), when it switches to a low-resistance state. This discharges the capacitor C_t via the triac's gate terminal, ensuring rapid turn-on.

Operational amplifiers

The final few pages of this chapter are devoted to operational amplifiers (opamps) and comparators. Both are linear integrated circuits (ICs) composed of many components, active and passive, formed in a single die of silicon. Their manufacture uses the same range of processes used in the production of discrete semiconductor devices, namely photo-lithography, diffusion, ion implantation by bombardment, passivation and oxide growth, metallization and so on. Although an opamp or a comparator may contain dozens of components, mainly active ones, it is here considered as a single active component. At one time this might have been considered a cavalier attitude, but opamps are now so cheap – often cheaper than a single small-signal bipolar transistor or FET – that such an approach is entirely justified. Indeed, in the frequency range up to one or two hundred kilo-hertz they are much simpler and more convenient to employ than discrete devices, which they furthermore outperform. The need for them existed long before the IC opamp was available. They were therefore produced first with discrete semiconductors and passive components, and later in hybrid form – transistor chips and miniature passive components mounted on an insulating substrate encapsulated in a metal or plastic multi-lead housing. And even before that there were operational amplifiers built with thermionic valves.

Figure 3.12a shows the usual symbol for an opamp. It has two input terminals and one output terminal. The *non-inverting* input terminal (NI or +) is so called because if it is taken positive with respect to the other input terminal, the output voltage will also move in a positive direction. Conversely, if the *inverting* input terminal (I or −) is taken positive

with respect to the other input terminal, the output will move in the inverse or negative direction. Ideally, the operational amplifier has no offset voltage: that is to say that if its input terminals are at the same voltage, the output terminal will be at zero volts, as shown in Figure 3.12b. Furthermore, the output should respond only to differential inputs, i.e. voltage differences between the two input terminals, so as the wiper of the potentiometer in Figure 3.12b is moved away from the central position towards either supply rail, the output voltage should remain unaffected.

A voltage variation common to both inputs, as in Figure 3.12b, is called a *common mode input*. If a 1 V common mode input results in a 1 mV change in output voltage, the *common mode rejection ratio* (CMRR) is described as 60 dB or 1000 : 1; this would be a very poor performance for a modern opamp. An opamp's rejection of common mode AC signals is poorer than for DC voltage changes, and is progressively worse the higher the frequency. Ideally, the common mode input range – the range of common mode voltage for which the CMRR remains high – would extend from the negative to the positive supply rail voltage. In practice, it often only extends to a point 2 or 3 volts short of each rail voltage, though with many opamps, including the CA3140 and the ubiquitous LM324, it does extend to and include the negative supply rail. This feature is becoming more common as many recently introduced opamps are designed for single-rail working, i.e. using just a positive supply and 0 V ground. A few opamps have a CM input range extending right up to the positive rail, the LF355 being one example, though this is only a typical, not a guaranteed, characteristic.

The earliest commonly available integrated circuit opamp, the 709, could exhibit an unfortunate behaviour known as *latch-up*. If the common mode input voltage exceeded the limits of the specified range, the sense of the amplifier's gain could reverse. This effectively interchanged the I and NI input terminals, turning negative feedback into positive feedback and locking the circuit up with the output stuck 'high' or 'low' as the case may be. Naturally, the circuit designer arranged for the input normally to remain within the

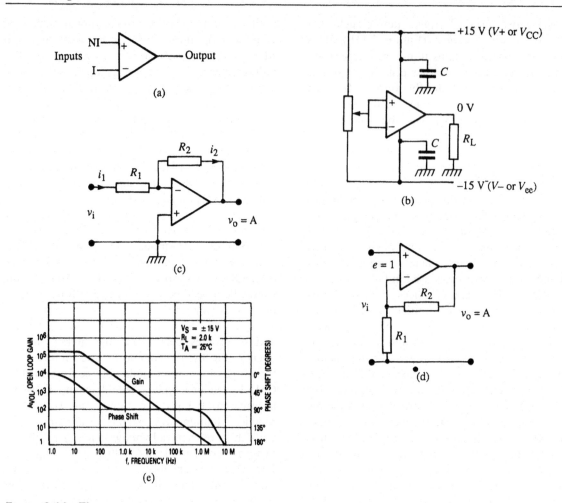

Figure 3.12 The opamp.
  (a) Symbol.
  (b) Ideal opamp before the application of feedback (see text).
  (c) The inverting connection.
  (d) The non-inverting connection.
  (e) Device TL081 gain and phase versus frequency, open loop (reproduced by courtesy of Motorola Inc.).

common mode range, but an unforeseen voltage spike capacitively coupled into the opamp's input circuit, or some other interference mechanism, could kick the circuit into the latched-up state. All modern opamps are arranged so that the gain does not reverse its sign when the CM range is exceeded.

Another important opamp parameter is the *power supply rejection ratio* (PSRR). A typical modern opamp would provide 100 dB PSRR at DC, but as with CMRR, this figure deteriorates with frequency. Modern opamps, like their thermionic forebears, are specified as having a certain minimum voltage gain at DC, though as with the other parameters, the gain falls with increasing frequency. Voltage gain is a particularly appropriate parameter for opamps with JFET or MOS input stages, as the input current for such types is negligible. However, bipolar opamps and their discrete semiconductor ancestors draw a small but finite input current – namely the base current of the transistors forming the input stage. In some cases this could be more significant than the input voltage, so the gain of discrete semiconductor opamps was often quoted as a *transresistance*, i.e. as so many volts change at the output per microampere change of input current. Modern bipolar opamps draw so little input bias current that they too are specified in terms of voltage gain.

Just as the input CM range would ideally extend from rail to rail, so ideally should the output range. Again, in practice, when driving a load connected to ground as in Figure 3.12b, most opamps will run out of steam when the output voltage is within one volt or so of the supply rail. Opamps with *complementary* MOS (CMOS) output stages can better this, reaching ever closer to the supply rail the higher the resistance of the load. Other characteristics of the ideal opamp of Figure 3.12b are very high voltage gain (even the venerable 709 managed 12000 minimum, 45000 typical), a low output resistance (50 ohms open loop is typical, reduced in practice by negative feedback), and a gain maintained constant up to as high a frequency as possible.

The very high gain of the opamp is the key to its usefulness. Imagine it connected as in Figure 3.12c, where for clarity the necessary power supplies and highly advisable local decoupling capacitors have been omitted. Whatever the output voltage v_o, the differential input voltage between the two input terminals will be exceedingly small owing to the amplifier's enormous gain. Therefore the two inputs must be at virtually the same input voltage, namely 0 V. Since the NI terminal is grounded: the inverting terminal can thus be described as a 'virtual earth'. If it is further assumed that the amplifier's transresistance is very high, i.e. the input current it draws is negligible, then $i_1 + i_2 = 0$, so $i_1 = -i_2$ and actually flows in the direction shown. So $v_o = i_2 R_2 = -i_1 R_2$ whence $v_o/v_i = -R_2/R_1$. So the gain of the complete circuit is defined by the ratio of the two resistors, provided only that the opamp's voltage gain is exceedingly high and that the input current it draws is negligible compared with i_1. Naturally enough, Figure 3.12c is known as the *inverting connection* and, bearing in mind that in this case the inverting terminal is a virtual earth, the input resistance of the circuit is just R_1. You can have any gain you like, from less than unity upwards, by

selecting the appropriate ratio of R_2/R_1, providing it falls well short of the *open loop gain A* of the opamp, i.e. the gain measured with $R_2 = \infty$ and $R_1 = 0$.

Remember that the opamp manufacturer usually quotes a typical value for A and also a minimum value, say one-third of the typical, but no maximum value. You can easily see what this means in terms of gain accuracy. Take the *non-inverting connection* of Figure 3.12d, for example.

The voltage at the junction of R_2 and R_1 equals $v_o R_1/(R_1 + R_2)$. Denote the fraction $R_1/(R_1 + R_2)$ of the output voltage fed back to the input by the symbol β. Also, assume that the output voltage is numerically equal to A, so that the opamp's input voltage e is just unity, as indicated in Figure 3.12d. Then by inspection,

$$v_i = \left(\frac{R_1}{R_1 + R_2}\right)A + 1 = \frac{AR_1 + R_1 + R_2}{R_1 + R_2}$$

and $v_o = A$. So

$$\frac{v_o}{v_i} = \frac{A(R_1 + R_2)}{AR_1 + (R_1 + R_2)}$$

and dividing top and bottom by $R_1 + R_2$ gives

$$\frac{v_o}{v_i} = \frac{A}{AR_1/(R_1 + R_2) + 1} = \frac{A}{1 + A\beta}$$

This is the gain with *negative feedback*, i.e. when the fraction β of the output is subtracted from the input. If $A\beta$ is very much greater than unity then the denominator approximately equals $A\beta$, and so the gain simply equals $1/\beta$. Suppose you want a stage gain of $\times 100$ or 40 dB, then you might make $R_1 = 100$ R and $R_2 = 9$ K9, whence $\beta = 0.01$. If the type of opamp being used had a minimum open loop gain (often called the large-signal open loop voltage gain A_{vol} or the large-signal differential voltage amplification A_{vd} and quoted in volts per millivolt in manufacturers' data sheets) of 10000 (or 10 V/mV), then the worst case gain will be

$$\frac{v_o}{v_i} = \frac{10\,000}{1 + (0.01 \times 10\,000)} = 99$$

or 1% low. Clearly the larger the gain required, the smaller β must be, and the larger A must then be to ensure that $A\beta \gg 1$.

In the inverting connection of Figure 3.12c one is not feeding back a fraction of the output voltage, but rather feeding back a current proportional to v_o and balancing this against a current proportional to the input v_i, so a different approach is appropriate. Equating the currents through R_1 and R_2,

$$\frac{v_i - 1}{R_1} = \frac{1 - (-A)}{R_2}$$

whence

$$v_i = \frac{R_1(1 + A)}{R_2} + 1$$

and of course $v_o = -A$. Hence, after a little rearrangement,

$$\frac{v_o}{v_i} = \frac{-A}{1 + (A + 1)/G}$$

where $G = R_2/R_1$. If $(A + 1)/G \gg 1$ then

$$\frac{v_o}{v_i} = \frac{-A}{(A + 1)/G} \approx -G$$

so G is simply the *demanded gain*, as derived earlier. Interestingly, in this inverting case, where one can choose R_1/R_2 so as to give a gain of less than unity, the condition $(A + 1)/G \gg 1$ may not be sufficient. In fact, A must be much greater than G or unity, whichever is the larger.

This analysis of opamp circuits has been in terms of ratios such as A and G and of symbols for variables such as v_i and v_o. Apart from stating v_i to be the input voltage, it has not been defined exactly, so it could apply equally well to a change of input voltage from one steady value to another, or to a continuously varying signal like a sine wave. In the latter case, however, the simple analysis above only applies exactly up to the highest frequency at which A is a real number, having no associated phase shift to provide a j component. In fact, for many opamps A begins to fall at quite a low frequency. This is due to a top-cut or low-pass characteristic deliberately built in to the opamp. Figure 3.12e shows characteristics typical of an opamp of this type, known as *internally compensated*. The open loop gain of such an opamp starts to *roll off* at a frequency in the order of 10 Hz and continues to fall at a rate of 6 dB per octave, associated with a 90° phase lag, until it reaches

unity. Beyond this unity-gain bandwidth, typically 3 MHz in the case of the TLO81, the rate of roll-off of most opamps increases to -12 dB per octave or beyond, associated with a phase shift of 180° or more. At these frequencies the negative feedback provided via R_1 and R_2 will have become positive. However, as the amplifier's gain is by then less than unity the circuit will be stable even in the case where $\beta_1 = 1$ (i.e. 100% feedback: $R_1 =$ infinity in Figure 3.12d).

When building or modifying existing circuits or developing new ones, it not infrequently happens that they don't work as expected. When trying to diagnose the problem, there is no real substitute for a thorough understanding of how they should work. So it's worth looking at the inverting opamp connection in more detail, remembering that only at very low frequencies is A_{vol} a real number with no associated phase shift.

Figure 3.13b shows the response of a typical opamp in the inverting mode with the feedback resistor R_f in Figure 3.13a open-circuit. If R_i is short-circuited (or the input impedance of the opamp is very high) then the voltage e at the inverting input is simply v_i. Taking this to be unity, then at very low frequencies the output will be numerically equal to A_{vol} which is typically 200 000 in the case of the opamp whose characteristics are illustrated in Figure 3.12e. Figure 3.13b presents the same information as the Bode plot of Figure 3.12e, but represented as an output vector whose magnitude and phase relative to the unity input vector varies with frequency in the way illustrated. Figures 3.15e and 3.16b both illustrate the open loop response of the amplifier, i.e. the response in the absence of feedback. In the vector diagram of Figure 3.13b, the unit input has been drawn to the left of the diagram's origin or ground reference point, so that v_o extends to the right at low frequencies, giving us the sort of circle diagram met earlier. Here, though, as already noted, at frequencies beyond the unity-gain frequency the phase shift increases to 180°, corresponding to a 12 dB per octave roll-off, resulting in the little pothook on the locus of v_0 as ω tends to infinity. Figure 3.13c shows the closed loop vector diagram for the inverting unity-gain connection, i.e. where $A = 1$. At

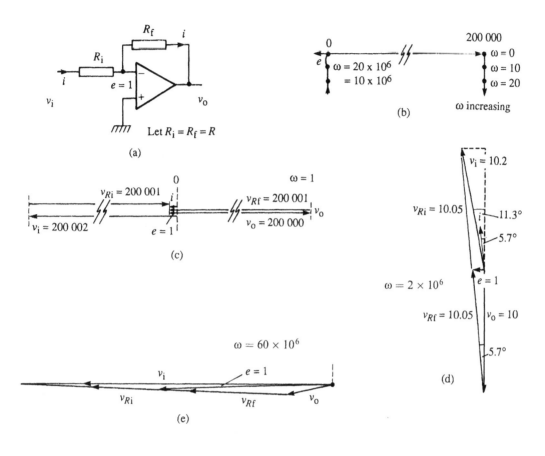

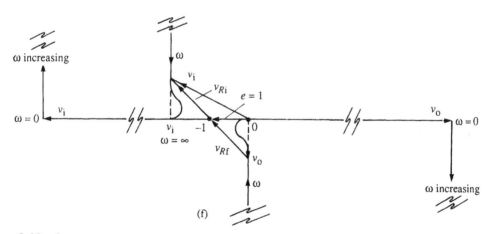

Figure 3.13 Opamp inverting connection.
 (a) Circuit.
 (b) Typical open-loop gain/frequency plot.
 (c–e) Closed loop vector diagrams for $\omega = 1$ (0.16 Hz), $\omega = 2 \times 10^6$ ($\sim$ 300 kHz), $\omega = 60 \times 10^6$ ($\sim$ 10 MHz).
 (f) Closed loop vector diagrams combined, showing frequency locus.

frequencies well below 10 Hz, A_{vol} has zero phase angle, so the vectors are all in line, in Figure 3.13c therefore they have been offset vertically for clarity. Note that for an A_{vol} of 200 000 the gain is just short of unity by one part in 100 000, since v_{Ri} (the voltage across the input resistor R_i) and v_{Rf} both equal 200 001, as shown. Figure 3.13d shows the situation at 300 kHz, where the open loop gain has fallen to 20 dB or ×10, again taking the voltage at the opamp input e as unity. The output voltage v_o lags $-e$ by 90° but, as there is still 20 dB of loop gain in hand, v_o only lags v_i by an additional 11.3° over the expected 180°. By *loop gain* is meant the excess of the opamp's gain at any given frequency over the demanded gain at that frequency. It can be seen from the vector diagram that although the open loop gain of the TL081 at 300 kHz is only ×10, the gain is only 2% short of unity. This is due to the 90° phase shift between e and v_o: without the phase shift, the vector diagram would look like Figure 3.13c and the gain shortfall would be 20%. Figure 3.13e shows the situation at 10 MHz, well beyond the unity-gain frequency. v_o now has almost 180° of excess phase shift, bringing it almost back in phase with e and v_i. Figure 3.13f summarizes the closed loop cases of Figure 3.13c, d and e, showing the loci of v_i and v_o relative to the inverting terminal input voltage e of unity. Note that the behaviour shown beyond the unity-gain frequency is a bit conjectural as it assumes that the opamp's output impedance is zero. In fact, in the case of the TL08I it is nearer 200 ohms, although at low frequencies this is reduced to a very low effective value by the application of negative feedback. However, at 10 MHz and beyond there is no loop gain to speak of, so the locus of v_o will not finish up actually at the origin as ω tends to infinity, but somewhere else close by. Figure 3.14a shows the general non-inverting connection. Figure 3.14b shows the corresponding zero-frequency vector diagram for a demanded gain of +2. If you compare it with Figure 3.13c you will find that it is identical except for the location of the zero-voltage or ground reference point, and the gain shortfall is likewise one part in 100 000 for an A_{vol} of 200 000. If, however, R_2 = zero, making

the demanded gain +1, then the actual gain is only 1 in 200 000 parts less than unity, since the loop gain $= A_{vol}$.

Opamps which are internally compensated for use at gains of unity upwards, such as the TL081, make useful general purpose opamps. If they are to be used for audio-frequency operations where any small DC offset is unimportant, then offset adjustment connections are unnecessary. Consequently only three pins per opamp are required, namely NI and I inputs and the output. So a standard 14-pin dual in-line (DIL) plastic package can accommodate four opamps, with the two remaining pins providing connections for the positive and negative supply rails (e.g. TL084, LM324 and MC3404). However, internally compensated opamps are not the best choice for applications requiring a gain substantially in excess of unity; the higher the required gain, the less optimum they prove. Figure 3.14c shows that at a gain with feedback (GWF) or demanded gain of ×1000 or 60 dB, the internally compensated TL081 family will provide a frequency response which is flat to only a little over 1 kHz. For this application the uncompensated TL080 would be a much more appropriate choice. Indeed, as Figure 3.14d shows, even the externally compensated μA709 opamp, dating from the 1960s, can provide 60 dB gain at a small-signal bandwidth of 300 kHz when appropriately compensated. Note that the rate of change of output voltage of an opamp is limited to some maximum rate called the *slew rate*, determined by the IC process used and the compensation components, internal or external. This slew rate limit results in the full-power bandwidth being substantially less than the small-signal bandwidth.

There is a half-way house between opamps internally compensated for unity gain and those completely uncompensated. This is the *decompensated* opamp, which is partially internally compensated and can be used without further external compensation, down to a specified gain. For example, the Motorola MC34085 quad JFET input opamp is compensated for A_{vcl} (closed loop gain or gain with feedback) of 2 or more, whilst the Signetics 5534 bipolar single low-noise opamp is compensated for gains of 3 or more.

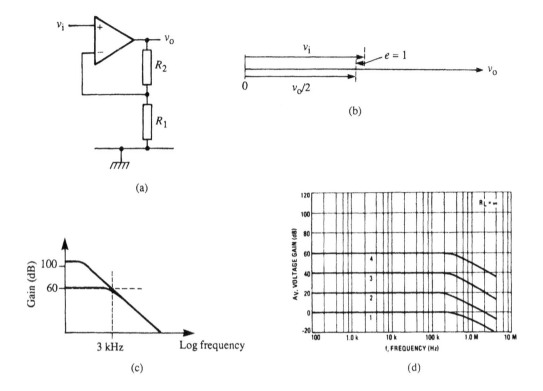

Figure 3.14   More about opamps.
  (a) Non-inverting connection. Demand gain $= 1/\beta = (R_1 + R_2)/R_1$.
  (b) For the case where $\beta = 0.5$ $(R_2 = R_1)$.
  (c) Gain/bandwidth plot for gain $= 1000$ (60 dB), for internally compensated opamp (see Figure 3.12e).
  (d) $G = 60, 40, 20$ and 0 dB for externally compensated opamp MC1709, with appropriate compensation components for each gain setting (reproduced by courtesy of Motorola Inc.).

A long-standing limitation on the IC opamp manufacturer's art has been the difficulty of producing high-performance on-chip PNP transistors. The traditional method was to use lateral PNP transistors, but these have a transition frequency f_T of only about 2 MHz, severely limiting the obtainable performance of the complete opamp. But some time ago now, National Semiconductor introduced the vertically integrated PNP (VIP) structure, leading to opamps such as the LM6161/64/65 family with a 300 V/μs slew rate, a 4.5 MHz power bandwidth and drawing only 4.8 mA supply current. The LM6161 is unity-gain stable, whilst the 6165 is stable for gains of ×20 or more.

Another exciting development is the current feedback opamp. Standard opamps have a high

input impedance at both the inverting and non-inverting terminals. However, when an opamp is used in the inverting mode, where the non-inverting terminal is tied to ground, the input impedance is simply equal to R_1 whilst when it is used in the non-inverting mode, the inverting input terminal is internal to the circuit, the input 'seeing' only the non-inverting input. The conclusion is that there is really no need in practice for a high input impedance at the device's inverting input.

Current feedback opamps capitalize on this relaxation and use a low-impedance inverting input. Without going into the details, this permits the opamp designer effectively to remove a pole in the frequency response. The result is that whereas the bandwidth of a regular opamp falls as you demand a higher set gain, owing to its

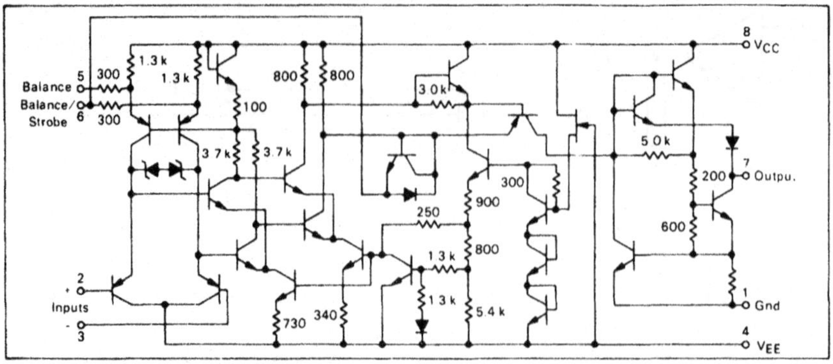

(a)

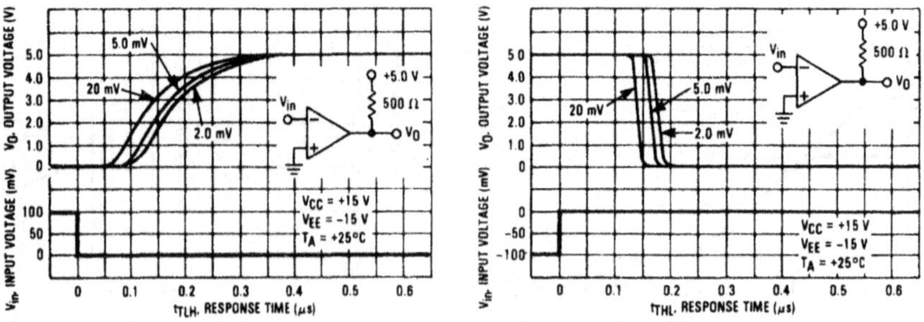

(b)

Figure 3.15
(a) Typical comparator.
(b) Performance (response time) for various input overdrives (LM111/211/311) (reproduced by courtesy of Motorola Inc.).

limited gain-bandwidth product, that of a current feedback opamp does not – at least to a first-order approximation. Current feedback opamps can therefore offer wider bandwidth at high gain, faster slew rates and shorter settling times than standard opamps, even decompensated ones. Nor is this premium AC performance achieved at the cost of sacrificing the DC characteristics. Input offset voltages and temperature coefficients as low as 300 μV and 5 μV/°C (typical) are available in a device with a 100 MHz bandwidth, making current feedback opamps an excellent choice for critical applications such as 12-bit two-pass sub-ranging flash analog-to-digital converters.

At the time of writing, current feedback opamps are available from five American manufacturers, including Analog Devices. We can expect to see more manufacturers offering such devices in the future.

Finally, in this section on opamps, mention must be made of *Norton* opamps such as the LM3900 and the MC3301. These are generally similar to conventional opamps but, although they have a differential input, it is ground referenced rather than floating, as in a conventional opamp. Consequently they are mainly used in single supply rail AC coupled circuits and less demanding DC comparator or logic applications.

Comparators

To finish this review of active components, a word about comparators. The sole purpose in life of a comparator is to signal as quickly and as accurately as possible whether the voltage at one of its inputs is positive or negative with respect to the other input. This represents one *bit* of information, and a comparator's output is designed to be compatible with transistor-transistor logic (TTL) and other logic families. One of the earliest comparators is the 710, and this is still one of the fastest general purpose comparators, only bettered by special purpose emitter coupled logic (ECL) compatible types. Later comparators such as the LM111 are suitable for use with $+15$ V and -15 V supplies as used by opamps, rather than the less convenient $+12$ V and -6 V rails required by the 710.

Although comparators, like opamps, have a differential input, high values of gain, common mode and power supply rejection, and even in some cases (see Figure 3.15a) input offset adjustment facilities, they are very different in both purpose and design. A comparator's response time is quoted as the time between a change of polarity at its input (e.g. from -95 mV to $+5$ mV or $+95$ mV to -5 mV, i.e. a 100 mV change with 5 mV overdrive) and the resultant change of logic state at its output (measured at half-swing), as illustrated in Figure 3.15b. Given suitable circuitry to interface between its rail-to-rail swing and logic levels, an opamp can be used as a comparator. However, its response time will generally be much longer than that of a purpose designed comparator. For the opamp is only optimized for linear operation, wherein the input terminals are at virtually the same voltage and none of the internal stages is saturated or cut off. Its speed of recovery from overdrive at the input is unspecified, whereas the comparator is designed with precisely this parameter in mind. Furthermore, whilst an opamp may be specified to survive a differential input voltage of ±30 V, it is not designed to be operated for long periods *toggled*, i.e. with any substantial voltage difference between the NI and I terminals. Extended periods of operation in the toggled state, especially at high temperatures, can lead to a change in input offset voltage in certain opamps, e.g. BIMOS types.

Reference

1. Current-feedback opamps ease high-speed circuit design. P. Harold, *EDN*, July 1988.

Questions

1. Define intrinsic silicon. What type of diode depends upon it for its action?
2. Which has the greater mobility in silicon, holes or electrons? What is the consequence for RF transistors?
3. What is the incremental or slope resistance R_d of a diode, in terms of the current I mA that it is passing?
4. In normal use, the peak-to-peak RF voltage across a varicap diode is kept small relative to the bias voltage. In what application is this not the case?
5. How does the forward base emitter voltage of a transistor, at a given collector current, vary with temperature? How is a transistor's f_T measured?
6. What is the approximate value of the low-frequency input resistance of a common emitter transistor when $I_c = 2$ mA?
7. What is r_{bb}' and what is its significance for the rise and fall times of a switching transistor?
8. What restrictions are there on the operating V_{gs} of a junction FET? What does FET action depend upon; majority or minority carriers?
9. Why is an opamp such as the TL080 preferred over an internally compensated type, such as the TL081, for an audio frequency amplifier with 60 dB gain?
10. What is the minimum gain which can be provided by an opamp operating in the non-inverting connection?

4 Audio-frequency signals and reproduction

Audio frequencies are those within the range of hearing – which is a definition vague enough to mean different things to different people. To the line communication engineer, everything has to fit within a 4 kHz wide telephony channel, whilst to the radio telephone engineer, audio frequencies are those between 300 Hz and 3 kHz. To the hi-fi enthusiast the definition means 20 Hz to 20 kHz, no less, even though only the young can hear sound waves of 15 kHz and above.

This chapter looks at the amplification of audio-frequency electrical signals and their transformation into sound waves. It also looks at the inverse process of turning sound waves into electrical signals and of recording and playing them back, and so will adopt the hi-fi definition of audio as 20 Hz to 20 kHz.

Audio amplifiers

Figure 3.3c shows a perfectly serviceable, if simple, audio-frequency (AF) amplifier. This is a small signal *single-ended class A* amplifier, meaning that it employs a single transistor biased so as to conduct all the time, with equal positive and negative current swings about the mean current. Of the 9.9 mW of power it draws from the supply, most is wasted, even when it is producing the maximum output voltage swing of which it is capable. But this is generally of little consequence since efficiency is only really important in a later power output stage, where one may well be dealing with many watts rather than a few milliwatts. The gain of the circuit of Figure 3.6c can be estimated without even putting pencil to paper, by making certain approximations:

1. The device's current gain is very large, so that the collector current virtually equals the emitter current.
2. Base/emitter voltage variations are due solely to the emitter current variations flowing through the transistor's internal r_e, where $r_e = 25/I_e$, i.e. the device's mutual conductance $g_m = (40 \times I_e)$ mA/V.
3. Collector voltage variations have negligible effect on the collector current.

Then at frequencies where the reactance of the 100 μF emitter decoupling capacitor is very low, a 1 mV input at the base in Figure 3.3c will cause a 40 μA change in emitter (collector) current, from approximation 2, since in this case $r_e = 25\,\Omega$ and so $g_m = 40$ mA/V. This will cause a collector voltage change of 40 μA × 3 K3 = 133 mV, or in other words

$$\text{voltage gain } A_v = \frac{\text{collector load resistance } R_c}{\text{internal emitter resistance } r_e}$$

Owing to the approximations, $A_v = 133$ is an optimistic upper bound, even assuming the input is provided from a zero-impedance source and the output feeds into a following stage with infinite input resistance. Now the input resistance of the circuit shown is approximately $h_{fe}r_e$, in parallel with the 33 K and 56 K bias resistors, say typically 3 K3 if $h_{fe} = 150$. So if the stage were the middle one in a long string of identical stages, then the gain per stage would simply be $h_{fe}/2$, since the collector current of one stage would be equally divided between its collector load resistance and the input of the following stage. This gives a gain for the stage of 150/2 and represents a more practical approximation than the figure of 133 derived above. The output impedance of the stage is approximately 3300 ohms, assuming the collector slope resistance is very high. Therefore the maximum power the stage can deliver will be obtained when the load resistance connected to the output, downstream of the 10 μF DC block, is also 3300 Ω (*see* maximum power theorem – Chapter 1). With its 1 mA standing bias current, the maximum current swing that the stage can handle without cutting off completely for a part of the half-cycle is 2 mA peak-to-peak (p–p). Of this, 1 mA p–p will flow in the load and 1 mA p–p corresponds (assuming a sine wave and no distortion) to $1/(2\sqrt{2})$ mA RMS. The power in a 3K3 load is then $W = I^2 R = \{(1/2\sqrt{2}) \times 10^{-3}\}^2 3300 \approx$ 0.4 mW, giving an efficiency at maximum output of $0.4/9.9 \approx 0.04$ or 4%.

Even if one discounts the power wasted in the bias resistors and the 2K7 emitter resistor, it is clear that a class A stage with collector current supplied via a resistor would result in a very inefficient power amplifier stage. The traditional solution to the problem of efficiency in single-ended class A output stages is transformer coupling: in certain circumstances a *choke feed* could be used instead (Figure 4.1a). Here, the circle represents an idealized power amplifier device, be it bipolar power transistor or power FET. When the alternating input signal v_i is zero, a constant standing current I_a flows through the device, defined by suitable bias arrangements (not shown). If the DC winding resistance of the choke L_c is much lower than the load resistance R_L, the DC current will all flow through the choke, as shown. However, any AC signal current i_a through the device will flow through R_L in preference to L_c, provided that the reactance ωL_c of the latter is much greater than R_L, where i_a represents the instantaneous value of the signal current, not its RMS value. This can be arranged to apply right down to the lower audio frequencies, less than 100 Hz say, by designing L_c to have a reactance X_L much greater than $2\pi 100 \times R_L$ – no simple task, as it happens. To obtain the required inductance coupled with a winding resistance much less than R_L a ferromagnetic core will be required. However, the choke is carrying a DC current, which will produce a flux in the core and could even saturate it. Therefore an airgap is introduced into the magnetic path, such as to increase the reluctance and set the DC component of flux at just one-half of the maximum usable value. But designing the choke is a little previous. First one must decide the value of R_L which in turn is determined by the available supply voltage V_s and the output power P_o which is required. Given V_s and P_o, then an approximate value for R_L can be simply calculated. From Figure 4.1a the maximum peak-to-peak current swing i_{pp} max cannot exceed $2I_a$, for I_a cannot fall below zero and, to avoid distortion, the positive and negative swings must be symmetrical. Likewise the voltage at the active device's output v_{pp} max cannot fall below zero and is thus limited to $+2V_s$ on the other half-cycle. Since the peak-to-peak current or voltage excursion of a sine wave is $2\sqrt{2}$ times the RMS value, the maximum output power available cannot exceed $P_o = (v_{pp}/2\sqrt{2})(i_{pp}/2\sqrt{2})$, given a load resistance $R_L = v_{pp}/i_{pp} = V_s/I_a$; whence $P_{o\,max} = v_{pp}^2/(8 \times R_L) = i_{pp}^2 R_L/8$. This power would actually be obtainable with an ideal device, as can be seen in Figure 4.1b, where a *load line* representing R_L has been superimposed on an idealized I/V plot. Note that the values of the parameter v_i shown are arbitrary incremental values relative to the quiescent no-signal condition. Note also that on the negative-going swings of v_i, when the current i is less than I_a, the stored energy in the choke causes the output voltage to rise above V_s, forcing a current $I_L = i_a - I_a$ through R_L.

In practice, the power device will have a finite bottoming voltage and its output characteristic will only approximate the ideal shown, as in Figure 4.1c. This will have two effects: first, the maximum available output power will be less than $v_{pp}^2/8R_L$; and second, the uneven spacing of the v_i lines will introduce distortion. For a given V_s a value of I_a and R_L would be chosen so as to maximize both the peak-to-peak voltage and current swings: this means choosing a load line to extend into the knee of the highest practicable i_a curve as shown. The maximum and minimum values of current $I_{a\,max}$ and $I_{a\,min}$ and of voltage V_{max} and V_{min} (as indicated in Figure 4.1c) then

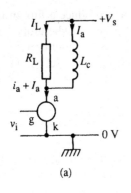

(a)

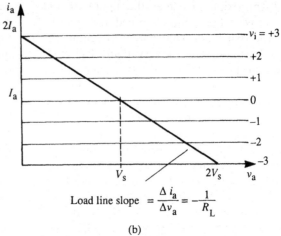

$$\text{Load line slope} = \frac{\Delta i_a}{\Delta v_a} = -\frac{1}{R_L}$$

(b)

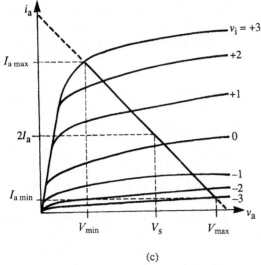

(c)

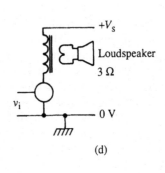

(d)

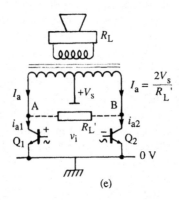

(e)

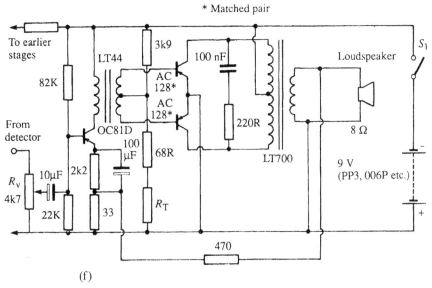

(f)

Figure 4.1 Audio amplifiers.
  (a) Choke fed single-ended class A stage.
  (b) Ideal output characteristics.
  (c) Practical output characteristics.
  (d) Transformer coupled class A output stage.
  (e) Transformer coupled push-pull class A amplifier.
  (f) Practical form of (e) as used in early transistor radios (operates in class B for battery economy).A temperature dependent resistor R_T may be included to compensate for variation of the output transistors' V_{be} with temperature. This avoids crossover distortion at low temperatures, and excessive battery drain when hot. The modest amount of negative feedback improves fidelity from shocking to poor.

define the load line and hence R_L and also the maximum power output. A little more power would be obtainable by increasing the peak-to-peak input v_i , but only at the expense of rapidly worsening distortion. Conversely distortion at full rated output could be reduced, at the cost of poorer efficiency, by designing the stage to have an actual maximum output considerably in excess of its nominal rated output. However, distortion can be more economically reduced by the application of negative feedback, as will appear later.

The choke feed arrangement of Figure 4.la is inconvenient in that usually R_L cannot be freely chosen: for example, in an audio power amplifier R_L would be a loudspeaker with a resistance of 3, 8 or 15 Ω typically. Thus the *transformer-coupled* arrangement of Figure 4.ld is more practical, enabling the speaker resistance to be transformed to whatever value of R_L is convenient, by suitable choice of transformer ratio. An upper bound on

the efficiency of a single-ended class A output stage is readily derived from Figure 4.1b. In the no-signal condition, dissipation in the power device is $V_s I_a$ watts; the output power is zero, and so is the efficiency. At full output power, $V_{pp} = 2V_s$, so the power in the load is $(2V_s)^2/8R_L = V_s^2/2R_L = V_s I_a/2$. Meanwhile, the supply delivers a sinusoidal current varying between zero and $2i_a$ at a constant voltage of V_s. So the average power drawn from the supply is still $V_s I_a$ watts. Thus the *efficiency* η (lower-case Greek letter eta) is given by

$$\eta = \frac{\text{useful power in load}}{\text{total power supplied}} = \frac{V_s i_a/2}{V_s i_a} = 50\%$$

The power in the load thus equals the power dissipated in the active device, and this seems to agree with the maximum power theorem, but does it? It was shown earlier that the maximum power theorem states that the maximum power will be obtained in the load when the load resistance equals the internal resistance of the source. Now the output slope resistance of the ideal active device in Figure 4.lb (the source resistance as far as R_L is concerned) is infinitely high, and still much higher than R_L even in the practical case of Figure 4.1c. The solution is quite simple: the value of R_L chosen in practice is that which will extract the maximum power from the source within the limits of the available voltage and current swings – for the output device is not an ideal generator in its own right, but only a device controlling the flow of power from the ideal voltage source V_s. As applied to the output of an amplifier, the maximum power theorem is in effect a 'maximum gain theorem'. If the drive voltage v_i is so small that the output voltage swing will not exceed $2V_s$ even with $R_L = \infty$, then it is indeed true that maximum power will be obtained in a load R_L equal to the output slope resistance of the device.

Figure 4.le shows a *transformer coupled push-pull* class A amplifier, sometimes described as class A *parallel* push-pull. The load resistance R_L is transformed to an equivalent collector-to-collector load resistance R_L', where $R_L' = 2V_s/I_a$. As transistor Q_1 swings from bottomed to cut-off, point A swings from 0 V ($2V_s$ negative with respect to point B) to $+2V_s$ ($2V_s$ volts positive with respect to B). Hence $V_{pp} = 4V_s$ and $I_{pp} = 2I_a = 2(2V_s/R_L')$. Maximum power in the load is thus $2V_s^2/R_L'$ watts, whilst power drawn from the supply is $4V_s^2/R_L'$ giving $\eta = 50\%$, the same result as for the single-ended class A stage. However, there are several advantages compared with a single-ended stage. First, for a given rating of transistor, twice the output power can be obtained, but most importantly there is now no net DC flux on the transformer core, so no airgap is needed. This makes it easier – and more economical – to provide a high primary inductance and low winding resistance. On the debit side, the two transistors should be a well-matched pair, and the circuit requires a balanced drive voltage v_i to provide the inputs needed by Q_1 and Q_2.

With a push-pull circuit like Figure 4.1e, efficiency can be increased by operating the devices in *class B* – an option not available with a single-ended stage as at Figure 4.1a. In class B the standing current I_a is reduced to a very low value, ideally zero. The stage now draws no current when $v_i = 0$, giving rise to the alternative name of *quiescent* push-pull. If you assume the same peak current through each transistor, namely $2I_a$ then the same value of R_L' is required for the class B stage as for the class A stage. The maximum output power for the class B stage then turns out to be the same as for the class A stage, but V_s supplies a pulsating current, namely a full-wave rectified sine wave of peak value $2I_a$. Now the average value of half a cycle of a sine wave is $2/\pi$ of the peak (a handy result to remember), so the power drawn from the supply V_s works out at $(8/\pi)(V_s^2/R_L')$ compared with $4V_s^2/R_L'$ for the class A stage. Thus the maximum efficiency at full output for a push-pull class B stage works out at $\eta = \pi/4$ or 78% against 50% for the class A stage. Of course, with practical devices, more like Figure 4.1c than b, the achievable efficiency will be less, say 70% at best; this gives rise to the rule of thumb that a class B output stage can supply an output power of up to five times the dissipation rating of each output transistor.

Matched devices are just as essential for a class B power amplifier as they are for a class A amplifier. However, the distortion in a class A amplifier will be low at small signal levels, since each transistor is operating in the most linear middle part of its current range. In a class B amplifier, on the other hand, each transistor is biased at one end of its current swing, i.e. is almost cut off. It is therefore difficult to avoid some crossover distortion (discussed later) and consequently some overall negative feedback (NFB) is always used with class B amplifiers. In the case of early transistor radios, this usually succeeded in improving the performance from shocking to poor.

A desirable economy in any audio power amplifier is the omission of the heavy, bulky and expensive output transformer. Figure 4.2 shows various ways in which this has been achieved. In Figure 4.2a a transformer is still employed but only for the drive signal, not the output. This arrangement is known as *series* push-pull, or even – confusingly – single-ended push-pull. A little later the arrangement of Figure 4.2b,

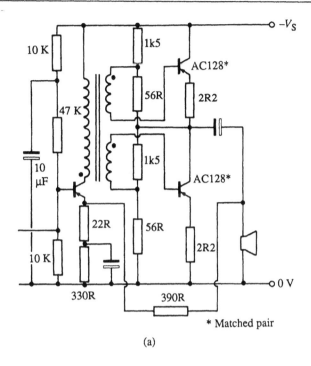

(a)

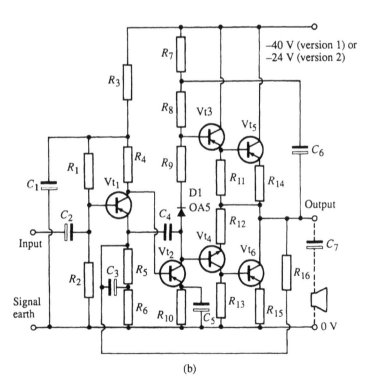

(b)

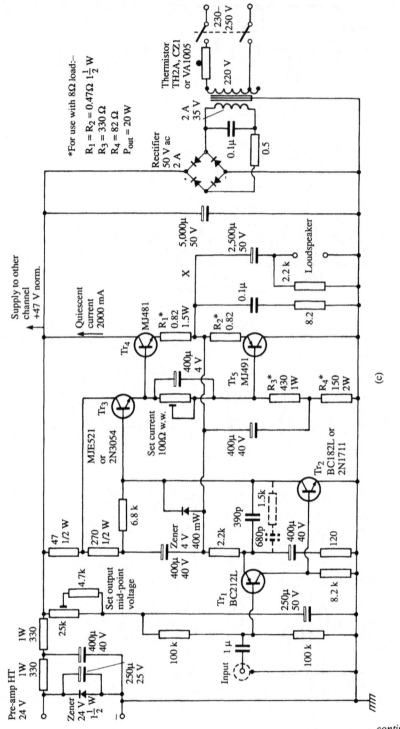

(c)

continued

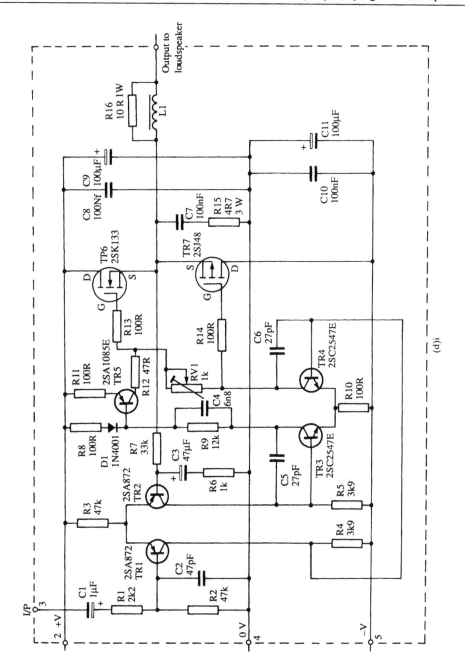

(d)i

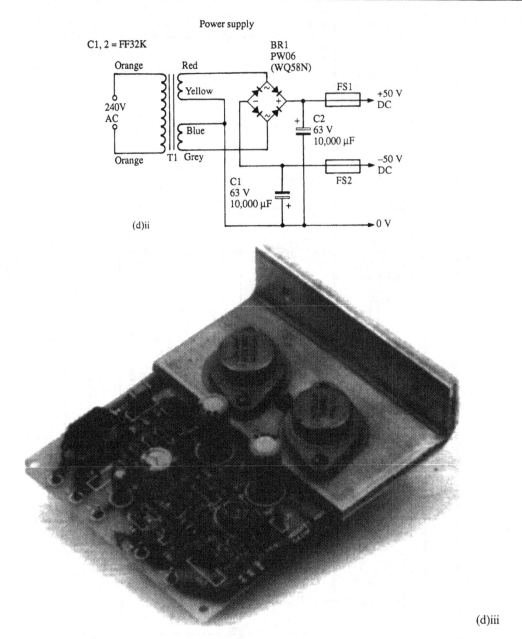

(d)ii

(d)iii

Figure 4.2   Audio power amplifiers.

(a) Series push-pull. Dot on coils indicates start of winding. All windings in same sense. Therefore all dot ends go positive together; dot voltages are in phase.

(b) Quasi-complementary push-pull.

(c) Full complementary push-pull amplifier. The dotted components (680 pF, 1.5 kΩ) can be added if electrostatic speakers are used.

(Parts (b) and (c) reproduced by courtesy of *Electronics and Wireless World*.)

(d) Complementary MOSFET high-power hi-fi audio amplifier. Power 150 W. Frequency response 15 Hz to 40 kHz. Total harmonic distortion 0.01% at 1 kHz. Input sensitivity 850 mV RMS for rated output. Damping factor 200 (8 Ω load). Main heatsink not shown. (Reproduced by courtesy of Maplin Electronic Supplies.)

known as *quasi-complementary* push-pull, became popular. This was because, at the time, silicon PNP high-power transistors were readily available whereas similarly rated NPN types with matching characteristics were unobtainable or very expensive, although lower-power NPN devices were common and cheap enough. Nowadays matched *complementary pairs* of bipolar power transistors are readily available (Figure 4.2c), whilst matched N and P channel audio power FETs are now commonly used in high-power audio amplifiers (Figure 4.2d). FETs have the advantage of a drain-current gate-voltage characteristic that provides a more gradual transition from the cut-off state to conducting than the corresponding collector-current/base-voltage characteristic of a bipolar transistor (compare Figures 4.3a and 3.2g). The trick with either bipolar or FET devices is to choose a small quiescent current such that the input/output characteristic of the stage is as linear as possible, this is known as *class AB* operation.

Practical circuits invariably incorporate measures to stabilize the quiescent current against temperature and supply voltage variations. Before looking at how NFB reduces the remaining distortion, here's an overview of what different sorts of distortion there are.

Distortion

The most noticeable and therefore the most objectionable type of distortion is due to *non-linearity*. Figure 4.3b shows, in exaggerated form, the sort of distortion which might be expected in a single-ended class A amplifier. Linear enough at very small signal levels, at maximum output the peak of the sine wave is more compressed on one half cycle than on the other. This type of distortion is called *second order*, because the transfer characteristic can be decomposed into two parts as shown: an ideal linear part and a parabolic component. Thus $v_o = A(v_i + kv_i^2)$, so if $v_i = V \sin(\omega t)$ then

$$v_o = A\{V \sin(\omega t) + k[V \sin(\omega t)]^2\} \qquad (4.1)$$

$AV \sin(\omega t)$ is the wanted amplified output, but we also get a term $Ak\{V \sin(\omega t)\}^2$. As the chapter on trigonometry in any maths textbook will tell you, $\sin^2\theta = (1/2)(1 + \sin 2\theta)$. So the unwanted term is

$$AkV^2 \sin^2(\omega t) = \tfrac{1}{2}AkV^2 + \tfrac{1}{2}AkV^2 \sin(2\omega t) \quad (4.2)$$

The term $(1/2)AkV^2$ is a small DC offset in the output, which is not audible and so need not concern us. However, if $\omega = 2765$ radians per second, making $AV \sin(\omega t)$ the note 440 Hz (concert pitch A above middle C), then the amplifier output will also include a component of amplitude $(1/2)AkV^2$ and frequency 880 Hz. Thus the second order distortion has introduced a second-harmonic component in the output which wasn't in the input. Figure 4.3b(iii) shows two sine waves, one twice the frequency of the other. You can see that if they are added, one peak will indeed be flattened whilst the other becomes even peakier, as shown in Figure 4.3b(i).

Assuming that $k = 0.1$ and that $V = 1$ corresponds to full rated-output, it follows from equations (4.1) and (4.2) that at full output the amplitude of the second harmonic is $(1/2)kV^2 = 0.05$ times the wanted fundamental, i.e. there is 5% distortion. At one-tenth of full output, where $V = 0.1$, the second-harmonic component is $(1/2)(0.1)(0.1)^2$ or 0.0005, representing only 0.5% distortion, or one-tenth of the previous figure. Expressing the same thing logarithmically this means that, for every 1 dB decrease or increase in v_i, the fundamental component of v_o falls or increases by 1 dB but the second-harmonic distortion component falls or increases by 2 dB, which explains why the distortion is worse in louder passages of music.

Now 5% of distortion on a single sine wave does not in fact sound at all objectionable, but you don't normally listen to isolated sine waves. The trouble begins when v_i includes many different frequencies at once, as for example in music. To get some idea of the problem, consider what happens when v_i consists of just two equal amplitude sine waves of different frequencies, say 1000 Hz and 1100 Hz. Then

$$v_o = A\left[\frac{V}{2}M + k\left(\frac{V}{2}M\right)^2\right]$$

where M stands for 'music'; in this case $M = \sin(\omega_1 t) + \sin(\omega_2 t)$, which has a maximum value of 2, giving a peak output voltage v_o of V as before, with $\omega_1 = 2\pi1000$, $\omega_2 = 2\pi1100$ radians per

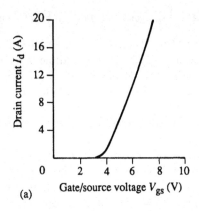

(a)

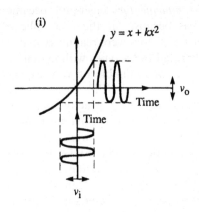

(i)

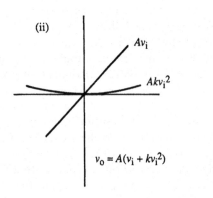

(ii)

$$v_o = A(v_i + kv_i^2)$$

(b)

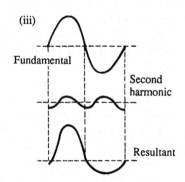

(iii)

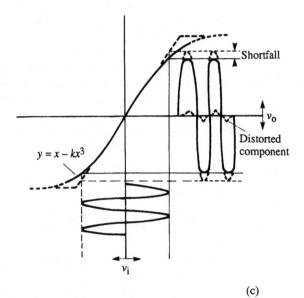

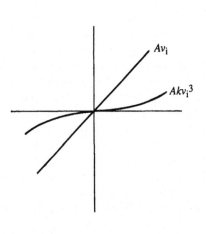

(c)

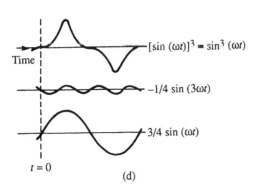

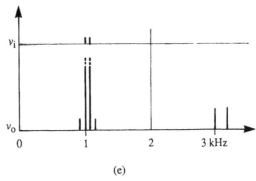

Figure 4.3 Even-order and odd-order distortion.
(a) Typical N channel MOSFET I_d/V_{gs} characteristic.
(b) Second-order distortion, typical of a single-ended class A amplifier.
(c) Third-order distortion, typical of a push-pull amplifier.
(d) Third-order distortion analysed.
(e) Third-order intermodulation distortion with two tones.

second. The distortion comes from the $k[(V/2)M]^2$ term, i.e. $k\{(V/_2)[\sin(\omega_1 t) + \sin(\omega_2 t)]\}^2$. Writing $(\omega_1 t) = A$ and $(\omega_2 t) = B$ for short, the expansion for

$$(\sin A + \sin B)^2 = \sin^2 A + \sin^2 B + 2 \sin A \sin B \tag{4.3}$$

is needed. The $\sin^2$ term has cropped up earlier, so you can tell from (4.3) that there will be components at the second harmonic of each tone, i.e. 2000 Hz and 2200 Hz. But there is also the 2 sin A sin B term. Now any handy maths textbook will tell you that $\cos(A + B) = \cos A \cos B - \sin A \sin B$ and $\cos(A - B) = \cos A \cos B + \sin A \sin B$. Subtracting the first from the second gives $\cos(A - B) - \cos(A + B) = 2 \sin A \sin B$. Thus v_o contains not only harmonics but also sum and difference frequencies of 2100 Hz and 100 Hz. These are not harmonically related and will therefore sound discordant. Clearly with the complex sounds of many different frequencies typical of music, be it pop or classical, there will be a host of discordant sum and difference products, giving the typically 'woolly' or 'muddy' sound of an early 78 RPM record or the soundtrack of a prewar film.

Figure 4.3c shows (exaggerated) the effect of the *third-order* distortion typical of a well-balanced class A push-pull amplifier. The transfer charac-

teristic is virtually linear at very small amplitudes, but at larger excursions the gain falls, being reduced by an amount proportional to the cube of the voltage swing. At some higher input amplitude than shown, the output will cease to increase altogether as the output devices bottom: but assume that over the range shown the distortion is purely third order. Then for a single tone as shown in Figure 4.3c,

$$v_o = A(v_i - kv_i^3)$$
$$= A\{V \sin(\omega t) - k[V \sin(\omega t)]^3\} \tag{4.4}$$

Now $\sin^3 \theta = (3/4) \sin \theta - (1/4) \sin 3\theta$, so there is a third-harmonic distortion term, but also a distortion component at the fundamental. The complete distortion term has been shown in Figure 4.3c, it represents the *shortfall* of v_o compared with what it would have been without the curvature of the transfer characteristic caused by the third-order term. This is shown to a larger scale in Figure 4.3d and also decomposed into its two constituent sine waves. Bearing in mind that it is multiplied by $-k$ in (4.4), the sin(ωt) distortion term indicates that the curvature of the transfer characteristic has reduced the fundamental output, as well as introducing a third-harmonic distortion term. Again, a few per cent of 1320 Hz tone when listening to the note A above middle C is not objectionable or even noticeable: it is the effect on complex sounds like music that is of concern. Substituting the two-tone signal previously denoted by M into (4.4) and turning the handle on the maths, you will find no less than half a dozen distortion terms, including components at the two input frequencies of 1000 Hz and 1100 Hz. These have been shown (exaggerated) in Figure 4.3e in the frequency domain rather than, as up to now, in the time domain. In this form the diagram indicates frequency and RMS amplitude but not relative phases of the components. The output contains not only the two input tones at 1000 and 1100 Hz and the third harmonic of each, but also third order *intermodulation products*. These are $2f_1 - f_2$ and $2f_2 - f_1$, in this case at 900 Hz and 1200 Hz. You will not be surprised to learn that (at least at modest distortion levels) 1 dB change in v_i results in a 1 dB change in the fundamental level in v_o but a 3 dB change in the distortion products. The intermodulation products, being close in frequency to the input tones from which they result, are very noticeable – even if not in principle more discordant than the sum and difference terms found in the case of second-harmonic distortion. Some writers have therefore claimed that a single-ended amplifier is preferable for the highest-fidelity applications to any push-pull amplifier. However, the reasoning is hardly convincing. It is true that if two identical single-ended stages are harnessed to work in push-pull, the remaining distortion will be purely odd order (mainly third). But an amplifier is not immune from third-harmonic distortion simply by virtue of being single ended: after all, if it is overdriven it will run out of gain on both positive- and negative-going peaks. At lower levels, the second-harmonic components of the two halves of the push-pull amplifier cancel out whilst the remaining third-harmonic distortion is simply that which is found in the two single-ended stages originally. It may be true that a single-ended amplifier with 0.2% distortion, all second order, is preferable to a similarly rated push-pull amplifier with 0.2% distortion, all third order, but that is not usually the choice. Given two of the class A single-ended stages, it is possible to design them into a push-pull amplifier of twice the power rating with virtually zero distortion, even at the higher rating.

On the other hand, many push-pull amplifiers do indeed leave something to be desired, it is true, for the above reasoning only applies to class A push-pull amplifiers, whereas the majority are class B designs. The problem here is *crossover distortion*. Figure 4.4a shows two stages with a curvature in one direction only (like Figure 4.3b), overlapped so as to provide an almost perfectly linear average characteristic giving an undistorted output (i). Next to it, (ii) shows the effect of reducing the small class AB standing current towards pure class B, i.e. moving the two halves of the transfer characteristic in the negative direction: a region of low gain appears at mid swing. Excess forward bias is no better, as a + movement in the characteristic shows. The increase in gain results in a larger output, due entirely to a region of excess gain at midswing, as shown at (iii). Now the distortion at full output in (ii) or (iii) may amount to only a fraction of a per cent and sound quite acceptable. But the same amount of crossover distortion will still be present at smaller outputs, representing a much higher, very objectionable, percentage distortion as shown in (iv). Thus, unlike class A amplifiers with primarily second- or third-harmonic distortion, with a class AB or B push-pull amplifier the percentage distortion in the output actually gets worse rather than better at lower output levels. Even though the quiescent bias may be optimized, there is no guarantee that the shape of the curvature at the origin, at the foot of each half of the characteristic, will be such as to give a linear average (shown dashed) through the origin. Even if it does, the quiescent setting may vary with temperature or ageing of the components; so, as mentioned earlier, NFB is employed to ameliorate matters. If efficiency were of no importance, the amplifier could be biased up to class A by moving the two halves of the characteristic in the positive direction so far that points A coincide with points B, eliminating crossover distortion entirely. Crossover distortion results in a comparatively sharp little wrinkle on each flank of the waveform, as in (ii). Being symmetrical there are no even-order distortion products; nor is the third-order component particularly large. However, many other odd-order products – fifth, seventh, ninth, eleventh and so on – are all present,

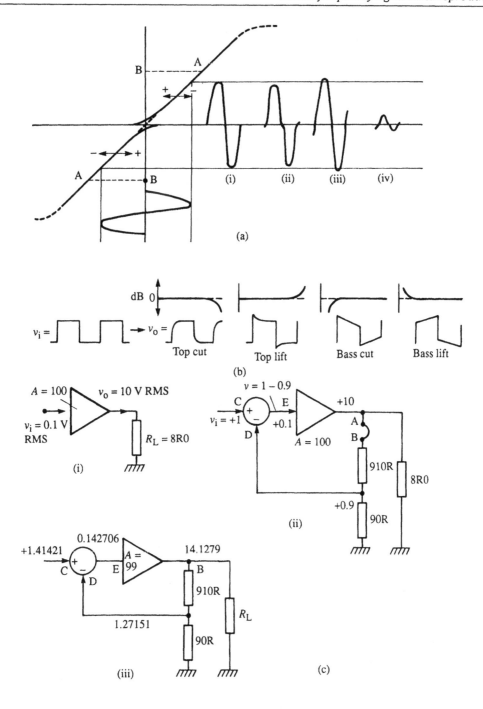

Figure 4.4
(a) Crossover distortion.
(b) Effect of tone controls (frequency distortion).
(c) Effect of negative feedback on distortion.

resulting in a multitude of intermodulation products as well as third, fifth, seventh etc. harmonics, which explains why even quite small levels of crossover distortion sound so objectionable.

The previous section has dealt at some length with non-linearity, which gives rise to waveform distortion: the next variety of distortion can be disposed of relatively easily. In fact we are so used to it and find so little to object to in it that it is frequently not thought of as distortion at all. I refer to *frequency distortion*. An amplifier may be perfectly linear in the sense that the ratio of output to input voltage is independent of volume, whilst being quite non-linear in terms of its frequency response. Indeed the whole purpose of tone controls is to enable the listener to adjust the degree of frequency non-linearity to suit his taste, especially as regards high and low notes using treble and bass controls, or in a more complex manner with a 'graphic equalizer'. These enable the ratio v_o/v_i, i.e. the gain A to be set to exhibit different values for different frequencies. Frequency distortion in a power amplifier is very small indeed: it is introduced, if and when required, in the early small-signal stages of the amplifier. A square wave is a convenient signal to show the effect of frequency distortion, since it is a signal containing many frequencies. However, as they are all harmonically related to its lowest or fundamental frequency, it has a constant determinate shape, unlike the constantly changing waveform of a piece of music. Figure 4.4b shows the effect of frequency distortion on a square wave. What is not so obvious is that the frequency distortion is accompanied by *phase distortion*. For example, not only does top cut reduce the amplitude of the higher frequencies in v_o compared with their relative level in v_i, it also retards them in phase. Like frequency distortion, phase distortion is not objectionable or even noticeable, at least in mono reproduction. To prevent it altering the spatial sound image in stereo reproduction, the tone controls are arranged to operate equally on both channels.

There are also other forms of distortion. One of these, transient intermodulation distortion (TID), was for long unrecognized, although it has attracted considerable attention for some years now. However, that topic is best left until after taking a look at the use of negative feedback to reduce distortion.

Negative feedback

NFB is applied around the main amplifier of, for example, a hi-fi system, principally to reduce the amplifier's distortion to a very low level. Strange as it may seem, there is no generally agreed figure – like 0.1%, for example – which may be taken as an adequate specification for all hi-fi amplifiers. It appears that the human ear is so discerning that distortion levels lower than this are noticeable, under favourable listening conditions, at least in the case of classical music. (In the case of pop, rock etc., limiters, volume compressors, fuzz boxes and other non-linear electronic musical 'aids' frequently render the question of distortion at the reproducing end of the chain quite immaterial.)

To see how NFB reduces distortion, take a concrete example: an audio power amplifier with a non-inverting open loop voltage gain $A_{vol} = 100$ and a 'shortfall' (Figure 4.3c) of 1% (without feedback) when supplying an output of 10 V RMS into an 8 ohm load, i.e. 12.5 W. If the load is a reasonably efficient loudspeaker in a domestic setting, this should provide enough sound to satisfy anybody, and probably too much for the neighbour's liking. Asume further that it is a class A amplifier, or a very well set up class B one, so that the distortion is entirely third harmonic, as in Figure 4.3c. Figure 4.4c summarizes the plot so far. Now add a voltage divider at the output to provide a signal whose amplitude is 9% of the output, i.e. a feedback fraction $\beta = 0.09$, and a network at the input which subtracts the feedback voltage from the input voltage, now applied at point C (Figure 4.4c(ii)). The voltage levels have all been marked in and, just to check that they all tie up, note that the gain-with-negative-feedback or closed loop gain

$$A_{vcl} = \frac{A_{vol}}{1 + \beta A_{vol}} = \frac{+100}{1 + (0.9 \times 100)} = 10$$

as in the diagram. But this isn't fair, you may argue, and I agree entirely; it assumes the amplifier is perfectly linear, whereas it has already been assumed to have 1% open loop shortfall at

10 V RMS output. It applies well enough at low levels where the percentage distortion is negligible, but at 10 V RMS output it is no longer a case of pure sine waves. Consider the situation at the positive peak of the output waveform: it should be 10 V RMS or 14.1421 V at the peak, when v_i instantaneously equals 1.414 21 V. But assuming a 1% shortfall (relative to a distortion-free characteristic, as defined in Figure 4.3c) then the voltages at various points round the circuit are actually as shown in Figure 4.4c(iii). If you want to derive them for yourself to check, take my tip and assume a nominal voltage at E of unity, so that the voltage at A is 99. It is then easy to fill in the rest and, finally, to proportion all the answers to a v_i of 1.414 21 V, i.e. 1 V RMS. Comparing (iii) with (ii), there is an actual output of 14.1279 V peak instead of $\sqrt{2}(10) = 14.1421$ or 99.9%; at E there is 0.142 706 V peak instead of 0.141 421 V peak or 100.91%. In other words the 1% open loop 'distortion' of the amplifier has been reduced by a factor of 10, the amount of gain sacrificed when the NFB was incorporated. Nine-tenths of the open loop 'distortion' can still be seen on the waveform at E; it is in the opposite sense, however. It thus predistorts the drive to the main amplifier, to push it relatively harder at the peaks than elsewhere.

Note that *shortfall* and *distortion* are not the same thing in normal parlance. Figure 4.3d shows that three-quarters of the shortfall is due to a component at the fundamental frequency which reduces the net fundamental, and only one-quarter is due to the third-harmonic component. Now a distortion meter measures the power at harmonic frequencies only and compares it with the output at the fundamental; it does not know about or measure the slight reduction in fundamental output. So one must not talk of an open loop distortion of 1% at 10 V RMS, when what is really meant is a shortfall of 1%.

If v_i in Figure 4.4c(ii) is set to zero, the link AB removed and a signal of 1 V RMS applied at B, the output will be −9 V RMS, the minus sign indicating that the output is inverted, since D is a subtract or inverting input. Thus the *loop gain* or 'gain within the loop' is ×9 or 19 dB, whilst the *gain reduction due to feedback* (A_{vcl} compared with A_{vol}

is ×10 or 20 dB: not so very different. The heavier the overall negative feedback, the closer the loop gain approaches the gain reduction due to feedback, and the two terms are often used loosely as though they were synonymous. Either way, you can see that the NFB has reduced the amplifier's third-harmonic distortion at 10 V RMS output from 0.25% to 0.025%, a very worthwhile improvement, whilst the 1% gain shortfall has been reduced in the same ratio. As noted earlier, the amplifier's open loop distortion appears, inverted, on the waveform at point E. It is often possible to disconnect an amplifier's NFB line and examine its open loop distortion directly. In some designs, however, the NFB is DC coupled and used in setting the amplifier's operating point; in this case the NFB cannot conveniently be disabled by just disconnecting it. The amplifier's open loop distortion can still be measured, though, simply by examining the signal in one of the early low-level stages. Assuming that these are very linear and that virtually all of the distortion occurs in the output stage, then one has only to examine the signal in a stage following the combination of the input and feedback signals. This can be very illuminating, not to say alarming. Consider an amplifier with 0.2% distortion at full rated output: doesn't sound too bad, does it? Suppose it has 40 dB of loop gain, though: one can deduce straight away that there is a massive 20% of open loop distortion, and confirm this by examining the signal in one of the early stages!

Heavy negative feedback can, then, considerably reduce the level of distortion of a test signal consisting of a single sine wave of amplitude within the amplifier's rating. However, NFB is far from the universal panacea for all hi-fi ills. Consider the amplifier response of Figure 4.3c. If the input is increased beyond that shown, the crushing of the peaks will gradually become more severe until they reach the flat dashed portion of the characteristic where no further increase is possible; the output devices have run out of supply rail voltage. Note that as Figure 4.3d shows, for every 1% increase in third-harmonic distortion there is a 3% reduction in gain. The result is a gentle sort of limiting, with gradually increasing distortion on overdrive. Imagine, however, that the same amplifier is improved with heavy negative feedback: the transfer characteristic will be considerably linearized, as shown long dashed in Figure 4.3c. As the input is increased, the output will rise *pro rata* with no gain reduction or distortion until the maximum possible amplitude is reached. Beyond this point the peaks are simply sliced off as with a sharp knife. Hard limiting of this sort involves high orders of distortion and, on programme material, severe intermodulation, resulting in a very nasty sound indeed. Nor is the problem limited to the output stage. Once the output clips, the feedback signal can no longer increase *pro rata* with the input signal; the loop gain has momentarily fallen to zero. With the amplifier now open loop, not only is the output stage overdriven, but in all probability earlier stages as well. This may shift their DC operating point, resulting in distortion even on smaller inputs following an overdriven transient, until they recover. This is one form of *transient intermodulation distortion* (TID).

In fact, heavy NFB can make severe demands upon the earlier stages even when the amplifier is not overdriven. As noted above, they have to handle an inverted version of the distortion caused by the output stage, superimposed upon the normal signal. The distortion includes harmonics at two, three, four etc. times the input frequency as well as sum frequency and higher-order components when the input consists of more than just one frequency. Consider a cymbal clash in a symphony, for example: it includes frequency components at up to 20 kHz and usually occurs when the rest of the orchestra is in full spate to boot. Thus the inverted distortion components which must be handled linearly by the earlier stages will include components up to 100 kHz or more, riding on top of an already full amplitude swing. This is a severe test for the penultimate stage which drives the output devices; if it fails the test the result is transient intermodulation distortion. This can result in discordant difference frequency components appearing which would not have been produced if a more modest degree of NFB had been employed. In fact, it is now generally agreed that the best fidelity is obtained from an amplifier with as little open loop distortion as possible, combined with a

modest amount of NFB, used well within its rating.

A course commonly adopted to ease the design problems of applying NFB is to limit the frequency range over which the full loop gain is applied. For example, suppose that the gain of the amplifier block in Figure 4.4c(ii) is rolled off by 20 dB, from ×100 to ×10, over the frequency range of 1–10 kHz. Over the same frequency range, the overall gain with feedback A_{vcl} will fall a little, but this can be easily corrected if desired with pre-emphasis in the preamplifier supplying v_i. At frequencies of 10 kHz and above the loop gain is now less than unity, so the opportunities for TID due to feedback are largely removed. It is, of course, true that harmonic distortion of signals at 10 kHz and above will not be reduced by feedback, but the harmonics will all be at 20 kHz or higher anyway, i.e. above the limit of audible frequencies. But why roll off the gain at all? The answer is that it will roll off of its own accord anyway, at some sufficiently high frequency, simply because practical amplifying devices do not have an infinite bandwidth. It is up to the designer to ensure that the gain rolls off in a controlled, reproducible manner, so that there is no excessive gain peak, or worse still, possibility of oscillation. The same criterion applies at the low-frequency end, if the amplifier has AC coupled stages, e.g. RC interstage coupling and an output transformer in a valved amplifier. Transistor amplifiers are usually DC coupled throughout in the forward path, and frequently the feedback path is also DC coupled, so stability considerations are limited to the high-frequency end of the spectrum.

It was noted in the previous chapter that an internally compensated opamp is designed with a *dominant lag* which rolls off all of the open loop gain before the phase shift due to other stages becomes significant. If, however, in a high-power audio amplifier you wish to retain say 20 dB of loop gain up to as high a frequency as possible, despite the earlier remarks about TID, then you might try using a higher rate of roll-off. Figure 4.5 illustrates this case, where the gain rolls off at 12 dB per octave. In Figure 4.5a the basic amplifier with an open loop gain A of ×100 has been shown as inverting so that the feedback voltage $A\beta$ at D can be simply added to

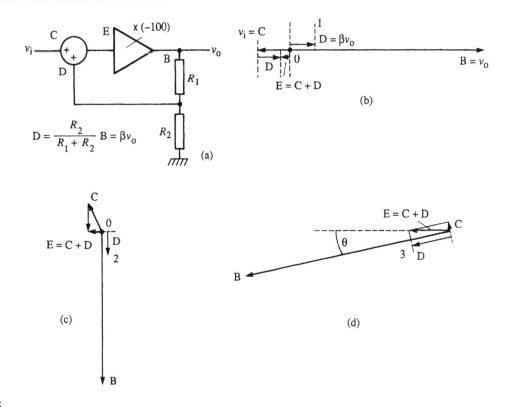

Figure 4.5

the input voltage v_i applied at point C. In Figure 4.5b the vector diagram for low frequencies (before the frequency response has started to roll off) shows that the amplifier input at E is simply the sum of v_i and the inverted feedback voltage: it is also of course in antiphase with vo at B and just one-hundredth of the amplitude. Assuming $\beta = 0.1$, then $A_{vcl} = 100/[1 + (100 \times 0.1)] = 9.09$, so $v_o = 9.09v_i$. In Figure 4.5c the two *coincident lags* are contributing between them a total of 90° phase shift between the voltages at E and B, and of course a corresponding 6 dB fall in the open loop gain A_{vol}. But despite the small phase shift between v_i and v_o (at points C and B) the gain is virtually unaffected - as was also found in the previous chapter in connection with internally compensated opamps. Figure 4.5d shows the situation at a much higher frequency, where the two lags have been running for some octaves and hence not only is the amplifier's phase shift very nearly 180°, but its gain is greatly reduced. Note that, for clarity, Figure 4.5d has been drawn to an expanded scale

relative to b and c. At the particular radian frequency ω shown in Figure 4.5d (call it ω_p), the magnitude of the feedback voltage at D is the same as the magnitude of the amplifier's input at E. As can be seen, v_i at point C is now very small compared with the voltage at D and hence also compared with the output; owing to the (almost) 180° phase shift, the negative feedback has become positive, resulting in a substantial gain peak at ω_p.

In Figure 4.6a the locus of the feedback voltage has been plotted as a function of ω, from zero to infinity. For convenience, v_o is not shown; it is just the same only $1/\beta$ times larger. Figure 4.6b shows the open loop gain (from v_i to the junction of R_1 and R_2) and the phase shift Φ, i.e. with the feedback connection broken. Points 1, 2 and 3 indicate the zero frequency, 90° corner frequency and unity loop gain frequency ω_p, corresponding to the vector diagrams of Figure 4.5b, c and d. Figure 4.6b also shows the closed loop gain v_o/v_i with its peak at the critical unity loop gain (0 dB

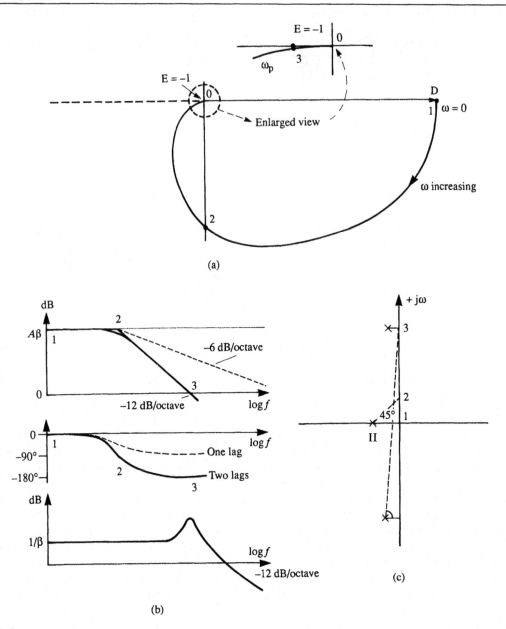

Figure 4.6

loop gain) point 3. Figure 4.6c shows the corresponding open and closed loop pole–zero diagrams. For open loop, there are two coincident poles (indicated by the roman numeral II) on the axis, each contributing 45° phase shift at point 2 on the $j\omega$ axis. For closed loop, point 3 indicates the frequency of maximum gain, the phase shift being the sum of the angles subtended at point 3 by each of the two closed loop poles, i.e. nearly 180°. Since Figure 4.6a is incomplete – showing the voltage at points D and E but not v_i at C – it can do duty for both the open and closed loop cases. Note that the closer the open loop phase shift to 180° at ω_p, the smaller the voltage v_i at C in Figure 4.5d, until at 180° the amplifier is capable of supplying its own input without any need for v_i at all! This corresponds to an infinite closed loop peak in Figure 4.6b and to closed loop poles actually on the $+$ and $-j\omega$ axis in Figure 4.6c. The result is a self-sustaining oscillation at ω_p; the amplifier is unstable. If the phase shift exceeds 180°, so that the locus in Figure 4.6 (a) passes to the north of the point $-1 + j0$, the amplitude will rapidly increase until limiting occurs, resulting in a non-sinusoidal oscillation. Since at each peak, positive or negative, the amplifier has to recover from overload, the frequency of the non-sinusoidal oscillation is lower than ω_p; this is called a *relaxation oscillation*. For stability, the locus must not pass through or enclose the point $-1 + j0$: this is called the *Nyquist stability criterion*.

The upshot of this exercise is the conclusion that two coincident lags is perhaps not the ideal way to control the high-frequency performance of a high-power audio amplifier. Of course one could argue that if the gain peak comes well above 20 kHz, what does it matter? In the case of an audio-frequency amplifier there will be no components of v_i at that frequency anyway: you could even add a simple low-pass filter ahead of the amplifier to make sure. Now if the amplifier were perfectly linear without NFB, there might be something in this argument – but then if it were we would not need NFB anyway. However, it was shown earlier that when an amplifier has significant open loop distortion, the net input at E includes harmonic components, representing the difference between the input voltage and the feedback sample of the output voltage. So even if v_i were band limited to 20 kHz, the input to the amplifier at E is not: one had better think again.

Figure 4.7 shows one way of reducing the gain peak and thus enhancing stability: it is a method frequently used in linear regulated power supplies. The forward gain A has one of the two lags cancelled at a frequency lower than ω_p, i.e. it is a *transitional lag* (see Figure 2.6). The effect on the open loop Nyquist, Bode and pole–zero diagrams is shown in Figure 4.7. It increases the phase margin at the critical unity loop gain frequency (Figure 4.7a), with the locus approaching infinite

frequency with a lag of only 90°. A similar effect can be achieved by leaving both of the lags in the forward gain A to run indefinitely, but bridging a capacitor C across R_1 in Figure 4.5a such that $CR_1 = 1/\omega_4$. This arrangement, which modifies both the frequency response and the phase response of the feedback voltage, is a simple example of a *beta network*. In high-performance wide band linear amplifiers such as are used as submerged repeaters in frequency division multiplex (FDM) telephony via submarine cables, quite complex β networks are called for. Once the amplifier designer had done his bit and produced the best possible amplifier, the specialist β network designer would take over and produce the optimum β network, taking into account the frequency response, phase, linearity and noise requirements.

An alternative way of ensuring amplifier stability is to roll the gain off at 10 dB per octave, corresponding to a maximum loop phase shift of 150°. The Nyquist and open loop Bode and pole–zero diagrams are shown in Figure 4.8, from which it can be seen that the 10 dB per octave is approximated by alternating between 6 and 12 dB per octave. The result is a 30° phase margin at ω_p: constructing a vector diagram like Figure 4.5d with $\theta = 30°$ will quickly convince you that the peak is limited to approximately twice the amplitude of the low-frequency response, or just $+6$ dB.

I have tended to concentrate on the use of NFB to reduce distortion, but it should be clear that it will tend to reduce any difference between the input and the feedback sample of the output voltage, due to whatever cause. Thus, for example, if any noise, hum or other extraneous signal is picked up in the earlier stages of the amplifier, its level at the amplifier's output, relative to the wanted signal, will be reduced in proportion to the loop gain. In addition, the NFB will reduce the amplifier's output impedance, again in proportion to the loop gain.

Having looked at class A, AB and B audio power amplifiers, a word now about *class D* amplifiers (which enjoyed a brief period of popularity) might not come amiss. The earlier analysis of amplifier efficiency showed that the class B amplifier was considerably more efficient

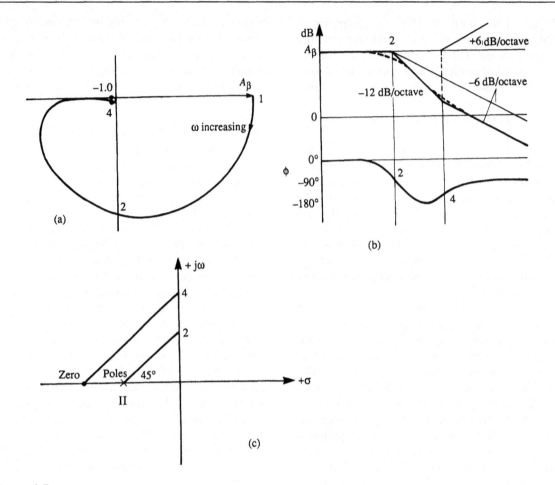

Figure 4.7

than a class A amplifier, mainly because the transistors only conducted heavily when the voltage across them was lowest. The class D amplifier pushes this philosophy to the limit: each output transistor is either bottomed or cut off. Figure 4.9 shows the scheme. With no input signal, the output transistors conduct alternately, applying a supersonic square wave to the inductor: the result, if the inductor has a high reactance at the switching frequency, is that only a small triangular magnetizing current flows. When an input signal is present, the *mark/space ratio* is modified so that the upper transistor conducts proportionately longer than the lower, for up to 100% of the time at the positive peak of a full output amplitude sine wave, and vice versa for negative peaks. The

effect of the inductor is to integrate or smooth the current pulses, resulting in a sinusoidal current at the signal frequency flowing through the loudspeaker's voice coil, together of course with a small supersonic ripple component.

The output stage of Figure 4.9b has the attraction of simplicity, but there will be some switching loss as, with bipolar devices, the turn-on of one transistor will be quicker than the turn-off of the other. In Figure 4.9a the on drive to one device can be delayed relative to the off drive to the other, thus positively avoiding any conduction overlap. Of course, with this arrangement, as soon as one device turns off the output voltage will fly off in the direction of the other rail, so the catching diodes shown will be necessary. In fact, they are necessary

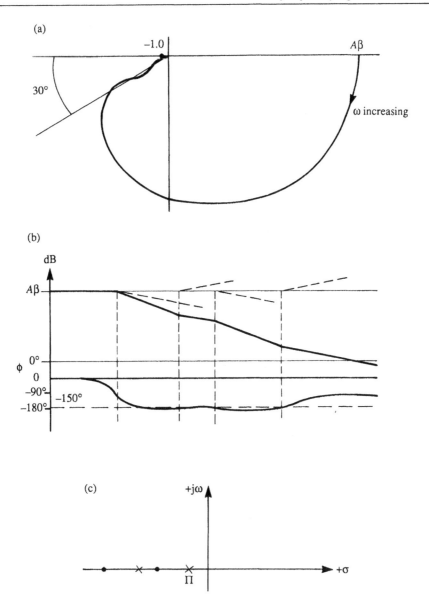

Figure 4.8

anyway since the magnetizing current is out of phase with the voltage, reversing sign half-way through each half-cycle under no signal conditions, as shown in Figure 4.9d.

The full circuit of Figure 4.9c operates as follows. Weak positive feedback via R_{12} round Tr_2, Tr_3, Tr_4 and Tr_5 causes them to act as a latch with a small

hysteresis. Overall NFB is applied from the output via R_8 to the input stage Tr_1 which operates as an integrator due to C_2. The collector voltage of Tr_1 always moves in such a direction as to drive the latch via R_5 to switch back in the opposite direction. The switching frequency is determined by the slew rate of the integrator and the amount of hysteresis of the

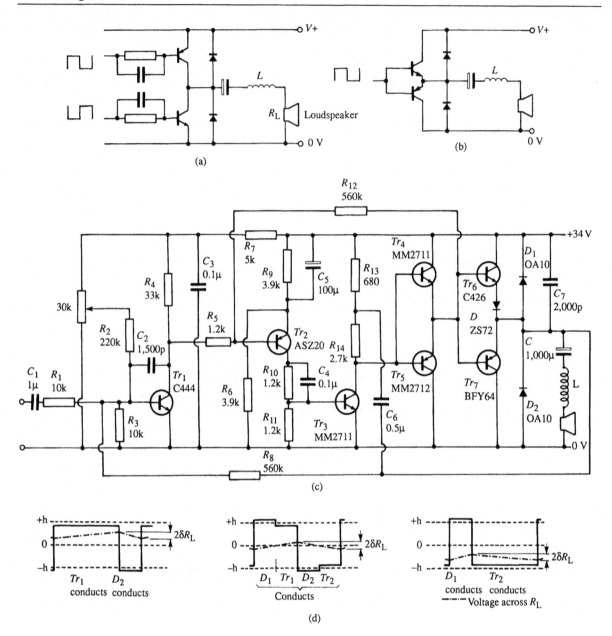

Figure 4.9 Class D audio amplifiers.
  (a) Common emitter class D output stage.
  (b) Common collector class D output stage.
  (c) Complete class D audio amplifier.
  (d) Output current and voltage waveforms for positive, zero and negative output levels.
(Parts (c) and (d) courtesy of *Electronics and Wireless World*.)

latch: about 50 kHz sounds plausible. The inductor L is 100 µH or more, the higher the inductance the higher the efficiency, but if it is too large the amplifier's response will no longer be reasonably flat up to 20 kHz. The circuit shown provided 5 W output with less than 0.25% distortion at all volume levels.

One problem with this type of amplifier is the generation of audible difference tones due to intermodulation between the input signal and the switching frequency under certain circumstances. This could be minimized by raising the switching frequency (a simple matter with modern high-speed semiconductors) and using a different type of pulse width modulator, maybe the scheme is worth resurrecting after all. That is more than can be said for the *sliding bias* amplifier, which was used in very early models of transistorized car radio. A single output transistor was used, for reasons of economy: not only were power transistors expensive but, being germanium, they had a top junction temperature of around 75°C. Recalling that in a class A amplifier the dissipation in the transistor(s) is lowest at full output, in the single-ended class A sliding bias amplifier, the transistor was normally biased at a fraction of the standing current required to cope with full output. A part of its drive signal was rectified and smoothed, and used to increase the bias in sympathy with the size of the signal. Thus the device always draws just enough current to cope with the present size of the signal without distortion – or at least that's the theory!

Loudspeakers

The reduction in amplifier output impedance resulting from the employment of NFB was mentioned earlier. In amplifiers employing heavy negative feedback, e.g. 40 to 50 dB of loop gain, the output impedance is reduced to a very low level – a fraction of an ohm[1,2]. This is generally reckoned to be a good thing since it increases the damping effect on a moving coil loudspeaker, thus discouraging the cone from flapping around at frequencies where it has a mechanical resonance. The result should be a smoother, more level frequency response. However, NFB can only

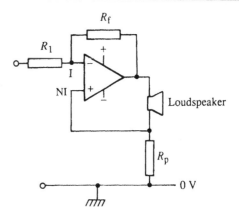

Figure 4.10 Use of positive feedback to achieve zero output impedance.

reduce the output impedance to a very low value, it can never actually make it zero. In the world of hi-fi marketing, superlatives are the order of the day, so it would look impressive to be able to claim an output impedance of zero, making the amplifier appear to the speaker as an ideal voltage source. This can be achieved by the judicious application of a little current derived *positive feedback* (PFB). Figure 4.10 shows the scheme. If the loudspeaker is disconnected, there is no current through R_p and hence no positive feedback. On connecting the loudspeaker, the output voltage (in the absence of PFB) would fall, even if only fractionally (thanks to the NFB), owing to the amplifier's residual output impedance. However, in Figure 4.10 the current through the very low-value resistor R_p causes a voltage drop which, fed back to the amplifier's non-inverting terminal, causes the output voltage to rise slightly. This completely offsets the internal voltage drop across the amplifier's output impedance, effectively providing a zero output impedance. Indeed, one could increase the PFB even further and cause the amplifier to exhibit a modest negative output impedance, if desired, say $-8\,\Omega$, cancelling out the voice coil resistance and resulting in a near infinite damping factor, at least at low frequencies where the reactance of the voice coil is negligible.

Audio amplifiers, whether hi-fi or otherwise, are usually rated to drive a nominal *load impedance,* such as a 4, 8 or 15 ohm loudspeaker. The vast

(a)

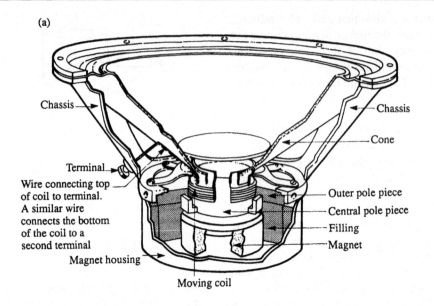

Chassis

Chassis

Cone

Terminal

Wire connecting top
of coil to terminal.
A similar wire
connects the bottom
of the coil to a
second terminal

Outer pole piece

Central pole piece

Filling

Magnet

Magnet housing

Moving coil

(b)

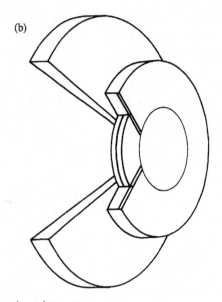

Figure 4.11
(a) Vital elements of a moving coil loudspeaker unit.
(b) Goodmans loudspeaker unit (cutaway), showing the moving coil, the magnet and pole pieces, and the chassis.
(Reproduced from *Beginners' Guide to Radio*, G. J. King. Heinemann Newnes 1984.)

majority of reproducers are *moving coil* speakers whose impedance, at least at low frequency, is resistive and equal to one of the three common values just mentioned. Figure 4.11 shows the

principles and construction of a moving coil loudspeaker. The voice coil is located in a powerful magnetic field. Current flowing in the voice coil causes the latter to be subject to a mechanical force at right angles to both the current and the magnetic field, i.e. along the axis of the coil. Thus the alternating audio-frequency currents cause the voice coil and cone assembly to vibrate in sympathy, turning the electrical signal into an acoustic one. A lightweight paper cone of small size results in a relatively high speaker efficiency at middle and high frequencies, but with negligible output at lower frequencies.

Such a speaker is appropriate in a small transistor portable radio. For hi-fi work a much larger speaker with an extended bass response and a much stronger, stiffer cone is preferred. The stiffness of the cone reduces its tendency to vibrate in various 'bell' modes at different frequencies, resulting in a smoother frequency response with less *coloration*. However, the much greater weight of the larger, stronger cone results in a less efficient speaker – except of course at the lower frequencies that the smaller speaker was incapable of reproducing. Considerable ingenuity has been exercised over the years to develop loudspeaker systems capable of fairly realistic reproduction of all sorts of music, from symphony orchestra to cathedral organ. For the latter, a bass response right down to 32 Hz or better still 16 Hz would be ideal, although few domestic living-rooms are large enough to do justice to such a speaker system, or even to accommodate it conveniently, whilst the frequency range of broadcast or recorded music (CDs excepted) extends only down to 50 or 30 Hz respectively. Details of bass reflex enclosures can be found in many standard works[3]. The non-resonant acoustic line is another design capable of providing extended bass response, as is the exponential horn loaded loudspeaker. The latter is very efficient, owing to the excellent matching of the air column's acoustic impedance to the cone of the speaker: this in turn results in a relatively small peak-to-peak excursion of the cone even at low frequencies, minimizing distortion (such as caused by the coil leaving the linear area of gap flux) and intermodulation. However, these three types of speaker enclosure are all large if dimensioned for response down to 50 Hz or lower. With so many people living in 'little boxes', the demand is for ever smaller loudspeaker enclosures; but the laws of physics are inexorable and such enclosures are not capable of radiating a satisfactory level of distortion-free sound at low frequencies, despite various palliatives.

One popular palliative is the loudspeaker with a long-throw voice coil and very flexible roll surround supporting the outer edge of the cone. The theory is that greater bass radiation can be achieved in a smallish box by enabling the cone to move back and forth at low frequencies with a relatively long throw. Clearly if the magnetic gap and the voice coil were the same length (in the axial direction) the coil would quit the flux entirely at each extreme of movement, resulting in gross distortion. But if the coil is much longer than the magnet's airgap (or vice versa, but the former is the cheaper), distortion due to this cause is avoided. Further, if the cone is made very strong (i.e. heavy), not only will break-up resonances be avoided, but its efficiency at mid-frequencies will be depressed to match that at low frequencies. With modern amplifiers capable of delivering tens or hundreds of watts, the low efficiency is of little consequence.

Long-throw loudspeakers can produce some unpleasant effects. Consider, for example, organ music. The high notes are radiated from a cone which is travelling back and forth an inch or more if a powerful low note is also sounding. This results in frequency modulation of the high note by the low note, owing to the *Doppler effect*. By using separate speakers – a woofer and a tweeter – to reproduce the low and the high notes, the effect can be diminished. It can be reduced even further with a three-speaker system of woofer, barker and tweeter, or bass, middle and treble reproducer. Figure 4.12 shows a very simple two-way crossover network. The low-pass filter for the woofer and the high-pass filter to keep bass out of the tweeter each cut off at an ultimate attenuation rate of only 6 dB/octave. This is barely adequate in practice, where 12 or sometimes 18 dB/octave would be used, but it shows the principle. Note that the arrangement is a *constant resistance* network, so that the amplifier sees a purely resistive load of its rated value, at all frequencies – or would if the loudspeaker impedance were constant.

In practice the impedance of a loudspeaker is far from constant, rising steadily with frequency once the reactance of the voice coil becomes appreci-

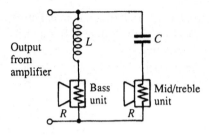

Figure 4.12 Constant resistance crossover circuit.

If $L/C = R^2$, the amplifier 'sees' a load resistance of R at all frequencies. At the crossover frequency f_c, where $f_c = 1/[2\pi\sqrt{(LC)}]$, each speaker receives half the total power.

able. This results in its drawing less current and hence receiving less drive power at higher frequencies. However, this is offset by a number of factors, including the speaker's greater directionality at higher frequencies, resulting in the energy being beamed forward rather than widely dispersed. Thus the moving coil loudspeaker produces a generally acceptable performance as far as the listener is concerned, though a graphical plot of its acoustic output as a function of frequency invariably looks rather like a cross-section of the Alps. Speakers also produce distortion and intermodulation due to non-linear and Doppler effects, but hi-fi enthusiasts readily tolerate high percentage levels of distortion in a loudspeaker which they would not dream of countenancing in an amplifier. It may be that the distortion produced by the acoustomechanical deficiencies of a loudspeaker is less objectionable than that produced by the electronic deficiencies of an amplifier, as some people maintain. But a more likely explanation is simply that loudspeaker manufacturers do not quote distortion figures for their products whereas amplifier manufacturers do.

Whilst the majority of loudspeakers are of the moving coil variety (first introduced by Kellogg and Rice in the 1930s) *piezoelectric* tweeters are now quite common. Much less common is the *electrostatic* loudspeaker, in which a large diaphragm, forming one plate of a capacitor, vibrates as a whole, providing a large radiating surface with very low inertia, as in the Quad electrostatic loudspeaker. This results in a very clean sounding output with very little distortion, intermodulation, coloration or 'hangover' after transients. Unfortunately the large size and high price of electrostatic speakers restrict them to a rather limited market.

Signal sources

Having dealt first – for no very good reason – with the back end of the audio chain, it's time to turn from power amplifiers and loudspeakers to signal sources such as microphones, pick-ups and tape recorders.

With the exception of purely electronic music, programme material – be it speech or music – originates from performers and the sound waves are turned into electrical signals by one or more *microphones*. The latter are called *transducers* since, like pick-ups and loudspeakers, they convert mechanical vibrations into electrical signals or vice versa. In fact, the *moving coil* microphone is very like a tiny loudspeaker in construction and produces an audio-frequency output voltage from its 'voice coil' in response to the movements of the cone occasioned by sound waves incident upon it. Sometimes two such microphones are mounted in a common mounting with their axes at right angles. They thus respond principally to sounds coming from left front or right front as the case may be, producing separate left and right channel components of a stereo signal. Moving coil microphones typically have an output impedance of 300 to 600 ohms and a sensitivity, at 1 kHz, of about -73 dB relative to 1 V per microbar. They are often supplied fitted with an internal 10:1 ratio step-up transformer providing an output of about -53 dB at an impedance of about 50 kΩ. Condenser microphones are also widely used. These work on the same principle as the electrostatic loudspeaker and, like the latter, early models needed a separate DC polarizing voltage. Nowadays this is furnished by an *electret,* in effect a capacitor with a positive charge trapped on one side and a negative on the other; it is thus the electrostatic equivalent of a permanent magnet. As with moving coil microphones, the frequency response of an electret condenser microphone

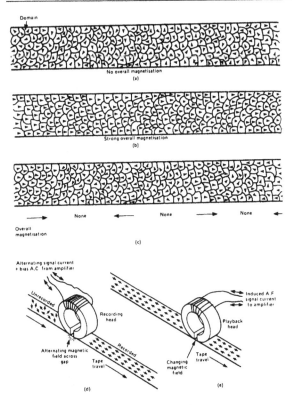

Figure 4.13 Magnetic tape recording.
(a) Unrecorded tape.
(b) Saturated tape.
(c) Recorded tape. In (a) to (c) the size of the domains is very much exaggerated; arrows represent the direction of the magnetization.
(d) Recording.
(e) Playback.

can be very wide, typically 50 to 16 000 Hz for a good quality type, and an FET preamplifier is usually built in. This is operated by a single miniature button cell and provides an output at 1 kHz of about −60 dB relative to 1 V/μbar.

The signal from a microphone may be broadcast direct or recorded on magnetic tape for later use. *Magnetic recording tape* consists of thin plastic, coated with finely divided iron oxide or other suitable magnetic powder. Once pins have been picked up with a magnet, they will tend to stick together even in the absence of the magnet. This effect is called *remanent magnetism* and is the basis of *tape recording*. Figure 4.13 shows diagrammatically how a recording head records the signal

onto the tape. The relation between the record current and the remanent magnetism is highly non-linear, so a high-frequency (50 kHz to 100 kHz) bias current is added to spread the recording to the linear parts of the characteristic (Figure 4.14). The same head can be used for *playback*, though the best machines use separate heads for record and playback, as each can then be optimized for the job it has to do. This also allows monitoring from the tape, the almost simultaneous playback which assures that the recording is going well. Another head uses the AC bias current at a much higher power to erase previous recordings.

The earliest tape recorders recorded just one track using the full width of a $\frac{1}{4}$ in wide tape moving at 30 in or 15 in per second. As tapes and recorders improved, it became possible to use speeds of $7\frac{1}{2}$ and even $3\frac{1}{4}$ in per second and to use half-track heads. One set of heads was used, recording along one half of the width of the tape; the take-up and supply reel were then interchanged (turning the tape reel over), so the same heads could then record or play a second track along the other half of the tape. Quarter-track heads are now commonly used, permitting two tracks to be recorded in each direction. The two tracks can be used simultaneously for the left and right channels of a stereo signal, or, on many machines, either can be used alone to provide four mono tracks per tape. The two tracks are then called A and B rather than left and right. Because of the interleaving of tracks in each direction, a four-track recording cannot be played on an older two-track machine (unless two of the four tracks – say the B track in each direction – were left blank), although a two-track recording can be played on a four-track A/B machine.

Mono cassette recorders use half-track heads in the same way as older half-track reel-to-reel recorders. With stereo cassette machines the two tracks of the stereo signal are on the same half of the tape, and both are erased at the same time by the erase head. (As in reel-to-reel recorders, the erase head is energized during recording, and the tape passes over it before reaching the record head.) Thus four-track mono use is not available, but mono cassettes can be played on stereo cassette decks and vice versa. Of course, the resultant

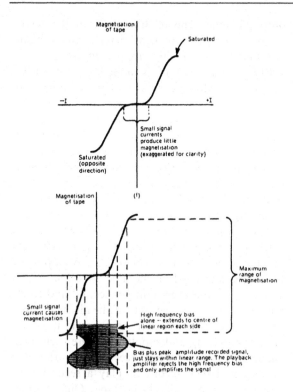

Figure 4.14 Use of high-frequency bias.
(a) No bias.
(b) With bias.

sound is always mono, except when a tape re-corded in a stereo machine is played in a stereo machine.

With the low tape speed ($1\frac{7}{8}$ in per second) and the narrower tracks on $\frac{1}{8}$ in tape, it is difficult to make the background noise from a cassette recorder as low as on LPs, reel-to-reel recorders or FM radio. Various *noise reduction* techniques have therefore been proposed, of which Dolby B is widely used in domestic cassette recorders. This substantially improves the signal/noise ratio at high frequencies (where background hiss is most noticeable) during quiet passages of music. The more sophisticated Dolby A system is used by broadcasting authorities and other professional users. Dolby B works by sensing the level of the upper frequencies in the audio signal and, if they are below a certain level, boosting them on record by up to a maximum of 10 dB. The reverse process is carried out on playback. Thus on attenuating

the pre-emphasized low-level high-frequency components to restore them to their proper level, the tape hiss on playback is reduced by 10 dB relative to a non-Dolby recorder. On the other hand, if the high-frequency components of the signal are of large amplitude, no emphasis is applied during recording, avoiding any possibility of distortion. On playback the hiss is not attenuated, but is not noticed as it is masked by the high level of the treble part of the signal. The Dolby system for professional applications works by dividing the audio spectrum into a number of bands and applying the above principle to each independently. Thus an improvement in the effective *dynamic range* of an analog tape recorder is obtained across the whole audio spectrum. The dynamic range is simply the volume range in decibels separating on the one hand the highest signal that can be handled without exceeding the recorder's rated distortion level, and on the other the no-signal noise level on playback.

The Dolby system is only one of a number of schemes designed to increase the dynamic range (lower the effective noise floor) of an analog tape recorder. For example, the Philips Signetics NE57IN is a 16-pin dual in-line (DIL) plastic IC dual compander in which each channel may be used independently as either a dynamic range compressor or an expander. Each channel comprises a full-wave rectifier to detect the average value of the input signal, a linearized temperature compensated variable gain cell block, and an internally compensated opamp. Features include a dynamic range in excess of 100 dB, provision for harmonic distortion trimming, system levels adjustment via external components, and operation from supply voltages down to 6 V DC. This versatile component can be employed for dynamic noise reduction purposes in analog tape recorders, voltage-controlled amplifiers, filters and so on.

Increasingly nowadays, analog recording methods (LPs, reel-to-reel and cassette tape recorders) are being replaced by digital recordings such as compact disc (CD) and the RDAT or SDAT digital tape formats. Fourteen-bit sampling at 44 K samples/second can provide a 20 kHz bandwidth with an 84 dB dynamic range, more than adequate for the highest domestic hi-fi standards,

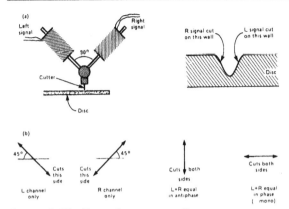

Figure 4.15   Recording a stereo disc.
(a) Principle of recording two channels in one groove.
(b) Motion of the cutter.
(Reproduced from *Electronics Questions and Answers*. I. Hickman. Newnes 1982.)

while sampling devices with 18, 20 or even 22 bit resolution, giving dynamic ranges in excess of 100 dB, are available. On the pop scene, however, dynamic range is not an important consideration. Consequently, on the cheaper sort of cassette tape recorder, a record-level control can be dispensed with entirely. All programme material is automatically recorded at maximum level, giving rise to virtually zero variation in volume level on replay, which fits in well with the output of pop records and local radio stations. On better quality cassette and reel-to-reel tape recorders, a record-level control and a volume unit (VU) meter or peak programme-level meter (PPM) are provided, permitting optimum use of the dynamic range of the recording medium – ferric, chromium dioxide or metal tape.

The other common programme source for sound broadcasting is the *disc* or gramophone record. Historic recordings on 78 RPM records and early 'long-playing' $33\frac{1}{3}$ RPM discs (LPs) were originally recorded monophonically; these are usually recorded or 'transcribed' onto tape to avoid further wear on the irreplaceable originals. Modern LP records are stereophonically recorded, as indicated diagrammatically in Figure 4.15. For playback a lightweight pick-up is used, the 'needle' or stylus being a conical artificial sapphire or a diamond (with the tip of

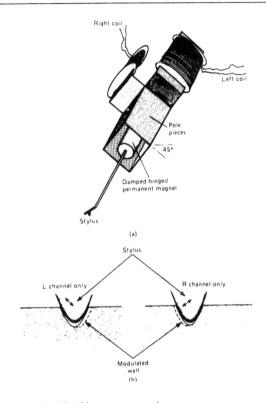

Figure 4.16   Playing stereo discs.
(a) Moving magnet cartridge.
(b) Motion of the stylus.
(Reproduced from *Electronics Questions and Answers*. I. Hickman. Newnes 1982.)

the point rounded), mounted on the end of a short arm or lever. In one inexpensive type of pick-up, the lateral movements of the stylus lever, as it traces the waggles in the grooves, bend a piece of piezoelectric material (e.g. barium titanate), producing an output voltage which is applied to an amplifier. Figure 4.16 shows a magnetic pick-up for stereo records, and indicates how it responds separately to the left and right channel signals, recorded one on each wall. When the same sound is present in both channels (or on a mono record), the stylus movement is purely horizontal.

A useful feature of the cheaper ceramic pick-up is that the output voltage is almost independent of frequency, for a given amplitude of groove deviation. This is not so with magnetic pick-ups, whether of the moving magnet, moving coil or variable reluctance variety. In fact, when gramophone records are made, the frequencies above about 2000 Hz are boosted and those below 500 Hz down to 50 Hz are attenuated relative to middle frequencies. The former improves the replay signal/noise ratio because the replay equalization restores the high-frequency level of attenuation, which at the same time reduces the objectionable hiss of high-frequency noise. The latter ensures that the amplitude excursion of the recording stylus is restricted to a value that avoids the need for excessively wide groove spacing when large amplitude low-frequency signals are being recorded. The low-frequency balance is likewise restored by the equalization circuit on playback. The playback characteristic is shown in Figure 4.17b; the recording characteristic is the inverse of this. As a result, the magnitude of the groove modulation corresponding to a given magnitude of the signal to be recorded is more or less independent of frequency, apart from the slight step in the 500–2000 Hz region – a convenient result for the ceramic pick-up already described. With magnetic pick-ups of all varieties, the output is proportional to the rate of flux cutting and hence, for a given amplitude of groove deviation, to frequency. This is restored to a flat response by an equalization circuit such as that shown in Figure 4.17a.

In tape recording, the playback process is again one of flux cutting the turns of a coil; so, for maximum recorded level on the tape, the replay head output voltage increases with frequency, at least at lower frequencies. Above a certain frequency, however, determined mainly by tape speed but also by head design and the nature of the magnetic coating on the tape, the response begins to fall again, giving a playback characteristic as a function of frequency for maximum recorded level, before equalization, as in Figure 4.18a. In part, the problem is that the higher the frequency the shorter the length of each recorded half-wavelength on the tape, so the magnetic areas tend to demagnetize themselves. The required playback characteristic is simply the inverse of the Figure 4.18a characteristic, i.e. a 6 dB per octave bass boost and a rather more vigorous treble boost

above the frequency of maximum output, to salvage an octave or two before the roll-off becomes too severe to cope with.

Now not only does the disc replay characteristic of Figure 4.17b equalize the output of a magnetic pick-up back to a level frequency response, but the phase characteristic shown also restores the original relative phases of all the frequency components of the recorded signal. The same may be said for the 6 dB octave bass boost used for tape playback, but compensation for the high-frequency loss is a different case entirely. The fall in high-frequency response on tape playback is due to factors such as tape self-demagnetization and head gap losses, and these are not 'minimum phase' processes, i.e. the attenuation is not accompanied *pro rata* by a phase shift. Consequently, the top lift on playback – together with pre-emphasis applied during recording (Figure 4.18b) – introduces phase shifts between middle- and higher-frequency components which were not present in the original signal. These are usually of no consequence for speech or music, but often had dire consequences in the early days of home computers, when storing programs on a cassette recorder, causing errors on replay during program loading.

Whether an audio signal originates from a high-grade studio microphone, a gramophone pick-up or a tape replay head, it will be of low amplitude and will require considerable amplification to bring it up to a usable level. A low-noise *preamplifier* is employed to raise the level of the signal, which is then, if necessary, equalized by an appropriate disc or tape equalization circuit. Low-noise audio preamplifier stages may use discrete components, e.g. a low-noise bipolar transistor such as a BC109C or a FET. However, frequently nowadays a low-noise opamp is employed, such as an NE5534 ($3.5\,\text{VnV}/\sqrt{\text{Hz}}$ noise) or OP-27 ($3.2\,\text{nV}/\sqrt{\text{Hz}}$) (both general purpose low-noise single opamps), an LM381 low-noise dual audio preamplifier for stereo tape or magnetic pick-up cartridge, or an HA12017 low-noise low-distortion single audio preamplifier.

Having now considered the entire audio chain from input transducer (microphone, disc or tape) right through to the output transducer (loudspeaker or headphones) – a word about those

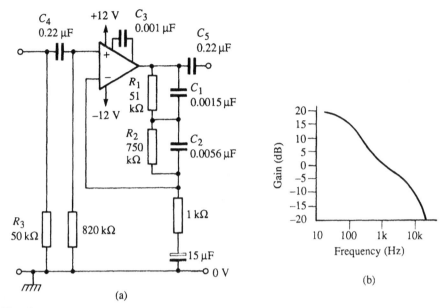

Figure 4.17   Equalization.
(a) The reactive elements in the feedback circuit cause the gain to fall (feedback increasing) with increasing frequency.
(b) Approximate response.

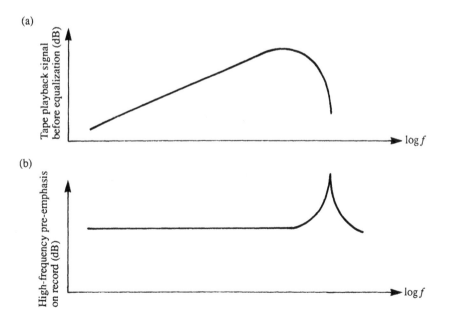

Figure 4.18

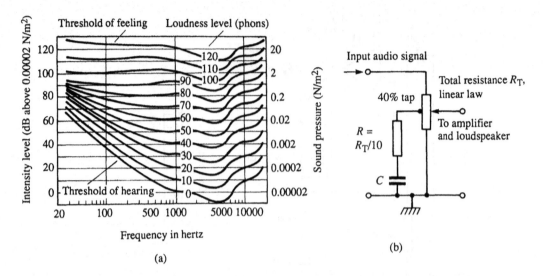

Figure 4.19
(a) Fletcher-Munson equal loudness contours.
(b) Compensated volume control provides bass boost at low levels, at frequencies below about $f = 1/2\pi CR_T$. Additional taps and CR shunt networks may be used to maintain compensation over a wider range of volumes.

important items, volume and tone controls. The operation of a *volume control* has already been covered in Chapter 1, so it will not be mentioned further – except to say that its exact location in the audio chain is not a trivial matter. The later it is placed (i.e. the nearer the large-signal back end) the less chance of it contributing noise to the output, but the greater the possibility of overload in the last stage preceding the volume control – and vice versa if it is located nearer the front. Likewise, *tone controls* must be suitably located in the audio chain. Simple bass and treble boost and cut circuits have been covered in Chapter 2. The more sophisticated type of control is discussed in a later chapter.

In some amplifiers and music centres, and even in the better class of transistor portable radio, the volume and tone controls are designed to interact; this is called a *compensated volume control* or loudness control. Figure 4.19a shows the reason for employing this arrangement. At high levels of sound, the apparent loudness – measured in phons – is more or less directly related to the level of acoustic power regardless of frequency, except for a region of greater sensitivity at around 4000 Hz. At lower volume the sensitivity of the ear

falls off at both low and high frequencies, but particularly at the former. Thus with a normal volume control, reproduction of music sounds distinctly thin and cold at lower listening levels. A compensated volume control reduces the level of the low frequencies less rapidly as it is turned down, approximating to the curves of Figure 4.19a. Thus reproduction at lower levels of volume retains the warmth and richness of the higher volume levels.

Figure 4.19b shows one way of arranging the required characteristic. Instead of the usual logarithmic track, the volume control is a linear one with a tap at 40%. This point is shunted to ground via a series combination of C and R. At middle and high frequencies, R is in parallel with the lower section of the volume control, giving it an approximately logarithmic characteristic. At lower frequencies where the reactance of C is high, the law is still linear. Thus as the volume is turned down from maximum, the bass is attenuated less than the middle and upper notes, giving the required characteristic.

Conventional rotary or slider potentiometer types of volume control can become noisy in operation or even intermittent, especially in

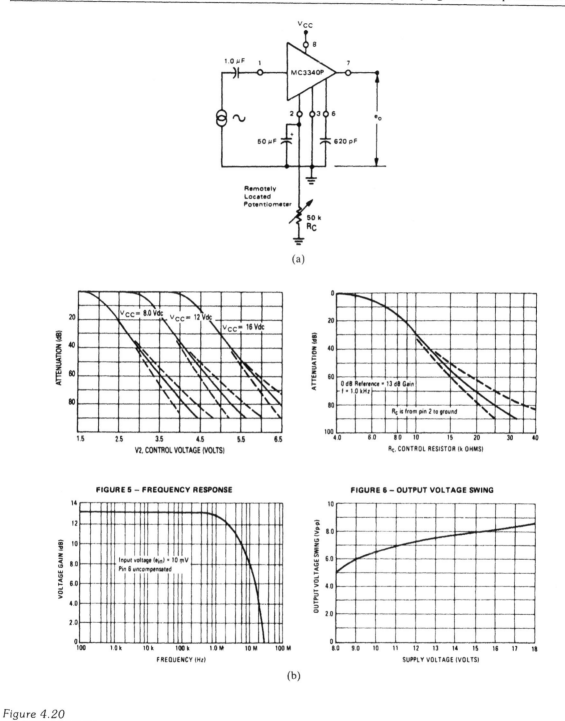

Figure 4.20
   (a) Typical DC 'remote' electronic volume control (MC3340P).
   (b) Typical attenuation, frequency response and output voltage swing characteristics
($V_{cc} = 16$ V DC, $T_A = +25°$C unless otherwise noted).
(Reproduced by courtesy of Motorola Inc.)

lower priced equipment, while the wiring to volume controls can be subject to hum pick-up or other forms of interference. In some audio equipment any possibility of such problems is eliminated by using an *electronic volume control*. This is an integrated circuit comprising either a variable gain amplifier or a voltage-controlled attenuator. The user's volume control then consists of a variable resistor or potentiometer which varies the voltage or current at the control input of the electronic volume control IC: this pin can be heavily decoupled since the control input is simply a DC level. The IC can be located on the main printed circuit board, avoiding any long audio signal leads, whilst the volume control can be located on a remote front panel at the end of wires as long as need be. A typical example is the MC3340P (see Figure 4.20).

References

1. Ultra-Low Distortion Class-A Amplifier. L. Nelson-Jones, p. 98, *Wireless World*, March 1970.
2. 15–20W Class AB Audio Amplifier. J. L. Linsley Hood, p. 321, *Wireless World*, July 1970.
3. Bass reflex – acoustical phase inverter – vented baffle, p. 845, *Radio Designers' Handbook*, F. Longford Smith, lliffe and Sons Ltd. 4th edition 1953.

Questions

1. The gain of a small-signal transistor amplifier, driven from and feeding similar stages, can often be approximated as $h_{fe}/2$. Explain how.
2. Derive from first principles the maximum theoretical efficiency of (i) a push-pull class A output stage, (ii) a push-pull class B output stage.
3. In a stereo reproduction system, why is the tone control always designed to operate equally on both channels?
4. It is not possible to open the feedback loop of a particular power amplifier, yet it is desired to measure its open loop distortion. Is this possible, and, if so, how?
5. Explain why and how a transitional lag is often used in the design of an amplifier with heavy overall negative feedback.
6. Name and describe two different types of distortion to which moving coil loudspeakers are subject.
7. In tape recording, is the high-frequency bias *added to* the audio signal, or *modulated by* the audio signal?
8. Describe the operation of the Dolby® B system. What level of high-frequency noise reduction is achieved on quiet passages?
9. Describe the considerations that influence how early or late in the chain the volume control in a high fidelity reproducing system should be placed.
10. Explain the operation of a compensated volume control.

5 Passive signal processing and signal transmission

Leaving aside the physical transportation of written messages (this includes everything from the runner with a message in a forked stick, through carrier pigeons, to the Post Office), there are three main ways of transmitting information between widely separated locations. These are communication over wire lines, that is telegraphy and telephony; wireless communication, that is radiotelephony, radiotelegraphy, sound and vision broadcasting; and optical communication. Of these, by far the oldest is optical communication. It includes the semaphore telegraph, heliograph and smoke signals for daylight use, and beacon fires or Aldis lamps for use after dark. However, these are all examples of 'broadcast' light signals even if, as with the heliograph and Aldis lamp, there is an important element of directionality. But optical communication is also the newest form, thanks to high-grade optical fibres or 'light pipes' and electro-optic transducers such as light emitting diodes (LEDs), lasers and photodetectors. These potentially offer huge bandwidth systems and, in conjunction with digital voice communications, are rapidly emerging as the new backbone of national trunk telephone networks. These will carry both voice and data traffic in digital form over an integrated services digital network (ISDN).

Practical line communication dates from 1837, the original telegraphic systems preceding telephony by about 40 years. Line communications form a major topic of this chapter. Practical communication by means of radio waves, on the other hand, dates only from Marconi's experiments of 1895, following Hertz's demonstration in 1888 of the actual existence of electromagnetic waves (these had been predicted by Maxwell in 1864). They have now assumed at least equal importance with line communication, and many services originally provided by the latter are now carried by the former. Intercontinental telephone circuits are carried by radio via satellites, and now by fibre optic submarine cables, supplementing the various transoceanic submarine telephone cables. These cables themselves were only installed in the years following the Second World War; prior to that transoceanic cables carried only telegraphy, the few intercontinental telephone circuits being carried by HF radio. Radio communication in all its forms is so important a topic that it is dealt with at length separately in a later chapter.

Transmission line communication and attenuation

It was found in the earliest days of line communication that signals were not transmitted without impairment. This took the form of *attenuation*: the signals were much weaker at the receiving end of a long line than at the sending end. Furthermore, higher-frequency signals were found to be more heavily attenuated than low-frequency signals, restricting signalling rates on very long telegraph lines to an impractically slow speed. Indeed, on the first transatlantic telegraph cable the effect was so severe as to render it useless: the impatient backers insisted on higher and higher sending end voltages in an attempt to get the signals through, with the result that the cable burnt out. The famous Scottish physicist William Thomson was called in and his analysis showed that the original design of cable was doomed to failure due to its enormous capacitance. He further showed that although the capacitance could not be substantially reduced, its effect could be compensated by the addition of an

appropriate amount of distributed series inductance. In 1866 a new transatlantic cable incorporating his ideas was completed and proved him
right. The backers got a cable that worked, and
William Thomson got a knighthood and later a
barony as Lord Kelvin of Largs.

Figure 5.la shows a simple telegraph system
suitable for the transmission of information by
Morse or one of the other telegraph codes. The
detector could be a buzzer or a bulb (either
connected directly or via a sensitive relay and
local battery), or an 'inker' to record the 'marks',
e.g. dots and dashes, on a moving paper tape. In
Figure 5.lb a very short section of the *two-wire line*
is shown, together with its equivalent circuit. Each
of the two wires will have series resistance and
inductance: furthermore, there will be capacitance
between the two lengths of wire and, at least in
principle, there may be a resistive leakage path
between them also. This is indicated by two
conductances $G/2$, where G is the reciprocal of
the leakage resistance between the wires. A two-
wire system as shown in Figure 5.lb is usually
balanced, that is to say that at any instant the
voltage on one line will be as many volts positive
with respect to earth as the voltage on the other is
negative, and vice versa. The advantage of this
arrangement is that if interference is picked up on
the line, either by magnetic coupling or electrostatic (capacitive) coupling, this will appear as
equal in-phase voltages on the two wires, and it
is called longitudinal noise, noise to ground or
common mode noise. The interference is a 'push-
push' signal, unlike the wanted balanced or 'push-
pull' signal, which is called a transverse, metallic or
normal mode signal. The receiving end circuitry is
usually arranged to respond only to the latter,
whilst rejecting the former. This is easily achieved
by coupling the signal at the receiving end through
a 1:1 ratio transformer whose primary is
'floating', i.e. its centre tap is *not* grounded.
Alternatively, the receive end may use the
common mode rejection provided by an active
circuit incorporating operational amplifiers.

One of the commonest forms of common mode
noise is mains power frequency hum. Pick-up is
usually localized and can be minimized by using a
twisted pair. An alternative cable design uses one

conductor to screen the other, providing good
isolation from electrostatic pick-up but less protection from unwanted magnetic coupling. This is
called a *coaxial cable* and is illustrated in Figure
5.1c. Here one of the two conductors is tubular
and completely surrounds the other. The inductance and resistance of the outer are smaller than
those of the inner, so the line is usually represented
by the unbalanced equivalent circuits shown. One
is a π section, being directly equivalent to the balanced π or 'box' section of Figure 5.lb; the other is
a T section, which is the unbalanced equivalent of
the H or balanced T section (not shown). Of
course, neither the T nor the π circuit is really
equivalent to the cable, where the impedances R,
L, C and G are continuously distributed, not
lumped as shown. But at frequencies where the
phase shift through the section is very small
compared with a radian, the equivalent circuit
behaves exactly like the real thing.

Now owing to the resistive components R and G
per unit length, the signal will be attenuated more
and more the further it travels along the line,
until – if the line is long enough – the energy fed
in at the input is virtually all dissipated before
reaching the other end. The line is thus effectively
infinitely long, so that whatever happens at the
receiving end can have no effect on the sending
end; it will be impossible to tell from the sending
end whether anything is connected to the receiving
end or whether it is open- or short-circuited. To
get the maximum signal energy into the line, it is
necessary to drive it from a matched source, so one
needs to know the infinite line's *input impedance*
Z_0. It turns out that

$$Z_0 = \sqrt{\left(\frac{R + j\omega L}{G + j\omega C}\right)} \qquad (5.1)$$

which has the dimensions of ohms and is in general
a function of frequency. This is inconvenient, since
to match it one would need a generator whose
source impedance also varied with frequency. At
zero frequency, $Z_0 = \sqrt{(R/G)}$. At very high
frequencies, where $j\omega L \gg R$ and $j\omega C \gg G$,
$Z_0 = \sqrt{(L/C)}$ (see Figure 5.ld). If now
$L/R = C/G$, Z_0 is independent of frequency: this
is called the *distortionless condition*. In practice this

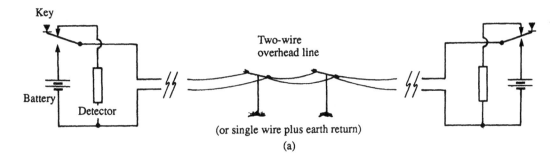

(a)

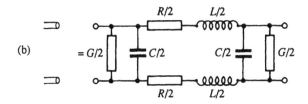

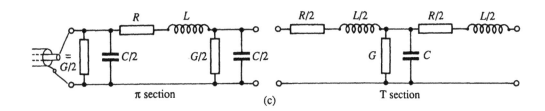

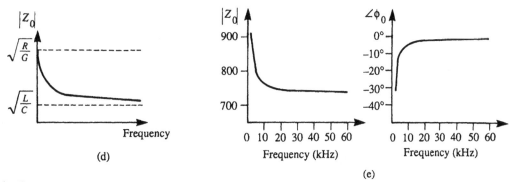

Figure 5.1   Transmission lines.
(a) Simple telegraph system.
(b) Two-wire line: balanced π equivalent of short section.
(c) Coaxial line: unbalanced equivalents of short section.
(d) Variation of characteristic impedance Z_0 of a line with frequency.
(e) Frequency and phase distortion for an air line.

condition is not met, since the conductance G between the two conductors is usually insignificant, so that all the loss is due to R. G could be artificially increased by bridging resistors across the line at intervals, but that would double the attenuation – which doesn't seem a good move. Fortunately, the dilemma only exists at low frequencies, as Figure 5.1e, relating to an air line, shows clearly. Nowadays audio-frequency air lines are only used for short connections, such as the last few tens of metres of a domestic subscriber's telephone line, from an overhead distribution pole to the house. In a multipair underground cable, such as from the distribution pole to the local exchange, the two wires of each pair are twisted together; the result is that the capitance C between them is very much larger, whilst the inductance is much lower, than for an air spaced line. Consequently, at audio frequencies $j\omega C \gg G$ but $R \gg j\omega L$, so that here Z_0 varies with frequency since (approximately) $Z_0 = (R/j\omega C)$. The frequency distortion arising from the variation of Z_0 can be largely removed by artificially increasing the inductance per unit length of the line. This is achieved, on twisted pair lines which are long enough to make such measures necessary, by inserting inductors in series with the conductors at intervals: these are called *loading coils*. Clearly, a loaded balanced line closely approximates to the equivalent circuit of Figure 5.1b. The loading results in Z_0 being nearly constant with frequency.

However, in the absence of loading, variations in Z_0 are not the only cause of signal distortion. To understand why, one must look at the other secondary line constant, the *propagation constant*, denoted by γ. This, like Z_0, is determined by the primary line constants R, L, C and G per unit length. The propagation constant is a complex number which describes the variation of amplitude and phase of a sinusoidal signal along the length of the line. Owing to the losses, the amplitude of the signal will decrease exponentially along the line. If the sending end voltage is 0.707 V RMS at some frequency ω rad/s, the peak sending end voltage E will be just unity. The peak voltage at any distance x metres along the line can then be simply expressed as $e^{-\alpha x}$ where α is the *attenuation constant* per unit length. Furthermore, the signal energy will

propagate along the line at a finite velocity: this cannot exceed the speed of light but is not much less, at least in the case of an air spaced line. Hence the peak voltage at some point down the line will occur a little later in time than at the send end: the greater the distance x, the greater the delay or phase lag. This can be expressed by $e^{-j\beta x}$ where β is the *phase constant* per unit length of line. The *wavelength* λ in metres along the line, i.e. the distance between points having a phase difference of 360°, is simply given by $\lambda = 2\pi/\beta$, assuming that β is expressed as radians per metre. So the voltage vector at any point x along the line relative to the sending end voltage E_s, and phase can be expressed in polar $(M\angle\phi)$ form as

$$E_x = E_s e^{-\alpha x} e^{-j\beta x} = E_s e^{-(\alpha + j\beta)x} = e^{-\gamma x} E_s$$

where the complex number γ is the propagation constant. A little tedious algebra which can be found in textbooks ancient[1] and possibly modern (although transmission lines was always a scandalously neglected subject) enables one to express γ in terms of the primary line constants R, L, C and G per unit length. It turns out that

$$\gamma = \surd[(R + j\omega L)(G + j\omega C)] \qquad (5.2)$$

The term β necessarily varies with frequency but, if the magnitude of the resistive terms is much less than that of the reactive terms, then

$$\gamma = \alpha + j\beta \approx 0 + j\omega\surd(L/C) \qquad (5.3)$$

i.e. the attenuation is negligible. The *velocity of propagation* v is given by $v = f\lambda$, i.e. frequency times wavelength, but $\lambda = 2\pi/\beta$ so

$$v = \frac{2\pi f}{\beta} = \frac{\omega}{\beta} = \frac{1}{\surd(L/C)} \qquad (5.4)$$

Now at last one can see why the first transatlantic telephone cable was a failure. Telegraph signalling speeds are down in the frequency range where R is significant, and, with the unloaded coaxial structure, L was small, C was large and Z_0 was consequently very low compared with R. Further, the distortionless condition was not met as G was very small. So there were three separate problems:

1. The characteristic impedance Z_0 varied with

frequency, so that the source could not be matched at all frequencies.

2. The attenuation constant α increased with frequency attenuating dots more than dashes.
3. The phase constant β was not proportional to ω, so that different frequency components travelled at different speeds; dots and dashes arrived in a jumble, on top of each other.

Resistive line sections: pads

At radio frequencies the reactive primary line constants C and L predominate, so that Z_0, α and β ought to be constant; but of course things are never that simple. In particular, just because $j\omega L \gg R$ and $j\omega C \gg G$, that doesn't mean that α should be negligible, only that, like Z_0, and ω/β, it should be constant with frequency. Unfortunately even that is wishful thinking, for the resistance of a piece of wire is not constant but rises with frequency due to *skin effect*. This effect is due to inductance and results in the current being restricted to the outer part of the wire's cross-section as the frequency rises, eventually flowing entirely in a very shallow layer or 'skin'. For this reason the wire of high-frequency coils and RF transformers is often plated with silver, or even gold. (Gold is a poorer conductor than silver, but the surface of the latter can tarnish badly. The high-frequency losses due to skin effect in silver can then exceed those in gold.) Furthermore, in a coaxial cable with solid dielectric, even if the dielectric constant is reasonably constant with frequency, there will be an associated series loss resistance which increases with frequency; hence foamed dielectric or partially air spaced feeders are used at UHF. The theory to date has only allowed for a shunt loss G, such as might be experienced at the supports of an open wire line on a wet day, not for a loss component in series with the capacitance C. In short, the secondary line 'constants' Z_0 and γ can vary with frequency, even if the primary line 'constants' really are constant, which in any case they aren't. Nevertheless, expressions (5.1) and (5.2) for Z_0 and γ are exceedingly useful and can take us a long way.

Consider for example a very short length of line in which R and G have deliberately been made

large. If the length l is only a centimetre or so, L and C will be very small; so Z_0 and α will be constant and determined by R and G, while β will be negligible, up to a few hundred megahertz. What you have in fact is an attenuator or *pad*. Its $Z_0 = \sqrt{(R/G)}$ and (β being negligible) $\gamma \approx \alpha = \sqrt{(RG)}$. Discrete components can be used as in Figure 5.2a and, provided the attenuation α is small, the total series resistance and shunt resistance can be very simply related to R and G as shown. For this to apply, the attenuation must not exceed a decibel or so. Now α defines the attenuation of a length of line in exponential terms. The loss in a line of length x where the attenuation constant is α per unit length is αx, i.e. $E_x/E_s = e^{-\alpha x}$. If αx is unity, the output is just e^{-1} times the input voltage or 0.368 per unit. This is an attenuation of 1 neper, whereas we normally work in decibels. An attenuation of 1 decibel implies an output voltage of 0.891 times the input – always assuming that, as here, one is working with a constant value of Z_0 throughout. Eight 1 dB pads in series would provide just less than 1 neper of attenuation, nine pads a little more; in fact 1 neper $= 8.686$ decibels, since $(0.891)^{8.686} = 0.368$. Clearly you could obtain a large attenuation – say 20 dB or 2.3 nepers – by connecting the required number of 1 dB pads in series, simulating a lossy line, but it is obviously much more economical to use just three resistors in a single T or π pad. However, the R and G elements can then no longer, by any stretch of the imagination, be regarded as continuously distributed, so the design formulae change. To stick with α in nepers, use the design formulae of Figure 5.2b. To work in decibels, start by calculating the input/output voltage ratio N from the required attenuation of D dB. Thus $N = 10^{D/20}$, and substituting N in the formulae in Figure 5.2c gives the required resistor values. Note that if the voltage (or current) ratio is very large, then (1) the coupling between input and output circuit must be very small, and (2) looking into the pad from either side one must see a resistance very close to R_0 even if the other side of the pad is unterminated. For if very little power crawls out of the far side of the pad, it must mostly be dissipated on this side. Thus, when N is very large, (1) R_p in a T

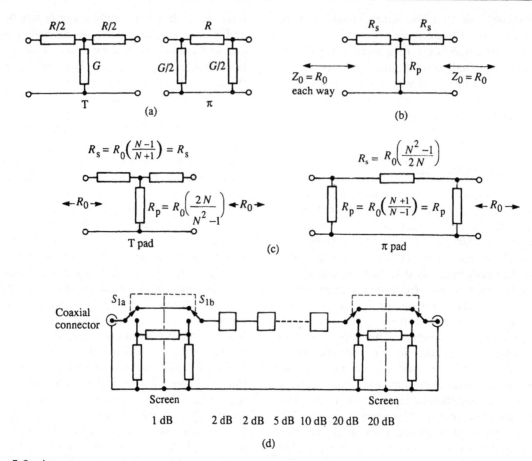

Figure 5.2   Attenuators.
(a) Lossy line section used as an attenuator. $Z_0\sqrt{(R/G)}$ and $\alpha = \sqrt{(RG)}$, only true for small α (in nepers: see text).
(b) Attenuator design in exponential form: $R_s = R_0 \tanh \alpha/2$, $R_p = R_0/\sinh \alpha$, true for all α (in nepers).
(c) Attenuator design in terms of input/output voltage ratio N: attenuation $D = 20 \log_{10} N$ dB.
(d) 0–60 dB attenuator with 1 dB steps.

circuit must be almost zero, and R_s in a π circuit almost infinity, and (2) R_s in a T circuit will be fractionally less than R_0, and R_p in a π circuit fractionally higher than R_0. In fact, as can be seen from Figure 5.2c, the R_s in a T circuit is the reciprocal of the R_p in a π circuit (in the sense that $R_s R_p = R_0^2$) and vice versa, for all values of N.

So that's how to design a pad of any given characteristic impedance, for any given attenuation, though in practice it will only be necessary to do so if a non-standard value of attenuation is required. Normalized resistor values for all common pad values are tabulated in Appendix 3. But what is the point of networks that serve no

useful purpose other than wasting power? There are many reasons, such as adjusting the level of the signal in, for example, a TV camera video signal path: if the level is too high and the gain of the preceding amplifier is not adjustable, a suitable attenuating pad can be employed. Or, when the level of a signal is being measured, it can be attenuated until it is equal to some standard test level such as 0 dBm, i.e. one milliwatt. The level of the original signal is then $+D$ dBm, where D dB is the attenuation of the pad. Similarly, the gain of an amplifier can be measured by attenuating its output until the level change through the two together is 0 dB – no net gain or loss. The gain in decibels then simply equals the attenuation of the pad. However, measurements like this presuppose that the attenuation of the pad can be set to any known desired value at will whilst keeping Z_0 constant. Whether using a T or a π pad, to achieve this, it would be necessary to adjust three resistors simultaneously, following a different non-linear law for series and shunt resistors. Such attenuators have been produced, but a much more common solution is to use a series of fixed pads and connect them in circuit or bypass them as necessary as shown in Figure 5.2d. This type of attenuator, properly constructed with miniature switches and resistors, would be usable up to VHF. Screens would not be necessary for the lower attenuation steps, but they would be essential for the 10 and 20 dB steps to prevent stray capacitance shunting R_s and thus reducing the pad's attenuation at very high frequencies. The switches might be manually operated toggle switches, or might be operated in sequence as required by cams on a shaft to provide rotary knob operation. The electrical design would be such as to ensure that with 0 dB selected, the internal wiring exhibited throughout a characteristic impedance Z_0 equal to R_0, usually 50 Ω.

Another use for pads is to enable accurate measurements to be made where two different impedance levels are concerned, without errors (or the necessity for corrections) for mismatch losses. For example, one might want to measure the gain at various frequencies of an amplifier with 75 Ω input and output impedances, using 50 Ω test equipment. A 50 Ω to 75 Ω mismatch pad – it should really be called a matching pad, as it is an anti-mismatch pad – would be used at the amplifier's input and another pad (the other way round) at its output. The design equations for a mismatch pad are given in Figure 5.3a. Note that in this case N is not the input/output voltage ratio but rather the square root of the input/output power ratio, since $R_{0\,in}$ does not equal $R_{0\,out}$. D can be chosen to be a convenient value such as 10 dB, making $N = 10^{10/20} = 3.162$. Allowing for both pads, the gain of the 75 Ω amplifier will then actually be 20 dB more than the measured value.

If using the above set-up to measure the stop band attenuation of a 75 Ω filter, the extra 20 dB

loss of the pads would limit the measuring capability undesirably. In this case it would be better to use minimum loss pads. Figure 5.3b shows that for a 1.5 : 1 impedance ratio such as 50 and 75 Ω, the minimum loss is about 6 dB. Whatever the impedance ratio, the minimum loss pad will turn out to be a two-resistor L type network. That is, in Figure 5.3a, if $R_2 > R_1$ then R_A in the T pad will be zero or R_C in the π pad will be infinity, whilst the remaining series and shunt resistors of course work out the same using either the T or the π formula. There is no reason why you shouldn't work out the resistor values for a mismatch pad with less than the minimum loss: 0 dB, say, would be a very convenient value. The difficulties only arise at the practical stage, for one of the resistors will turn out to be negative.

Reactive line sections: delay lines

The basic transmission line equations (5.1) and (5.2) were used to see under what conditions the line is distortion free and under what conditions it can be represented by a string of sections with the series and shunt impedances lumped: this turned out to be that α and β per section must both be small. Then, looking at purely resistive sections and extending this to the case where α is large, led to different design equations. Now let's look at purely reactive sections ($\alpha = 0$), both distributed and short lumped (β small), and then extend this to the case where β is large.

Starting with the case where β is small, (5.3) showed that $\beta = \omega\sqrt{(L/C)}$ and (5.4) that the velocity $v = 1/\sqrt{(L/C)}$ was independent of frequency. Thus different frequency components of a non-sinusoidal waveform will all arrive unattenuated at the end of a line as in Figure 5.4a, and in the same relative phases as at the sending end. For any component of frequency ω, the phase delay will be $\beta = \omega\sqrt{(LC)}$ radians per unit length, i.e. linearly proportional to the frequency. So at 0 Hz, the phase shift per unit length of line or per section (T or π) will be 0°. You might think *prima facie* that if the output is in phase with the input, the delay through the section must be zero – but not so. From (5.4) $v = 1/\sqrt{(LC)}$, and this is independent of frequency ω. So the time taken to travel

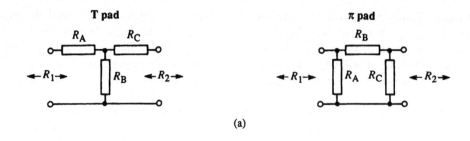

T pad   **π pad**

(a)

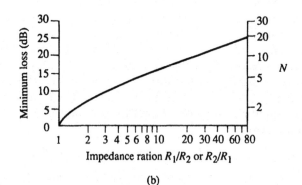

(b)

(a) Mismatch pads.

$$R_B = 2 R_0 \frac{N}{N^2 - 1}$$

$$R_A = R_1 \frac{N^2 + 1}{N^2 - 1} - 2R_0 \frac{N}{N^2 - 1}$$

$$R_C = R_2 \frac{N^2 + 1}{N^2 - 1} - 2R_0 \frac{N}{N^2 - 1}$$

$$\left.\begin{array}{l}\end{array}\right\}$$ T pad
$R_0 = \sqrt{(R_1 R_2)}$

$$R_B = \frac{R_0}{2} \frac{N^2 - 1}{N}$$

$$R_A = R_1 \frac{N^2 - 1}{N^2 - 2NS + 1}$$

$$R_C = R_2 \frac{N^2 - 1}{N^2 - 2(N/S) + 1}$$

$$\left.\begin{array}{l}\end{array}\right\}$$ π pad
$R_0 = \sqrt{(R_1 R_2)}$
$S = \sqrt{(R_1/R_2)}$

(b) Minloss pads.

Figure 5.3

unit distance is $1/v = \sqrt{(LC)}$ seconds, however low the frequency. It's true that as ω becomes infinitesimal, the phase $-\beta$ at the output of the unit length or section relative to the input becomes infinitesimally small also. But ω is the signal's rate of change of phase ϕ with t, so ϕ at the output only advances through β to $0°$ at an infinitesimal speed. The time delay $\sqrt{(LC)}$ seconds is in fact

sometimes referred to as the *DC delay*, being the limit of β/ω as ω tends to zero. This is illustrated in Figure 5.4b. The limit $d\beta/d\omega$ is usually called the *group delay* and is the reciprocal of the velocity v. For a uniform lossless transmission medium, v is independent of frequency, i.e. the group delay is constant. The group delay of free space is 3.33×10^{-8} seconds per metre, i.e. the reciprocal

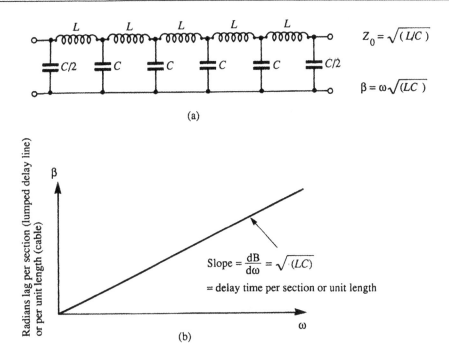

Figure 5.4
  (a) Delay line.
  (b) DC delay and group delay.

of the free space velocity of radio or light waves. The wave velocity in a balanced open wire transmission line, such as the feeder used to connect a high-power HF transmitter to a balanced antenna, is about 2.94×10^8 m/s or 98% of the speed of light. In a coaxial cable with solid dielectric the wave velocity is in the range 60 to 70% of 3.0×10^8 m/s, whilst for an underground telephone cable with lumped loading it may be 0.15×10^8 m/s or thereabouts, only one-twentieth of the speed of light.

In all these examples the delay experienced by the signal is incidental and of no importance, but in many applications delay is introduced into a transmission path deliberately. For example, by splitting the output of a transmitter and feeding it to two spaced antennas via different length feeders, so that the radiation from one antenna is out of phase with the other, the radiated power can be concentrated in a desired direction. For this purpose the delay can be provided by ordinary coaxial cable. For other purposes a rather longer delay

may be required and the length of cable needed can then become an embarrassment. So special delay cable may be employed; this is constructed so as to have an unusually high value of β. The time delay per unit length can be increased by increasing L and C per unit length, e.g. by using a spirally wound inner conductor in place of a straight wire. A typical application is to delay the output of an oscilloscope's Y amplifier on its way to the deflection plates. This enables one to see on the screen the very same rising edge which triggered the sweep. In early oscilloscopes, when a delay line was provided at all, it was composed of discrete LC sections and was bulky and expensive since a large number of sections was required to provide an adequate constant signal delay, even on an oscilloscope with a bandwidth of only 50 MHz, then the state of the art.

Discrete section delay lines are still used in many applications, a typical example being the tapped delay lines employed to permit time alignment of the various component signals of a colour TV

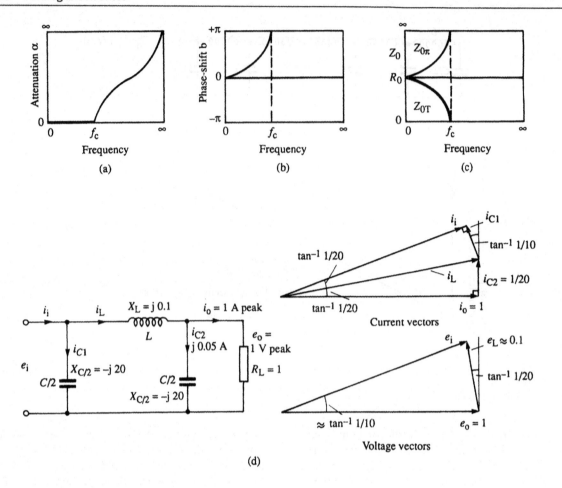

Figure 5.5   Delay line or simple low-pass filter: variation with frequency of
(a) attenuation
(b) phase shift
(c) characteristic impedance
(d) Operation at $f_c/10$.

waveform. As in the oscilloscope application, for this purpose also it is essential that all frequency components are delayed in time by the same amount, i.e. that the phase delay per section is strictly proportional to frequency corresponding to a constant group delay. With a delay line consisting of a series of π sections as in Figure 5.4a this is approximately so only when ω is much smaller than $2/\sqrt{(L/C)}$. As ω approaches $2/\sqrt{(L/C)}$, the phase shift per section increases more rapidly with frequency as shown in Figure

5.5b, indicating an increasing group delay. Beyond $2/\sqrt{(LC)}$ the phase shift remains constant at π radians per section, whilst the attenuation increases. Thus a delay line composed of discrete lumped LC sections behaves as a low-pass filter, in contrast to a line (such as coaxial cable) where L and C are continuously distributed. A lumped delay line will thus distort a complex signal unless its cut-off frequency is much higher than the highest frequency components of appreciable amplitude in the signal waveform.

Filters

An important aspect of passive signal processing is filtering. Filters are used to limit the bandwidth of a signal, e.g. before applying it to a transmission system. For example, the bandwidth of voice signals in a telephone system must be limited to fit within the allocated 4 kHz channel of a frequency division multiplex (FDM) group. (A *group* is a block of twelve 4 kHz telephone channels multiplexed to occupy the frequency range 60–108 kHz.) The same requirement to eliminate energy at 4 kHz and above applies equally to baseband voice signals which are applied to a digital modulator in a pulse code modulation (PCM) multiplexer, for application to an all-digital telephone trunk circuit. In these cases, a low-pass filter is required. Other applications require a high-pass filter, which passes only components above its cut-off frequency, or a band-pass filter, which passes only frequencies falling between its lower and upper cut-off points. Also met with are the band-stop filter – which passes all frequencies except those falling between its lower and upper cut-off frequencies – and the all-pass filter. The latter has no stop band at all, but is employed on account of its non-constant group delay, to compensate delay distortion introduced by some other component of a system: it is often called a phase equalizer.

The frequency response of the ideal *low-pass filter* would be simply rectangular, i.e. it would pass all frequencies up to its cut-off frequency with no attenuation, whilst frequencies beyond cut-off would be infinitely attenuated. In a practical filter, the stop band attenuation only rises gradually beyond the cut-off frequency. Furthermore, in practice, the pass band does not display a sharp corner at the cut-off frequency as shown in Figure 5.5a, since for convenience one uses a fixed value of R_0 for the terminations, where $R_0 = \sqrt{(L/C)}$. However, just as the group delay is constant only for values of $\omega \ll 2/\sqrt{(LC)}$, the input impedance of an LC π section is only equal to $\sqrt{(LC)}$ with the same limitation, rising to ∞ at the 'cut-off' frequency ω_c, i.e. at $2/\sqrt{(LC)}$ Hz, and becoming reactive above this frequency. It is

worth looking at this in a little more detail. Figure 5.5d shows the situation in an LC low-pass π section at a frequency well below cut-off, where $|X_L| = 1/10$ of the load resistance and $|X_C| = 10$ times the load resistance. Everything else has been normalized to unity for simplicity, working back from the output. From the current vector diagram, $i_L = 1 + j0.05$. This enables $e_i = e_o + e_L$ to be marked in on the voltage vector diagram, and hence $i_i = i_L + i_{C1}$ can be marked in on the current vector diagram. As you can see, $e_i = e_o$ and $i_i = i_o$ to a very close approximation, so the attenuation is zero and the phase delay is $\tan^{-1} 1/10 = 5.7°$ (one-tenth of a radian). Note, however that i_L, the current in the inductor, is slightly greater than i_i and i_o. This is because of a small additional component of current circulating round the LC circuit; this represents stored energy and is the Achilles heel of LC filters. It is responsible for the group delay increasing as the cut-off frequency is approached. Meanwhile you can see that the input impedance is resistive when R_L (the load resistance of the section) is the geometric mean of the impedance of L and C, i.e. $\sqrt{(X_L X_C)} = R_L =$ characteristic impedance Z_0. Hence

$$Z_0 = \sqrt{\left(j\omega L \frac{1}{j\omega C}\right)} = \sqrt{\left(\frac{j\omega L}{j\omega C}\right)} = \sqrt{(L/C)}$$

(5.5)

If both L and C have some associated loss resistance (series resistance R in the case of L and shunt conductance G in the case of C), the section's input resistance will still look purely resistive if both L and C have the same loss angle, i.e. $\tan^{-1}(R/\omega L) = \tan^{-1}(G/\omega C)$. If you redraw the vector diagrams to a larger scale with X_L replaced by $R + j\omega L$ and X_C replaced by conductance G in parallel with the reactance of C, you will find that i_i is in phase with e_i and Z_0 becomes $\sqrt{[(R + j\omega L)/(G + j\omega C)]}$, the distortionless condition for a cable which was quoted earlier. However, now e_i is greater than e_o, even at very low frequencies, showing that the filter section has a finite pass band loss. This is usually assumed to be negligible in filter design,

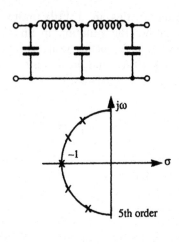

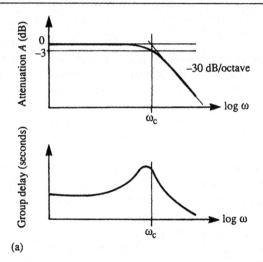

(a)

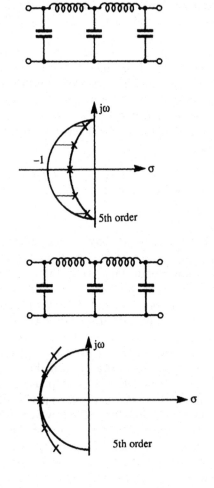

(b)

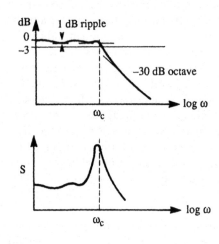

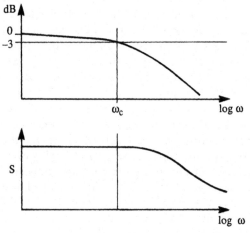

(c)

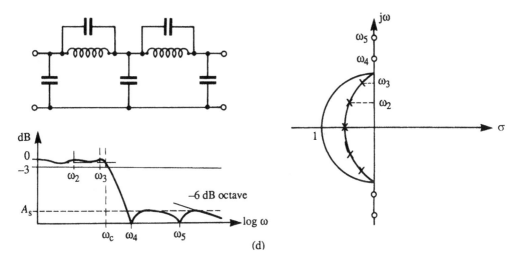

Figure 5.6 Passive low-pass filters.
  (a) Butterworth.
  (b) Chebyshev.
  (c) Bessel.
  (d) Elliptic.

but becomes important in high-order filters with sharp cut-off, e.g. Chebyshev and elliptic types.

The variation of Z_0 over the pass band was studied by Zobel,[1] who derived a filter design procedure which minimized the problem and at the same time permitted a much more rapid increase in attenuation at frequencies just above ω_c. This is achieved by introducing *finite zeros*, that is to say frequencies above ω_c where the attenuation becomes infinite, in addition to the infinite attenuation occurring at infinite frequency. Zobel filters use one or more basic *constant K LC* sections (like a lumped delay line) together with specially modified *m derived sections*, *m* being a parameter between 0 and 1, terminated with *m* derived half-sections. Terminating half-sections with $m = 0.6$ result in the value of Z_0 remaining within $+4\%$ and -9% of $\sqrt{(L/C)}$, the low-frequency value, right up to $0.9\,\omega_c$. However, whilst the resultant filter corresponds more closely to the ideal 'brick wall' filter shape, the group delay variations in the pass band of an *m* derived filter are substantially worse than in the prototype constant *K* filter.

There is some real virtue in a filter which presents a good match over the greater part of its pass band, since the response of such a filter will be less affected by deviations from R_0 in its driving source and load impedances. Nevertheless, *m* derived filters are now of purely historic interest, modern filter design taking account only of a filter's insertion loss versus frequency. The *insertion loss* (or gain) of any two-port network, passive or active, is defined as the resultant change of level at the receiving end of a measuring system when the network is introduced into the system's transmission path such that the network is working between resistive source and load terminations, usually $50\,\Omega$. Often the network is inserted between two $50\,\Omega$ 10 dB pads to ensure that the source and load are very close to the ideal, whilst if the network has a nominal impedance other than $50\,\Omega$ then mismatch pads may be used as described earlier. Modern filter design reflects various characteristic pole–zero plots which very conveniently describe the performance of both passive and active filters. Many different basic types are used, of which the following are some of the most important (see Figure 5.6).

The *Butterworth* or maximally flat amplitude filter is a good general purpose design (Figure 5.6a). It has a flat pass band; a reasonably fast cut-off, especially if a high-order design (many sections) is used; and a group delay characteristic which may be acceptable (in those cases where it is important) if a fairly low-order design is adequate.

The *Chebyshev* filter provides a sharper corner at the cut-off frequency; the price paid is small ripples of attenuation in the pass band and a severely degraded group delay characteristic. As can be seen from the pole–zero plot in Figure 5.6b, the poles of a Chebyshev filter are on an ellipse displaced to the right relative to those of a Butterworth, which are equally spaced around a semicircle in the $-\sigma$ half of the diagram. Thus attenuation due to the pole on the $-\sigma$ axis sets in earlier, the response being held up at higher frequencies by one or more pole pairs which have a higher Q (nearer the $j\omega$ axis) than the corresponding pairs in a Butterworth. The resultant pass band ripples are due to mismatch loss as the filter's impedance varies with frequency. The mismatch loss depends upon the reflection coefficient ρ (lower-case Greek letter rho); so, for example, a ripple depth of 0.1 dB corresponds to 15% reflection, whilst $\rho = 25\%$ corresponds to 0.28 dB ripple depth. Note that for an even-order filter the 0 Hz response occurs at the trough of the ripple, so the filter must be mismatched. This is achieved by working the filter between unequal source and load impedances, resulting in mismatch loss at 0 Hz, whilst at the peaks of the ripple it acts as an *impedance transforming* filter, matching the different source and load impedances to each other. At frequencies well into the stop band, the Chebyshev filter has the same rate of cut-off as a Butterworth, i.e. 30 dB per octave for the five-pole filters shown in Figure 5.6. The 'faster' cut-off of the Chebyshev is in reality just an earlier cut-off, its stop band characteristic approaching the 30 dB/ octave asymptote from above rather than, like the Butterworth, from below. Projected back, the asymptote cuts the 0 dB level well down the pass band, rather than at the corner (compare Figure 5.6b and a). Like the Butterworth and Bessel filters, the low-pass Chebyshev design is an *all-pole* filter, the number of poles equalling the number of reactive components. Indeed, these three filters all have the same circuit diagram, the only difference being in the component values.

An important point to bear in mind when comparing a Butterworth design with Chebyshev or elliptic filters is the different definition of cut-off frequency. For low-pass filters with a flat pass band amplitude response, such as Butterworth and Bessel, the upper limit of the pass band is taken as that frequency at which the attenuation has risen to 3 dB. For Chebyshev and elliptic filters, the cut-off frequency is the highest frequency at which the attenuation equals the attenuation at the troughs of the ripples: this is clearly shown in Figure 5.6.

The *elliptic* (or Caur) filter (Figure 5.6d) has both ripples in the pass band like a Chebyshev filter, and finite zeros in the stop band like an *m* derived filter. Like the Chebyshev filter, it is not suitable for use with complex signals requiring a constant time delay for all frequency components, i.e. where waveform preservation is important. The elliptic filter design, with the aid of finite zeros in the stop band, offers the fastest cut-off of any *LC* filter. However, each of these zeros is bought at the price of 12 dB/octave reduction in the final rate of cut-off. It is thus convenient to use a design with an odd number of poles, not only on account of its equal source and load impedances, but also to ensure that eventually the attenuation increases to infinity rather than remaining indefinitely at A_s. The *inverse hyperbolic* filter has a stop band like the elliptic filter but a flat pass band, i.e. the ripple depth is 0 dB.

If it is desired to limit the bandwidth of a square wave or pulse train whilst introducing the minimum of waveform distortion, then a *Bessel* filter, with its maximally linear phase and flat group delay, is a better choice, even though the rate of attenuation increase in the stop band is much lower than for other filter types (Figure 5.6c).

If *L* and *C* are interchanged as in Figure 5.7a, then a *high-pass* filter results. It shows infinite attenuation at 0 Hz; the attenuation then falls until, above ω_c the pass band has zero attenuation. If the series elements are series tuned circuits whilst the shunt elements are parallel tuned circuits, a *bandpass* filter response is produced (Figure 5.7b). Interchanging the position of the series and parallel tuned circuits gives a *band-stop* response. Both band-pass and band-stop filters of this type are narrow band de-

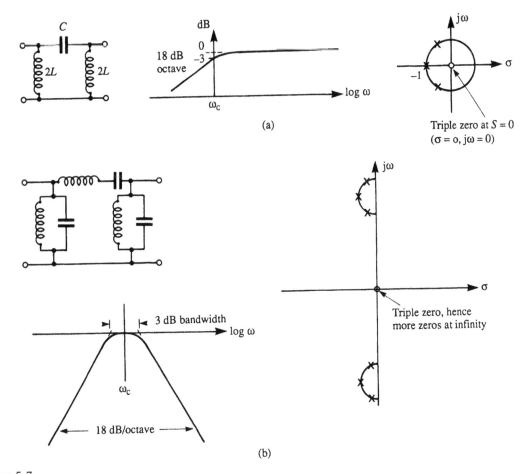

Figure 5.7
(a) Three-pole high-pass filter.
(b) Six-pole band-pass filter (three-pole low-pass equivalent).

signs, suitable for a small percentage bandwidth only, say half an octave at the very most. For very small percentage bandwidths (less than 1%), crystal resonators are used in place of resonant *LC* circuits. For pass bands of ±20% or more of the centre frequency, series connected high- and low-pass filters with overlapping pass bands offer a more satisfactory alternative. Similarly, a wide band-stop response can be produced with parallel connected high- and low-pass filters with non-overlapping pass bands. When connecting different filters in series and more particularly in parallel, care must be taken to ensure that unwanted interactions between the filters' terminal impedances do not cause problems.

As with attenuators, the modern electronic engineer seldom needs to design a filter, be it low, high or band pass, from first principles. Tabulated filter designs are to be found in many books, of which References 3, 4 and 5 are examples. The designs are usually normalized to 1 ohm source and load and to 1 radian per second cut-off frequency. To convert the tabulated values to those for a 50 ohm filter, divide all *C* values by 50 and multiply all *L* values by 50; to convert to a cut-off frequency of e.g. 1 MHz, divide all the *L* and *C* values by $2\pi \times 10^6$.

An *all-pass* filter or *phase equalizer* has no stop band. It is employed where a specific non-constant group delay response is required to counteract undesired group delay variations in a filter or other network. Most textbooks do not even mention the *all-stop* filter[6], which is in any case little used, for obvious reasons.

References

1. *Handbook of Line Communication*, Volume 1, HMSO London 1947.
2. Theory and Design of Uniform and Composite Electronic Wave Filters. A. J. Zobel, *Bell System Technical Journal*, Vol. 2, no. 1, 1923. See also Transmission Characteristics of Electric Wave Filters; Ibid., Vol. 3, no. 4, 1924.
3. *Simplified Modern Filter Design*, P. R. Geffe, Iliffe 1964.
4. *Handbook of Filter Synthesis*. A. I. Zverev, 1967 John Wiley & Sons Inc.
5. *Reference Data for Radio Engineers* (chapter 8). Howard W. Sams & Co Inc, 6th edition, 1975.
6. Simplified Filter Design Routine, B. Sullivan, *Practical Wireless*, Vol. 65, no. 4, p. 28, April 1989.

Questions

1. State the distortionless condition for a transmission line, in terms of the fundamental line parameters of L, R, C and G per unit length. Coaxial cable type RG62A/U exhibits a capacitance of 48 pF per metre. Its characteristic impedance $Z_0 = 93\,\Omega$; what is its inductance per metre?

2. Considering an attenuator as a lossy transmission line where the phase constant $\beta = 0$, express the characteristic impedance and attenuation in terms of the fundamental line constants. Hence derive the values of R and $G/2$ for the resistors of a $50\,\Omega$ π pad where the amplitude of the voltage at the output is 90% of that at the input.

3. You are required to measure the response of a low-pass filter with a characteristic impedance of $75\,\Omega$, using a $50\,\Omega$ measuring set. Design suitable minimum loss pads for use at the input and output. What loss must be subtracted from the measured loss, at any frequency, to give the loss due to the filter itself?

4. What would be the (notional!) values of the two resistors of a $50\,\Omega$ to $75\,\Omega$ minimum loss mismatch pad if it were to have zero loss?

5. Calculate the phase delay of the output of 1 metre of the cable in question 1, relative to the input, at 100 MHz.

6. Define group delay. What is the group delay of (i) 1 metre of the cable in question 1? (ii) free space?

7. 'In a practical filter, the stopband attenuation only rises gradually beyond the cut-off frequency' (see earlier in the chapter.) Why does the attenuation not rise sharply at the cut-off frequency, as in Figure 5.5a?

8. What is a finite zero? How many finite zeros are there in a 5 pole Caur (elliptic) filter?

9. Describe and compare the relative merits and limitations of Bufferworth, Chebychev and Bessel filters. Why must an even-order Chebychev (or elliptic) filter of standard design work between unequal source and load impedances?

10. Design a $50\,\Omega$ 5 pole 0.1 dB ripple elliptic low-pass filter with a cut-off frequency of 16.9 MHz, by denormalizing the appropriate values given in Appendix 8.

Chapter

6 Active signal processing in the frequency domain

Chapter 2 looked at passive circuits whose response was frequency selective, and Chapter 3 discussed operational amplifiers. This chapter brings these two items together and looks at active filters and related topics. Active filters are preferred at audio frequencies since they enable all types of filters and phase equalizers to be realized with suitable combinations of resistors, capacitors and opamps. Thus the inductors that would be necessary for a passive filter can be dispensed with entirely; this is a real blessing since good quality high-value inductors, such as would be required for audio-frequency filtering purposes, are both bulky and expensive.

First-order active circuits

The least sophisticated filter, be it low pass or high pass, is the first-order or single-pole section. As noted in Chapter 2, this has a fixed shape providing a gradual transition from the pass band to a 6 dB/octave roll-off in the stop band. It was also noted that, to obtain exactly the calculated performance, a passive CR section should be driven from a zero source impedance and work into an

infinite load impedance. Opamps can be used to provide the required source and load impedances, avoiding the otherwise necessary step of designing finite source and load impedances into the network itself – which in any case may not be practicable if they are not themselves resistive, and hence not frequency independent. Figure 6.1 shows two opamps used as *buffer* amplifiers, providing the CR section with a low source impedance and a high load impedance. The output amplifier is used in the non-inverting mode to provide a high input impedance and is shown connected to provide unity gain, though it could perfectly well provide a gain of greater than unity if required. The input amplifier is also shown as non-inverting, providing a very high input impedance and keeping the section as a whole non-inverting.

Now a passive CR section buffered by a pair of opamps hardly constitutes an active filter: this description is generally reserved for a circuit where the frequency determining passive components arc built into an opamp's gain determining network. So can one design, for example, a single-pole low-pass active filter with very high input impedance and low output impedance, all with a

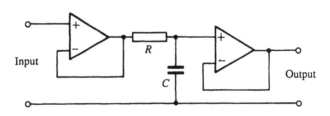

Figure 6.1 First-order low-pass circuit.

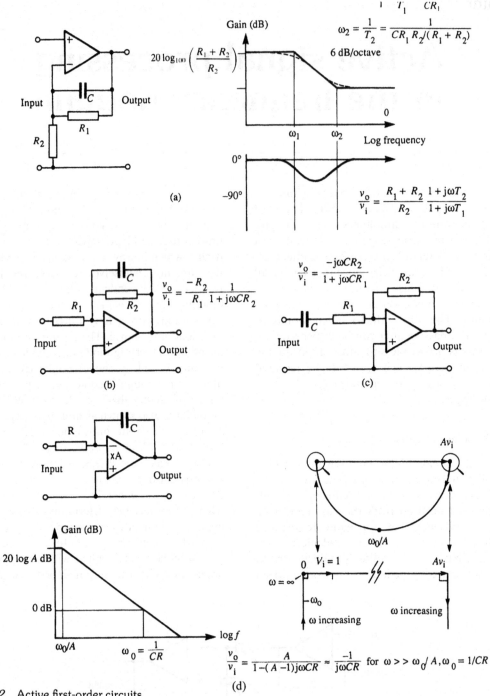

Figure 6.2   Active first-order circuits.
(a) Non-inverting transitional lag.
(b) Inverting low pass: $f_{3\,dB} = 1/2\pi CR_2$, low-frequency gain $= R_2/R_1$.
(c) Inverting high-pass: $f_{3\,dB} = 1/2\pi CR_1$, high-frequency gain $= R_2/R_1$.
(d) Almost ideal integrator (zero bias current and offset voltage).

single opamp? Figure 6.2a is probably the nearest you can get; it provides a low-frequency gain of $(R_1 + R_2)/R_2$, and this starts to roll off at a frequency of $1/2\pi CR_1$. However, as noted earlier, with the non-inverting circuit the gain can never be less than unity, so the circuit of Figure 6.2a provides a transitional lag response. If the gain is to roll off indefinitely beyond the corner frequency, one must use the inverting connection as in Figure 6.2b. Here there is an independent choice of low-frequency gain (set by R_2/R_1) and corner frequency (set by CR_2), but of course the input impedance is now just R_1 and the signal is inverted. Figure 6.2c shows the corresponding high-pass circuit.

For a response falling at 6 dB/octave over several octaves – say all the way from 20 Hz to 20 kHz – one could always use a passive lag. But with its unity response at frequencies below the corner frequency, the response at 20 kHz would be very small indeed. The problem does not arise with the active lag circuit of Figure 6.2b, as you can adjust the gain independently of the corner frequency. As R_2 becomes higher and higher, the corner frequency becomes lower and lower, and the pole moves closer and closer to the origin. But the unity-gain frequency is unaffected; it occurs at the frequency where the reactance of C equals R_1. When R_2 becomes infinite, the pole is (ideally) at the origin and the circuit is called an *integrator*: as the input frequency falls, the response rises at 6 dB/octave for ever more. In practice it is not possible to make a true integrator in this way, for the opamp's input current and offset voltage would cause its output to saturate at maximum or minimum voltage. Even without this practical difficulty, the gain can never exceed the opamp's open loop gain A, however low the frequency. So whereas for an ideal (inverting) integrator $F(s) = -1/j\omega CR_1$, for the circuit of Figure 6.2d $F(s) = A/[1 - (A - 1)j\omega CR]$. Likewise for the practical *differentiator* obtained by setting R_1 in Figure 6.2c equal to zero, the gain can only continue rising at 6 dB/octave until the opamp's open loop gain is reached. Note that in any case the open loop gain A of an opamp falls with frequency, so practical opamp differentiators are even less ideal than integrators.

Things really get interesting when we start to combine a first-order circuit – be it low pass or high pass – with one or more second-order circuits. However, before moving on to that stage, let's look at a first-order circuit with a finite zero, but not at the origin as in the case of a high-pass section. Figure 6.3a shows a passive low-pass CR circuit connected to the non-inverting input of an opamp. You will find the corresponding circle diagram and transfer function described in Chapter 2 (Figures 2.1, 2.5). The opamp's inverting input is connected to the junction of two equal resistors R between the circuit's input and output terminals. The heavy negative feedback will force the output to take up whatever potential is necessary to make the voltages at b and c, the opamp's input terminals equal. Further, since the input terminals draw negligible current, the same current must flow through both the resistors R, and the voltages across them must be equal and in antiphase with each other. This is shown in Figure 6.3b for a typical frequency somewhat below the corner frequency. Since the vectors db and ba are equal whilst ec and ca (the voltages across C and R_1) are in quadrature, ϕ_1 equals ϕ_2; and as the frequency rises from zero to infinity, ϕ_1 and ϕ_2 will both increase from zero to 90°. In fact the two resistors R act as a pantograph, drawing out a copy of the circle diagram at a scale of two to one as shown. Thus for unity input voltage v_i, represented by the vector ea in Figure 6.3b, v_o is also of unity amplitude but its phase, relative to the input, decreases from zero to $-180°$ as the frequency of the input rises from zero to infinite frequency. The frequency at which the phase shift is 90° is given by $\omega = 1/CR_1$. The circuit is called an *all-pass filter* or *phase equalizer*. The transfer function for a passive lag is, as found earlier, $F(s) = 1/(1 + sT)$, where T equals (in this case) CR_1. As this circuit contains but the one time constant, it is convenient to normalize the frequency to ω_c which equals $1/T$, so that $F(s) = 1/(s + 1)$ describes the voltage at the opamp's non-inverting input. Denoting the transfer function of the complete all-pass filter (APF) by $F(s)_{APF}$, you can describe it in terms of $F(s)$ by noting that it is twice as large and shifted to the left (see Figure 6.3b) by unity, since the input voltage is normal-

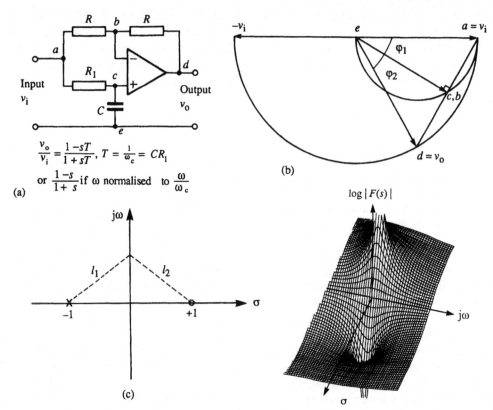

$$\frac{v_o}{v_i} = \frac{1-sT}{1+sT}, \quad T = \frac{1}{\omega_c} = CR_1$$

or $\frac{1-s}{1+s}$ if ω normalised to $\frac{\omega}{\omega_c}$

(a)

(b)

(c)

Figure 6.3 First-order all-pass circuit (phase equalizer).

ized like everything else. So

$$F(s)_{\text{APF}} = 2F(s) - 1$$

$$= \frac{2}{s+1} - 1 = \frac{2}{s+1} - \frac{s+1}{s+1} = \frac{1-s}{1+s}$$

This expression has a pole at $\sigma = -1$ due to the s in the denominator, but the numerator goes to zero at $\sigma = +1$; so the pole-zero diagram looks like Figure 6.3c, with a zero at $+1$. Moving up the $j\omega$ axis from zero frequency, the distances l_1 and l_2 to the pole and the zero increase but are always equal to each other. Since the magnitude of the response is inversely proportional to the distances to poles but directly proportional to the distances to zeros, the product remains unity. Also, as the output lag increases as we move anticlockwise relative to poles and clockwise relative to zeros, the phase increases twice as fast as does that at the non-inverting terminal of the opamp. So the pole–zero diagram predicts exactly the same behaviour as the vector diagram. Figure 6.3d shows a three-dimensional view of the pole–zero diagram but with $F(s)$ plotted on a logarithmic rather than a linear scale. The zero becomes an infinitely deep well, exactly like an upside-down pole, and symmetry shows the response on the $j\omega$ axis to be 0 dB (unity) at all frequencies.

If the right-hand R in Figure 6.3a is made progressively smaller, the vector db in Figure 6.3b shrinks and the zero in Figure 6.3c migrates further to the right. So at high frequencies the output is smaller than at zero frequency; the magnitude response looks like that of a transitional lag, but the phase finishes up at $-180°$ instead of returning to $0°$. As R tends to zero, the zero in Figure 6.3c and d migrates along the $+\sigma$ axis to infinity and the circuit becomes a simple first-order low-pass section. Conversely, if the right-hand resistor R is made progressively larger than its mate, the zero in Figure 6.3c migrates towards the origin and the output at high frequencies rises, since the 'pantograph' draws a larger and larger version of $F(s)$.

The all-pass filter is useful as a phase equalizer to correct phase distortion. For example, the high frequency attenuation associated with tape recording is not a minimum phase shift process, whereas the compensation applied in record and playback amplifiers is. The resultant phase distortion in the reproduced signal is of no consequence for audio signals but could cause errors when reading digital data from a diskette or other magnetic media.

Higher-order active circuits

Now on to second-order active circuits. Figure 6.4a shows an active second-order low-pass circuit called a *Kundert filter*. It is not the commonest active low-pass filter (LPF), but it behaves just like the more common Salen and Key filter (described later) and is particularly convenient for the purposes of analysis. It is clearly an LPF, for at zero frequency the capacitors can be considered open-circuit, so it is just a cascade of two unity-gain buffer amplifiers; whilst at infinite frequency the capacitors can be considered as short-circuits, preventing the passage of signals. What exactly it does in between depends on the values of the resistors and capacitors. To investigate this, Figure 6.4b shows the circuit 'unwrapped' by breaking it at point X in Figure 6.4a; the input voltage v_1, (from its ideal zero-impedance source) is set to zero, indicated by the ground point at Y in Figure 6.4b. We are left with a passive lead RC_1, and a passive lag RC_2, with buffer amplifiers. Clearly this circuit will not pass either very low or very high frequencies, whilst if the corner frequency of the lag or top cut is much lower than that of the lead or bass cut, there will always be a large attenuation through the circuit. If the two corner frequencies coincide, then at that frequency there will be a total of 6 dB attenuation through the circuit. If the corner frequency ω_{02} of the lag is higher than that of the lead ω_{01} as in Figure 6.4c, there will be a region between these two frequencies where the gain almost rises to unity. On the usual logarithmic scale of frequen-

cies as in Figure 6.4c, the low-pass and high-pass asymptotes are mirror images of each other while the phase responses are of exactly the same shape, passing through $+45°$ at ω_{01} in the case of the lead and $-45°$ at ω_{02} in the case of the lag. The frequency of minimum attenuation and zero phase shift occurs at the geometric mean of ω_{01} and ω_{02} call this frequency ω_0 as in Figure 6.4c. If you now connect the output of the circuit in Figure 6.4b back to its input, you have a circuit which can almost but not quite supply its own input at ω_0; it is no surprise therefore that, injecting a small signal at this frequency in series with R at point Y, it will be amplified considerably. Just how much, can be determined by analysing the circuit a little more formally. Starting at the output of the circuit in Figure 6.4a, one can write $v_o = v_3$ and

$$v_3 = v_2 \frac{1/j\omega C_2}{R + (1/j\omega C_2)} = v_2 \frac{1}{1 + j\omega T_2}$$

where $T_2 = RC_2$. Next, $v_2 = v_1$ and

$$v_1 = \left[(v_i - v_o) \frac{1}{1 + j\omega T_1} \right] + v_o$$

where $T_1 = RC_1$. The square bracket gives the part of the total voltage $(v_1 - v_o)$ across R and C_1, which appears across C_1; this plus v_o forms the input v_1 to the first opamp. Writing v_o in terms of the expression for v_3, and this in turn in terms of the expression for v_2 etc. gives

$$v_o = \frac{1}{1 + j\omega T_2} \left\{ \left[(v_i - v_o) \frac{1}{1 + j\omega T_1} \right] + v_o \right\}$$

Collecting all the v_o terms on one side and v_i terms on the other, then taking the ratio, gives

$$\frac{v_o}{v_i} = \frac{1}{(j\omega)^2 T_1 T_2 + j\omega T_2 + 1}$$

Now $(1/RC_1)(1/RC_2) = 1/(T_1/T_2) = \omega_{01}\omega_{02} = \omega_0^2$ say. Also, with a little shuffling, $T_2 = (1/\omega_0)\sqrt{(T_2/T_1)}$. So v_o/v_i can be rewritten as

$$\frac{v_o}{v_i} = \frac{1}{(j\omega/\omega_0)^2 + (j\omega/\omega_0)\sqrt{(T_2T_1)} + 1}$$

$$= \frac{1}{(j\omega_n)^2 + j\omega_n\sqrt{(T_2/T_1)} + 1}$$

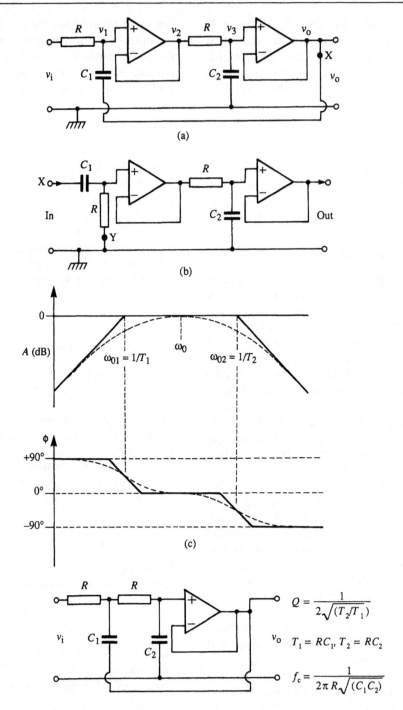

Figure 6.4   Some second-order low-pass sections.
  (a) The Kundert second-order low-pass filter section.
  (b) As (a), but with the loop 'opened out'.
  (c) Frequency response of (b) (see text).
  The Salen and Key second-order section.

where ω_n represents ω normalized to ω_0. More generally,

$$F(s) = \frac{1}{s^2 + s\sqrt{(T_2/T_1)} + 1}$$

This expression for $F(s)$ is identical to that we found in Chapter 2 for a series tuned circuit or two-pole passive low-pass filter (equation (2.6)). So it must behave in exactly the same way, and $\sqrt{(T_2/T_1)} = D = 1/Q$.

Figure 6.4d shows the better known *Salen and Key* active LPF. Here the lag circuit is not buffered from the lead, and in consequence $Q = 1/2\sqrt{(T_2/T_1)}$ as against $1/\sqrt{(T_2/T_1)}$ for the Kundert circuit. As a result of the square root, in the Salen and Key circuit the ratio of C_1 to C_2 for a given Q must be four times greater than in the Kundert circuit. Furthermore, in the latter there is a free choice of R and C for each time constant; it is not necessary to use the same value of R for both as shown in Figure 6.4a. By contrast, in the Salen and Key circuit, equal value resistors give the optimum arrangement.

The single passive lag circuit gives an ultimate rate of attenuation, well past the corner frequency, of 6 dB/octave. This is not a very sharp cut-off, although it could always be increased by cascading several sections: a cascade of three such sections would give 18 dB/octave, for example. The snag is that each would contribute its own 3 dB of attenuation at the corner frequency, resulting in considerable attenuation near the upper edge of the pass band and a slow initial rate of increase in attenuation in the stop band. A much better third-order filter results from combining a first-order circuit with a second-order circuit, since one can arrange the latter to have a peak in its response just below the corner frequency, to compensate for the roll-off of the first-order circuit. Choosing the right degree of peaking in the second-order circuit, obtains the sharpest possible corner frequency response without actually getting a peak of any sort in the pass band. This is called a *maximally flat* or *Butterworth* response, and you can derive the required value of D in the denominator of the second-order circuit without resort to the higher mathematics. All you need is algebra, with the aid of a useful result from the binomial theorem. In

case you never came across the binomial theorem or have forgotten it, it goes something like this. First, $(1 + x)^2 = 1 + 2x + x^2$, and this applies whatever the value of x. However, if x is very small compared with 1, then x^2 is very small indeed. For example, using δ (lower-case Greek letter delta) to indicate a number much less than one, if $\delta = 0.001$ then $1.001^2 = 1 + 2 \times 0.001 + 0.000\,001 \approx 1.002$. Conversely, the square root of $(1 + \delta) \approx 1 + \delta/2$. Also, since $(1 + \delta)(1 - \delta) = 1 + \delta^2 \approx 1$, then $(1 + \delta) \approx 1/(1 - \delta)$.

Armed with these results, let's see what happens to the attenuation of a third-order LPF at very low frequencies. The normalized response of the first-order section is $1/(j\omega_n + 1)$, and the magnitude of this is $1/\sqrt{(\omega_n^2 + 1^2)}$ by Pythagoras's theorem, since the j indicates that the ω_n term is at right angles to the 1. Writing δ for ω_n to indicate a very low frequency compared with the corner frequency of unity, then M_1 (the magnitude of the response of the first-order section) is $1/\sqrt{(\delta^2 + 1^2)}$ and, using the results we derived above, $\sqrt{(1 + \delta^2)} \approx 1 + \delta^2/2$ and $M_1 \approx 1 - \delta^2/2$. Thus at a frequency δ the response of a first-order section has fallen by $\delta^2/2$. By arranging that the second-order circuit has a response at δ of $1 + \delta^2/2$, the overall third-order response will be independent of frequency, at least at frequencies much smaller than the corner frequency.

The response of a second-order section is $1/[(j\omega_n)^2 + j\omega_n D + 1]$, where $D = 1/Q = \sqrt{(T_2/T_1)}$ for the Kundert circuit or $2\sqrt{(T_2/T_1)}$ for the Salen and Key circuit. Setting $D = 1$, this becomes $1/(-\omega_n^2 + j\omega_n + 1)$. At some low frequency δ the denominator becomes $(-\delta^2 + j\delta + 1)$, the magnitude of which is $\sqrt{[\delta^2 + (1 - \delta^2)^2]} \approx \sqrt{[\delta^2 + (1 - 2\delta^2)]} = \sqrt{(1 - \delta^2)} \approx 1 - \delta^2/2$ using the results of the binomial theorem. Then M_2 (the magnitude of the response of the second-order section) is $M_2 = 1/(1 - \delta^2/2) \approx 1 + \delta^2/2$, so $M_1 M_2 \approx 1$, i.e. the frequency response is level. Although the binomial approximations we have used themselves only hold for $\delta \ll 1$, the magnitude of the complete third-order response stays remarkably level, being only 0.5 dB down at $\delta = 0.707$ or half an octave below the corner frequency, as you can easily check with a pocket calculator. At $\omega_n = 1$ the response of the second-order section is unity since

$D = Q = 1$, i.e. for this value of damping the peak is entirely to the left of the corner frequency, as can be seen in Figure 2.8b for the curve $R = 1$, corresponding to $Q = 1$ for the active filter. Thus the complete third-order filter's response at the corner frequency is -3 dB, the same as the first-order section alone. However high the order of a Butterworth filter, the bandwidth is still defined as that frequency at which the response is -3 dB relative to the low- frequency response – or to the high-frequency response in the case of a high-pass filter.

Figure 6.5 shows the pole positions for the three-pole Butterworth filter. Note that the three poles are spaced around the semicircle at intervals of $180°/3$, the outermost poles being at half this angle from the jω axis. This is the general rule for a Butterworth filter; for example a two-pole maximally flat filter would have poles at $\pm 45°$. The required value of D is $2\cos 45°$ or 1.414; the response has no peak and is -3 dB at $\omega = 1$. Given the rule for the pole positions of a Butterworth filter, you can design any order filter with ease. For example, Figure 6.5c shows the pole positions for a fifth-order filter. In addition to a first-order section one needs two second-order sections – Salen and Key, Kundert, or any of the other second-order circuits you fancy – with D values of $2\cos 36°$ and $2\cos 72°$, or $D = 1.618$ and 0.618. At $\omega = 1$, $|F(j\omega)|$ collapses to $1/D$, so the two second-order sections have a response at the corner frequency of $20\log_{10} 1.618$ and $20\log_{10} 0.618$ or $+4.2$ dB and -4.2 dB. Of course you seldom need to design a filter, especially a commonplace type like the Butterworth, since the necessary C and R values are tabulated in numerous publications.[1] But getting inside a filter and taking it apart as I have just done will give you a clear idea of how it is supposed to work, and this puts you in a strong position when faced with one that doesn't.

State variable filters

Simply interchanging the C and R in each section of the low-pass filter of Figure 6.4a turns it into a high-pass filter with the same corner frequency.

The pole–zero diagram looks the same except that the two zeros that were at infinity are now at the origin. Turning now to the general purpose filter section *par excellence*, this is variously called the state variable or the biquadratic filter, depending on how the damping term D is organized. For simplicity, I may from time to time refer to all the versions as state variable filters (or SVFs for short). The SVF is perhaps a little prodigal in its use of opamps; it requires at least three where most second-order sections and some higher-order sections (e.g. Figure 6.5b) require only one. However, it does provide a choice of low-pass (LP), band-pass (BP) and high-pass (HP) outputs and, with the addition of just two or three resistors, all-pass and notch responses as well!

Figure 6.6a shows the circuit diagram and the relationships between the voltages at various points in the circuit, for generality using s in place of jω from the outset. To understand how it works, imagine the loop broken at the two points X and Y and R_Q disconnected. At 0 Hz the gain will be an enormous A^2 as shown in Figure 6.6b. However, at frequencies where the reactance of the capacitors C is small compared with A times the value of the resistors R, each of the two integrator stages will produce a 90° phase shift. So the locus of the output at the terminal LP as the frequency rises will be as shown, heading for a gain of unity with no phase shift when the reactance of C equals the value of R, where $v_{LP} = v_i$. If the loop were closed at X–Y the gain would be infinite and the circuit unstable. So one must apply some damping, which is shown in Figure 6.6a as coming from yet another opamp, but this is just to simplify the analysis at this stage. In Figure 6.6c is shown the closed loop vector diagram for the frequency $\omega = 1/CR$, i.e. at the normalized frequency of unity since $\omega_0 = 1/CR$. Each integrator provides a lag of 90°, but it looks like a lead since they are inverting integrators, owing to the opamps being used in the inverting connection. The additional input v_D applied via R_Q means that the circuit can no longer supply its own input; an external input v_i equal to $-v_D$ needed to close the vector diagram. If you sort out the various terms in Figure 6.6a and work out $F(s)_{LP}$, i.e. the ratio v_{LP}/v_i, you will find – surely

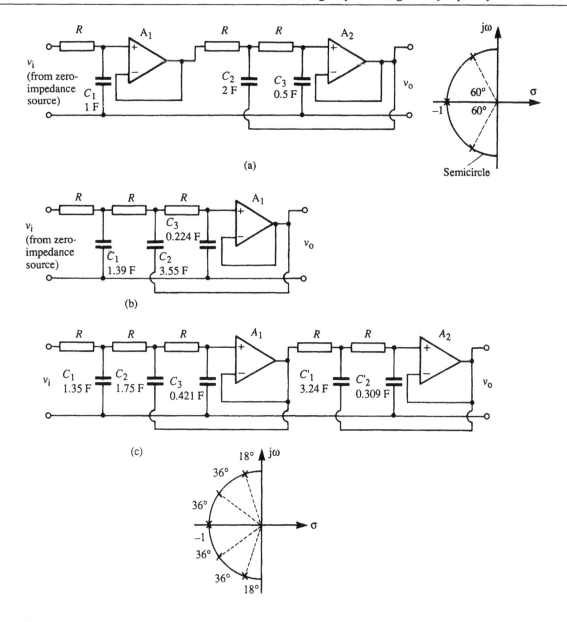

(a)

(b)

(c)

Figure 6.5 Butterworth low-pass filters.
(a) Third-order low-pass filter. Normalized to 1 Ω, 1 rad/s.
(b) Single-opamp version of (a).
(c) Fifth-order Butterworth low-pass filter.

no great surprise by now – that

$$F(s)_{LP} = \frac{-1}{s^2 T^2 + sT(1/Q) + 1}$$

where $Q = 1/D$. In the normalized case where $CR = T = 1$, $F(s)_{LP} = -1/(s^2 + sD + 1)$. The same sort of exercise will provide you with the results that $F(s)_{BP} = s/(s^2 + sD + 1)$ and $F(s)_{HP} = -s^2/(s^2 + sD + 1)$. Note that the three outputs at HP, BP and LP are always in

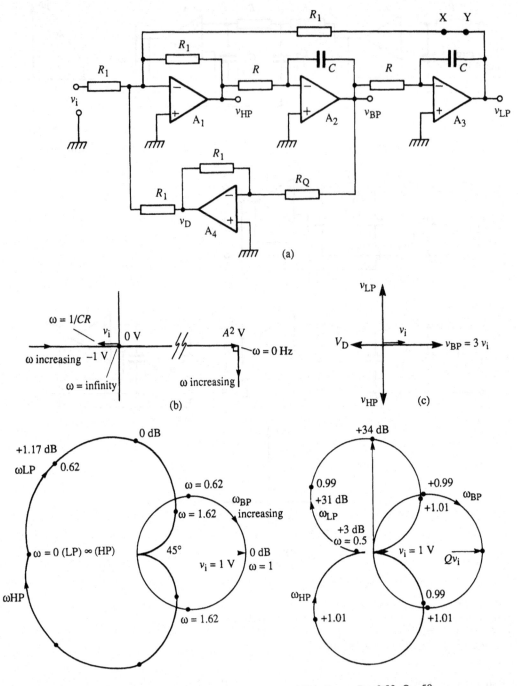

(a)

(b)

(c)

Low-*Q* case *Q* = 1, *D* = 1, slight peak

High-*Q* case *D* = 0.02, *Q* = 50

(d)

quadrature, but are only of equal amplitude as shown in Figure 6.6c at the corner or resonant frequency. How they vary with frequency is shown in Figure 6.6d and Figure 6.7. Notice that Q gives the magnitude of the output of a second-order section (for unity input) at the corner or resonant frequency, not the maximum output. This occurs, in the case of the low-pass output, at zero frequency for Q values of less than $1/2$, and at a finite frequency below the corner frequency for $Q > 1/2$. For higher and higher values of Q, the frequency of maximum response moves closer and closer to the corner frequency.

Since the HP and LP outputs are always in antiphase, combining them as in Figure 6.9a, there will be zero output at the resonant-frequency: a *notch* circuit. Adding in a suitably scaled contribution from the band-pass stage as in Figure 6.9b, the output phase will increase by 180° going

from low frequency to the resonant frequency, and by another 180° from there to a much higher frequency, while the amplitude will remain constant; this is an all-pass response. The higher the Q, the more rapid the phase change at the resonant frequency; in fact the picture looks like Figure 2.8b except that the limits are $+180°$ and $-180°$.

Figure 6.9a shows a commonly encountered version of the SVF. This has the advantage of economy, using only three opamps. Instead of inverting v_{BP} in A_4 (Figure 6.6a) and feeding a fraction $1/Q$ of it back to the inverting input of opamp A_1, a fraction of v_{BP} is fed to the non-inverting input of A_1 instead. If the fraction is one-third then Q is only unity, not 3: this is because the feedback from v_{LP} and v_{HP}, each via R_1, is effectively attenuated to one-third by the other resistors R_1 at the inverting input of A_1, whereas in Figure 6.6a the inverting input of A_1

Figure 6.6

(a) State variable filter.

$$v_{LP} = \frac{1}{j\omega T} v_{BP} \quad \text{or} \quad \frac{-1}{sT} v_{BP} : T = CR, s = \sigma + j\omega$$

$$v_{BP} = \frac{-1}{sT} v_{HP}$$

$$v_{HP} = -R_1 \left(\frac{v_{HP}}{R_1} + \frac{v_{BP}}{QR_1} + \frac{v_i}{R_1} \right)$$

Hence

$$\frac{v_{LP}}{v_i} = \frac{-1}{s^2 T^2 + sT(1Q) + 1}, Q = R_Q/R_1$$

In th normalized case where $T = 1$,

$$\frac{v_{LP}}{v_i} = \frac{-1}{s^2 + sD + 1} D = 1/Q$$

$$\frac{v_{BP}}{v_i} = \frac{s}{s^2 + sD + 1}$$

$$\frac{v_{HP}}{v_i} = \frac{-s^2}{s^2 + sD + 1}$$

(b) Open loop vector diagram at low-pass output: v_i short-circuited, link X–Y open, input at $X, R_Q = \infty$.

(c) Closed loop vector diagram at $\omega = \omega_o = 1/T, R_Q/R_1 = Q = 3$. Since all four resistors at the inverting input of A_1 are equal, the currents i_i, I_{HP}, i_{LP} and i_D are all directly proportional to v_i, v_i, v_{HP}, v_{LP} and v_D respectively. At $\omega = 1/CR, v_{HP} = -v_{LP}$, so i_{HP} and i_{LP} cancel out. Thus $v_i = -v_D = v_{BP}/Q$.

(d) Frequency responses, low and high Q.

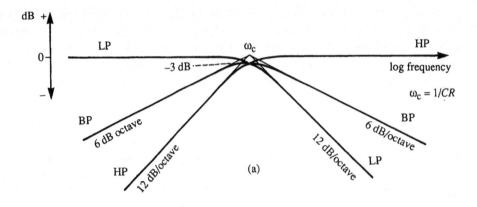

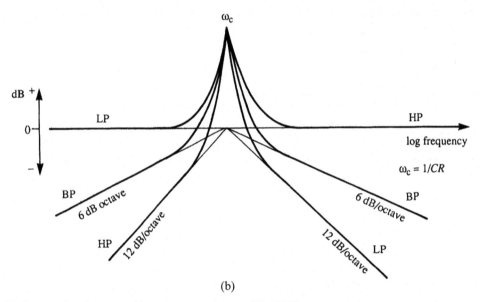

Figure 6.7   High-pass, band-pass and low-pass responses of the SVF.
(a) Low-Q response (asymptotes shown faint). Maximally flat case, $Q = 0.707, D = \sqrt{2}$, no peaking.
(b) High-Q response (asymptotes shown faint). As in (a), at all points the BP response lies between the other two.

is a virtual earth. Figure 6.9b shows another three-opamp second-order filter section, providing BP and LP outputs. This version is sometimes called the *biquadratic* filter, presumably because – like the APF of Figure 6.8b – it can provide a transfer function where both the numerator and the denominator are quadratic equations in s.

Further high-order filter design

The SVF provides an inherently high Q; indeed, it is infinite unless some deliberate damping is applied. With the Salen and Key and many other second-order circuits, on the other hand, high values of Q are only achieved with extreme component ratios (though there are ways round this). High values of Q are required in filters with a very sharp cut-off – the proverbial 'brick wall' filter for the second-order section with its poles nearest the jω axis. One can obtain such a sharp cut-off either with a Butterworth filter of very high order or, if willing to tolerate some ripple in the frequency response in the pass band, with rather fewer sections using a filter with a Chebyshev response.

Imagine a three-pole maximally flat filter such as in Figure 6.l0a, where the complex pole pair has been deliberately slid nearer to the jω axis; the resultant frequency response will have a peak as in Figure 6.10b. If you now make the single-pole stage cut off at a lower frequency by sliding the pole on the σ axis to the right, it will depress the peak back down to unity gain. The result is a dip in the response, as shown in Figure 6.10c. A fifth-order filter would have two dips, and if the poles are located on an ellipse, on the level at which they would have been in a Butterworth design, the peaks will all line up at unity gain and the troughs will all be the same depth. The more flattened the ellipse, the greater the ripple depth and the faster the rate of roll-off just beyond cut-off. However, the ultimate rate of roll-off is the same as a Butterworth filter, namely just $6N$ dB/octave, where N is the number of poles. The standard mathematical treatment of this *Chebyshev* filter is rather heavy going, and I haven't yet worked out a way to explain it with nothing more than algebra plus the odd complex variable, so you will have to take it on trust at the moment. As with the Butterworth design, the normalized component values for different orders of filter are tabulated in the literature, only there are many more tables for the Chebyshev design as there is a choice of ripple depth. Designs are commonly tabulated for ripple depths of 0.18, 0.28, 0.5, 1, 2 and 3 dB. If the first two ripple depths look odd, it is because they derive from designs for passive filters.[2] A ripple depth of 0.18 dB corresponds to a 20% reflection coefficient in the pass band at the troughs of the ripple, and 0.28 dB to 25%. With an odd number of poles, zero frequency corresponds to a peak of the ripple, i.e. to zero insertion loss in the case of a passive filter. With an even number of poles, zero hertz occurs at a trough of the ripple. Now in a passive filter composed (ideally) of purely reactive loss-free components, it is impossible to produce a finite insertion loss at 0 Hz in the filter itself. Therefore an even-order passive Chebyshev filter works between different design source and load impedances, the loss at 0 Hz being due to the mismatch loss. At the peaks of the ripple the filter acts as an impedance transforming device, matching the load to the source. With active filters this consideration does not arise, as the signal energy delivered to the load comes not from the input but from the power supply as dictated by the last opamp in the filter, which in turn is controlled by the earlier ones and ultimately by the input signal. This highlights the other major difference between passive and active filters: a passive filter passes signals equally well in either direction, whereas an ideal active filter provides infinite reverse isolation.

As with passive filters, so with active filters we can obtain a closer approach to the brick wall filter response by designing in *finite zeros* – frequencies in the stop band where the attenuation is infinite. A 'finite zero' – a finite frequency where the response is zero – has already appeared, see Figure 6.8a. If R_{HP} is made twice as large as R_{LP}, the frequency of the notch or zero would be higher than the resonant frequency ω₀. If you then adjust the circuit's Q to give a maximum response in the pass band of +0.5 dB relative to the 0 Hz response, this gives a 0.5 dB ripple second-order *elliptical* filter. Some writers call these Caur filters, whilst others simply regard them as a variant of the Chebyshev. Either way, you can see that sliding the notch down in frequency towards ω₀ whilst winding up the Q to retain the 0.5 dB ripple Chebyshev pass band shape, you can achieve as sharp a cut-off as you wish. There is just one snag: the closer the zero to ω₀, the less attenuation is left at frequencies beyond the notch – which only goes to prove that you don't get owt for nowt, as they say in Yorkshire.

Fortunately the maths required to work out the pole and zero locations for different orders of elliptical filters has all been done and the corresponding normalized component values for circuits have been tabulated; it is a little too heavy to wade through here, and you will find it in the many textbooks devoted entirely to active filters. Don't

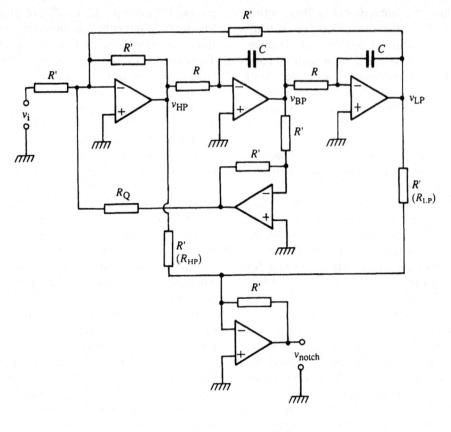

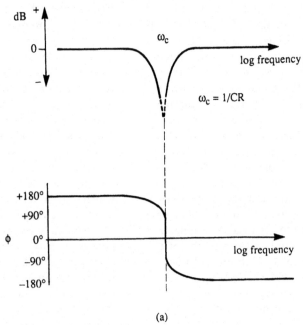

(a)

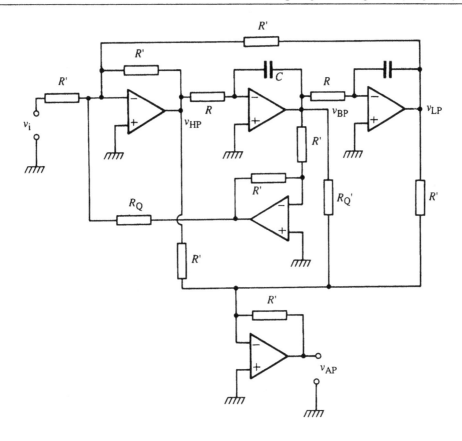

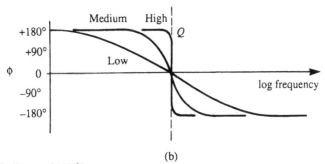

(b)

Figure 6.8   Notch and all-pass SVF filters.
(a) SVF notch filter, v_{HP} and v_{LP} are always $180°$ out of phase. At ω_c, $v_{HP} = v_{HP}$, therefore $v_{HP} + v_{LP} = 0$.
(b) SVF second-order all-pass filter (phase equalizer). If $R'_Q = R_Q$, amplitude response is unity at all frequencies. If not, then $|v_{AP}/v_i|_{\omega c} = R_Q/R'_Q$.

use the first elliptic filter design you come across, though; it might not be the most economical on components. The most economical sections are *canonic*, that is to say that they use only one *CR* or *time constant* per pole. So a canonic second-order (two-pole) section need only use two capacitors, each with an associated resistor, even if it also provides a pair of zeros. For zeros come 'for free',

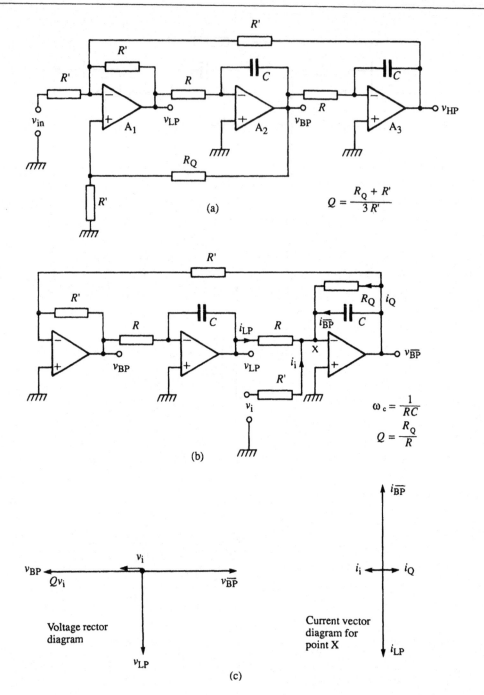

Figure 6.9   More second-order filter sections.

(a) Three opamp form of SVF.

(b) The biquadratic second-order section. If the biquadratic filter is tuned by adjusting the two resistors R, then Q is proportional to ω_c. This provides a constant bandwidth response. Owing to the input point chosen, the biquadratic filter has no high-pass output, but on-tune band-pass gain is independent of frequency.

(c) Vector diagrams at $\omega = \omega_c$ for the biquadratic section.

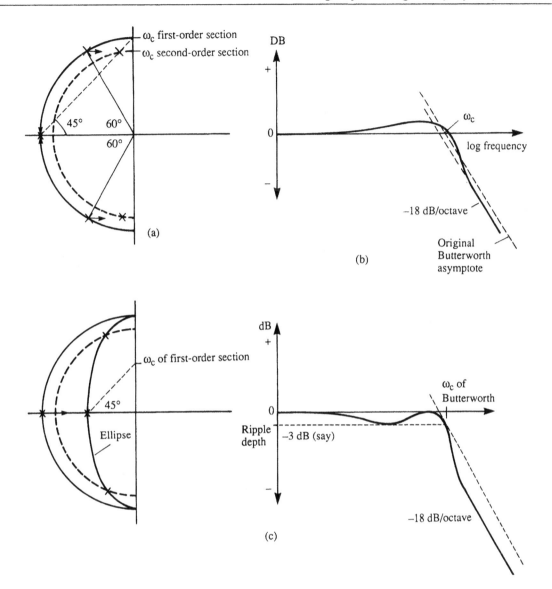

Figure 6.10   The Chebyshev filter (third-order example).
(a) Complex poles of Butterworth filter displaced to right.
(b) Resultant response.
(c) The Chebyshev response has the same ultimate rate of cut-off as Butterworth, but the peaking holds up the response at the corner, followed by a faster initial descent into the stop band.

even though they cost 12 dB/octave off the final filter roll-off rate. The tabulations for elliptic filters are even more extensive than for Chebyshev filters. For a Butterworth filter the only choice to be made is the number of poles, which is set by the normalized frequency f_s at which a given stop band attenuation

A_s is required. For a Chebyshev filter, the given f_s and A_s can be achieved with fewer sections at the expense of a larger pass band ripple, or with lower ripple but more sections. With the elliptic filter there are returns in the stop band as well as ripple in the pass band, so for a given order of filter and pass

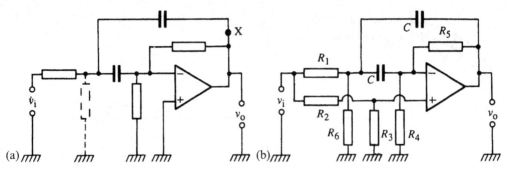

Figure 6.11   The SAB section and elliptic filters.
   (a) SAB (single active biquad) band-pass circuit, also known as the multiple feedback band-pass circuit.
   (b) SAB low-pass circuit with notch (finite zero-frequency response).

Design equations for first-order and SAB second-order sections in elliptic low-pass filters, normalized to $\omega_c = 1$ radian/second and $C = 1\,\mathrm{F}$ are as follows. For pole–zero pair $\Omega_1\sigma_1$ and Ω_2 (or $\Omega_3\sigma_3$ and Ω_4 etc., see Figure 6.12a) put $\Omega_p^2 = \Omega_1^2 + \sigma_1^2, \Omega_z = \Omega_2$ and 0 Hz stage gain $G = 1$. Then

$$R_5 = 1/\sigma_1$$

$$R_4 = R_5/(\Omega_z^2/\Omega_p^2 - 1)$$

$$R_6 = (R_4 + R_5)/[R_5^2\Omega_p^2 - 2(1 + R_4/R_5)]$$

$$R_1 = 1/(R_5\Omega_p^2 - 1/R_6)$$

$$R_2 = R_5$$

$$R_3 = R_4 \text{ (or choose } R_3 \text{, then } R_2 = R_3R_5/R_4)$$

If the circuit has an odd number of poles, then R' in first-order section is given by $R' = 1/\sigma_0$. Scale all results to required frequency and to convenient component values.

band ripple depth there is a further trade-off between f_s and A_s. For an even-order filter, the attenuation beyond the highest-frequency zero or f_∞ remains at A_s indefinitely (or at least until the opamps run out of steam), or rolls off at 12 dB per octave if you accept one less f_∞. In an odd-order filter, the odd pole on the axis ensures that the response at infinite frequency dies away to nothing, at a leisurely 6 dB/octave.

An important point to bear in mind, when comparing the responses of non-ripple filters such as the maximally flat amplitude Butterworth, maximally flat delay Bessel etc. with the responses of filters such as the Chebyshev and the elliptic, concerns the definition of *cut-off frequency*. For the former, the cut-off frequency is defined as the frequency at which the response is −3 dB, whereas for the latter it is defined as the frequency at which the attenuation last passes through the design pass-band ripple depth. If the ripple depth is

3 dB then the filters are directly comparable, but the response of say a 0.28 dB ripple Chebyshev low-pass filter with a 1 kHz cut-off frequency will not be 3 dB down until a frequency somewhat above 1 kHz.

To make an active low-pass elliptic filter one could use SVF sections for the second-order low-pass notch circuits, but a more economical filter results from a variant of the *single-amplifier biquadratic* (SAB) circuit.[3] Let's creep up on this ingenious circuit by stages. Figure 6.11a shows the SAB band-pass circuit; you can see it has no gain at 0 Hz if you imagine the capacitors open-circuit, and equally it has none at infinite frequency since the opamp's output is then effectively short-circuited to its inverting terminal. If you imagine the input grounded, the loop broken at point X and a signal inserted via the upper capacitor, you have a passive lead feeding an active (inverting) differentiator. At frequencies below its corner frequency the passive lead will produce nearly 90° phase advance, whilst the active differentiator will produce that much at all frequencies. So we have almost 180° of advance plus the inversion provided by the opamp, giving an open loop phase shift of nearly zero. If the open loop gain is unity when the phase shift is nearly zero, considerable peaking will result, i.e. the circuit will behave like a tuned circuit. Note that the resistor from the inverting input of the opamp to ground will have no effect as it is at a virtual earth.

Now look at the modified SAB circuit of Figure 6.11b. An attenuated version of the input signal is applied to the non-inverting input of the opamp; if the ratio of R_2 to R_3 is made the same as the ratio of R_5 to R_4, then the gain at 0 Hz will be unity, whilst the peak will be there as before. However, if R_6 and the other components are chosen correctly, there will be a frequency above the resonant frequency where the signal at the inverting input is in phase with the input and attenuated in the same ratio as the signal at the non-inverting input. Thus there is only a common mode input to the opamp but no differential input and hence no output – a notch, in fact. Figure 6.11 gives the design formulae for the second-order sections of an elliptic low-pass SAB filter, and for the first-order section if the design chosen has an odd number of poles. The design requires for each second-order section the σ and Ω (i.e. ω) values of the pole pair and the Ω value of the zero; the σ value of the zero is of course 0, as it is on the $j\omega$ axis. For an odd-order filter, the σ value for the single pole is also required. These values are all listed for various orders of filter with 0.1, 0.18 and 0.28 dB pass band ripple for various $f_s(2\pi\Omega_s)$ and $A_s(A_{min})$ combinations in Reference 2.

Figure 6.12 shows a five-pole elliptic low-pass filter, calculated using the design equations in Figure 6.11. The published tables of poles and zeros associate the lower-frequency zero with the lower-Q pole pair, as this results in less sensitivity of the response to component value tolerances. Note that although the gain in the pass band is between unity and -0.18 dB, the signal is attenuated before being applied to the non-inverting terminal of the opamp and then amplified by the

same amount. It is wise therefore when working with small signals to have some gain ahead of the filter, to keep the signal at the non-inverting terminal well clear of the noise level.

The higher-order filters discussed are used for critical filtering applications, where it is desired to maintain the frequency response of a system up to the highest possible value consistent with negligible response at some slightly higher frequency. A typical application would be the low-pass filtering of an audio signal before it goes into a sampling circuit of some sort, say the analog-to-digital converter in a compact disc recording system. In this instance, as is often the case with high-order filters, a fixed cut-off frequency is used. Where a variable corner frequency is required a very low-order filter is, fortunately, often perfectly adequate: a good example is the tone controls fitted in the preamplifier[4] section of a hi-fi amplifier (see Figure 6.13). These are usually first-order filters except for a high-frequency steep-cut filter, if fitted, which usually has one or two switched cut-off frequencies as well as an out-of-circuit position. The reason for this is that, to vary the cut-off frequency of a filter whilst preserving the shape of its response, it is necessary to vary the time constants associated with each and every pole in sympathy; thus, for example, a three-gang variable resistor is required for a three-pole filter. There are two-pole filter circuits which can be tuned by a single variable resistor, but remember that the transfer function of such a filter will contain a term in $T_1 T_2$, i.e. the product of two time constants. If the resistance associated with only one of these is varied, then a 10:1 ratio in resistance will provide only a $\sqrt{(10:1)}$ tuning ratio. Electronic tuning of high-order filters, e.g. those realized with the SVF circuit, is possible by using transconductance amplifiers or multiplying digital-to-analog converters in place of the resistors in the integrator sections.

For a number of years now another type of filter, the *switched capacitor* filter (not to be confused with the earlier and now little used *N*-path filter) has been available. The earlier types of switched capacitor filter, such as the MF10 introduced by Motorola and now second sourced by a number of manufacturers, are based on the SVF

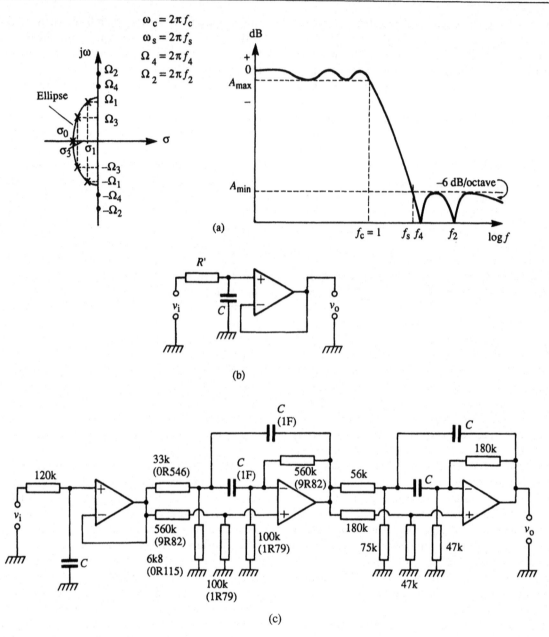

Figure 6.12   Five-pole active elliptic LPF.

(a) Five-pole elliptic filter. Unlike Butterworth type filters, the pole pair nearest the $j\omega$ axis actually lies outside the unit circle, to compensate for the effect of the nearest on-axis zero. Values for σ_0, Ω_1 etc. are read from Reference 2.

(b) First-order section, required when the number of poles N is odd. It is useful to include the odd pole on the $-\sigma$ axis, to ensure an ultimate roll-off at 6 dB/octave, as in (a). R' in ohms as in Figure 6.11, for $C = 1$ F, $\omega_0 = 1$ rad/s.

(c) Five-pole elliptic filter, A_{max} (ripple) $= 0.28$ dB, $A_{min} = 53.78$ dB, $\Omega_s = f_s/f_c = 1.6616$. Scaled to $f_c = 2.7$ kHz, giving a 3 dB bandwidth of about 3 kHz, designed as in Figure 6.11, using σ and Ω values from Reference 2. Reference 2 uses Ω to stand for normalized frequency in radians/second. In circuit, $C = 1$ nF. Resistors adjusted to nearest preferred value. Normalized values shown () for the highest Q section.

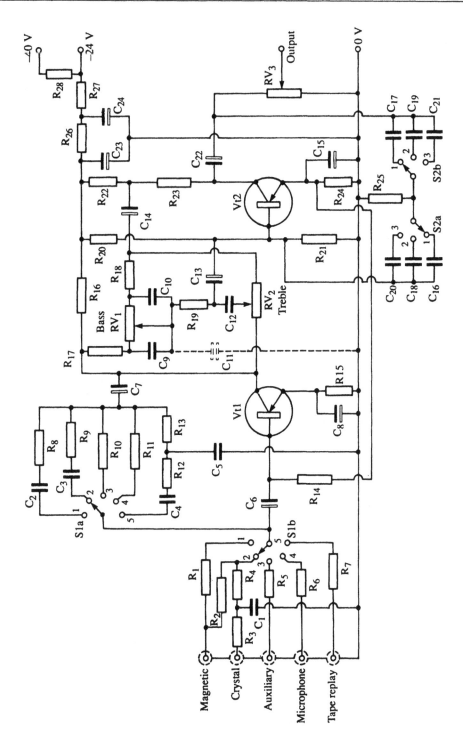

Figure 6.13 Preamplifier circuit including bass and treble controls. These each provide up to +12 dB (boost) and −16 dB (cut) relative to the central (flat) position of the control. This popular type of tone control stage was originally published by P. J. Baxandall in the October 1952 issue of *Wireless World*. (Reproduced by courtesy of *Electronics and Wireless World*)

configuration and can provide LP, HP, BP, notch and all-pass responses. These are analog sampled data filters, so the output is a stepwise approximation to the desired low-pass filtered version of the input signal, the steps being at 50 or 100 times the cut-off frequency. The steps can be effectively suppressed or rounded off by passing the output through a further fixed low-pass filter, which can often be a very simple low-order filter. The advantage of this type of filter is that it can be tuned by simply altering the necessary clock frequency, which is supplied to the filter at 50 or 100 times the required cut-off frequency. More recently, specialist analog IC manufacturers have introduced more complex filter ICs of higher order. For example, the XR-1015/1016 from EXAR Corporation is a seventh-order elliptic low-pass filter with the usual choice of ×50 or ×100 clock frequency. This filter may be used with clock frequencies between 1 kHz and 2.5 MHz, giving cut-off frequencies in the range 10 Hz to beyond 20 kHz.

References

1. Typical examples are: *Active Filter Design*. A. B. Williams, Artech House Inc, 1975. *Reference Data for Radio Engineers*, 6th Edition, Howard W. Sams and Co Inc. 1967.

2. *Handbook of Filter Synthesis*, A. I. Zverev, John Wiley and Sons Inc. 1967.

3. Sensitivity Minimisation in a Single Amplifier Biquad Circuit, P. E. Fleischer, *IEE Trans. Circuits Syst*. Vol. CAS-23, pp. 45–55, Jan 1976 (with 14 further references).

4. Transistor High-Fidelity Pre-Amplifier, R. Tobey, J. Dinsdale, p. 621, *Wireless World*, Dec. 1961.

Questions

1. Draw and explain the operation of a single opamp first-order low-pass filter with infinite (high) input impedance and zero (low) output impedance.

2. Design a single pole low-pass filter using a single opamp, having an input resistance of 100 kΩ, a passband gain of 35 dB and a cut-off frequency of 4 kHz.

3. In question 2, which would be the more appropriate, and why; an all bipolar opamp, or one with a FET or CMOS input?

4. An inverting opamp integrator circuit has a gain of unity at 1 kHz. What is the gain at (i) 10 kHz, (ii) 10 Hz? Ideally, as the frequency falls indefinitely, the gain rises indefinitely. What sets the frequency at which the gain ceases to rise?

5. Design a first-order all-pass filter as per Figure 6.3, to have 90° phase shift at 330 Hz. At what frequency will the phase shift be 135°?

6. Design a first-order all-pass filter to have a phase shift of 45° at the frequency where that in question 5 has a phase shift of 135°. If the two filters are cascaded, at what frequency will the phase shift be 315°? At this frequency, what is the time delay suffered by the signal?

7. What advantages does the Kundert second-order low-pass filter have over the Salen and Key circuit? Design a Kundert second-order high-pass filter with maximally flat response and a corner frequency of 100 Hz.

8. What is the disadvantage of achieving a third-order low-pass filter by cascading three first-order passive lags? Design a third-order Butterworth low-pass filter with a cut-off frequency of 3.3 kHz.

9. A state variable filter provides high-pass, band-pass and low-pass outputs. Describe how to combine these to implement (i) a notch filter, and (ii) an all-pass filter.

10. What is the attenuation at the design cut-off frequency of (i) a Butterworth filter, (ii) a 0.5 dB ripple Chebychev filter, and (iii) a 0.18 dB ripple elliptic filter?

7 Active signal processing in the time domain

This chapter looks at the processing of signals where the main emphasis is on waveform shape, and how it can be preserved or manipulated, rather than on the frequency response of the circuit and how it affects the signal. Some of the circuits described in the following pages are linear, whilst others are deliberately non-linear. The non-linearity may be provided by diodes, or by allowing an opamp to run out of its linear range, or by means of a comparator. The linear circuits may have outputs very different from their inputs. For example, in a constant current generator the sign and magnitude of the output current forced through the load is set by that of the input voltage. Again, in the frequency-to-voltage (F/V) converter the output voltage is proportional to the frequency of the circuit's input signal, regardless of amplitude or waveshape – and vice versa for its dual, the V/F converter.

Amplifier and multipliers

First, a look at the front end of the signal chain and at how a signal can be raised from (in many cases) a very low level to a level at which it can conveniently be processed, either on the spot or after being transmitted to a central location in, say, a manufacturing plant. Many signals start life as a very small voltage, only a few millivolts or even microvolts. Typical sources of such very small signals are thermocouples or platinum resistance thermometers for the measurement of the temperature of a furnace or processing oven, or resistive and piezoelectric strain gauges for the measurement of forces, accelerations or pressures. In a factory environment such signals often originate in

an electrically noisy environment and must be amplified to a usable level without contamination from undesired voltages, e.g. mains hum and harmonics of the mains frequency, often originating from phase-controlled rectifier motor drive or heater circuits. The unwanted electrical noise often takes the form of a common mode voltage, that is to say a voltage of equal magnitude with respect to earth or ground appearing on each of the two leads from the *transducer* – the device producing an electrical output corresponding to some physical parameter such as temperature, pressure or whatever. The problem is particularly acute where the transducer presents a high source impedance, e.g. a piezoelectric transducer, or micropipette electrodes used for physiological measurements in medical diagnostics. The instrumentation amplifier is a useful tool in such situations.

An operational amplifier exhibits a high degree of rejection of common mode signals at its two inputs, but when fitted with gain defining resistors the input impedance is high at the non-inverting input, i.e. in the non-inverting mode, but simply equal to the input resistor in the inverting mode. Unequal input resistance at the two input terminals is undesirable as it may cause conversion of any unwanted common mode signal to transverse or normal mode, which will then be indistinguishable from the wanted signal. The *instrumentation amplifier* is a combination of several opamps that provides a very high input impedance at both of its input terminals, which are 'floating' with respect to earth and respond only to normal mode signals. Figure 7.la shows the usual arrangement, although several other circuits provide equivalent performance. Each of the two amplifiers in the input stage

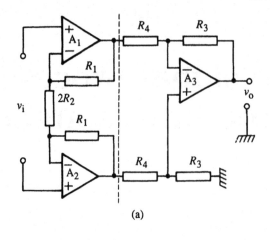

(a)

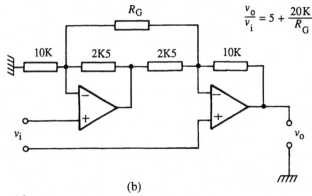

$$\frac{v_o}{v_i} = 5 + \frac{20K}{R_G}$$

(b)

Figure 7.1   Instrumentation amplifiers.

(a) Instrumentation amplifier with three opamps. Note that the resistors R_3 and R_4 form a bridge which is balanced to any common mode output from A_1 and A_2. In normal mode (push-pull, transverse), the input gain is R_1/R_2 and the output gain is R_3/R_4. In common mode (push-push, longitudinal), input gain is unity (0 dB) and output gain is zero ($-\infty$ dB), provided resistors are perfectly matched and A_3 is ideal.

(b) Two-opamp instrumentation amplifier.

is used in the non-inverting connection, the gain being defined by the feedback from the output to the inverting inputs in the normal way. Thus the input stage provides gain and impedance conversion but does not of itself reject common mode signals entirely; they are simply transmitted to the input of the output stage unchanged, i.e. at unity gain. If the input stage provides considerable gain to normal mode signals, for example 40 dB, then the input stage makes a useful contribution to the overall common mode rejection, in this case 40 dB. The rest of the rejection is provided by the output stage, by virtue of the way the signal is applied to it. The four resistors form a bridge which is balanced to common mode input signals to the output stage if the output voltage remains at ground potential. There is then no push-pull signal at the input terminals of the output opamp due to a common mode output from the first stage, provided the bridge is balanced. The output stage may also be used to provide gain and, provided that the ratio of the two resistors at the non-inverting input is the same as the ratio of the input and feedback resistors at the inverting input, the common mode rejection is maintained. Instrumentation amplifiers are available from many manufacturers and combine three opamps, together with the gain defining resistors, all in one

IC package. The early models were often of hybrid construction but today's designs are usually monolithic. (As indicated in Chapter 2, a hybrid is a module containing active devices such as opamps together with passive components, mounted on a substrate such as ceramic and encapsulated.) The AD524AD monolithic instrumentation amplifier from Analog Devices can be set for a gain of 1, 10, 100 or 1000 without any external components, or for intermediate gain levels with only a single external resistor. At unity gain it features 70 dB common mode rejection ratio (CMRR) and a 1 MHz bandwidth. Its input offset voltage of 250 µV maximum may be nulled with a preset potentiometer. The INA101HP monolithic instrumentation amplifier from Burr Brown offers an impedance of 10^{10} Ω in parallel with 3 pF at each input, 85 dB CMRR and a large output swing of ±12.5 V. The bias current at each input is only ±15 nA and the input offset voltage is trimmable to zero.

Isolation amplifiers are akin to instrumentation amplifiers with the additional feature of providing complete electrical isolation between the input and output circuits. The AD204JY from Analog Devices amplifies the input and uses it to drive a modulator. The resulting signal is transformer coupled to the output stage which converts it back to a replica of the input signal. Isolated power supplies for the input stage are provided by rectifying and filtering a 25 kHz transformer coupled supply. The device has a very high input impedance, a bias current of ±30 pA and a bandwidth of 5 kHz, and provides a continuous input/output voltage rating of 750 V RMS at 60 Hz or ±1000 V DC.

Instrumentation amplifiers are usually used at a fixed gain, set by internal or external resistors when the system of which they form part is manufactured. A requirement not infrequently arises for an amplifier whose gain is settable to any desired value by a control signal, either digital or analog. In the latter case, an operational transconductance amplifier can be used. Where digital control is needed, a number of different circuit arrangements are suitable. An early example used binary related resistors (R, $2R$, $4R$ etc.) switched by means of 2N3642 transistors

used in the inverted mode, connected so as to modify the gain determining feedback at the invert input of an opamp.[1] Another scheme, providing binary coded decimal (BCD) control of gain, uses CMOS analog switches to select the appropriate gain determining resistors.[2] Yet another scheme uses an AD7545 analog-to-digital converter's $R/2R$ ladder as the gain defining network for an opamp, using AC coupling to permit high gain without excessive offset voltage magnification.[3]

Nowadays, where a digitally programmable gain is required, a *programmable* amplifier (called a PRAM or PGA) is often used. These are available from many manufacturers. The 3606 series from Burr Brown are programmable gain instrumentation amplifiers whose gain is set with a four-bit binary word: the available gain values are ×1, ×2, ×4, ×8, ..., ×1024 at a bandwidth of 100 kHz at unity gain and 10 kHz at the highest gain settings. By contrast, for use with non-floating signal sources, the SC11310 programmable gain/loss circuit (PGLC) from Sierra Semiconductor has a single-ended ground referenced input and a −3 dB small-signal bandwidth typically in excess of 1.5 MHz when driving a 600 Ω load. It has nine gain control inputs, one selecting gain or loss and the other eight providing 256 gain/loss steps of 0.1 dB, i.e. a ±25.5 dB gain range. Given that distortion is typically 60 dB down on full output and that noise is very low indeed, the device provides a usable dynamic range of well over 100 dB.

Next, a look at some circuits which have already appeared in these pages. The *opamp integrator* in Figure 7.2a is identical to the inverting integrator described in Chapter 6, where, however, only its frequency response was considered. This, it turned out, increased indefinitely as the input frequency was made lower and lower, becoming (ideally) infinite at zero frequency. In Figure 7.1a, imagine that an input of +1 V is applied at the input at an instant when the output voltage is zero. The input current of 1 µA must be balanced by 1 µA current in the capacitor, assuming that the input impedance of the opamp is virtually infinite. The charge on the capacitor must therefore increase by 1 µC (one microcoulomb) per second; since the charge $Q = CV$, if $C = 1$ µF then the opamp's output

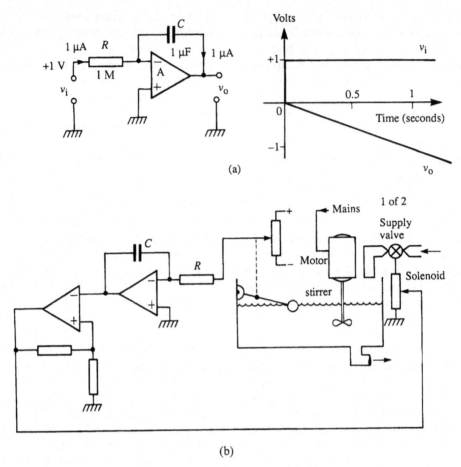

Figure 7.2
(a) Opamp integrator.
(b) Integrator as part of a process control system.

voltage must become 1 V more negative every second. This is true in the long term even if the input voltage is varying, provided that its average value is 1 V, for the charge on the capacitor will equal the average input current times the time for which it is flowing, i.e. the integral of the input current.

A comparator can be used to monitor the output voltage and, when it reaches a certain value, to control, say, the inlet valves to a liquid blending tank. Thus we have the basis of a simple *bang-bang servo* system. Figure 7.2b shows the arrangement, with the integrator and comparator controlling the inlet valves, to make up as and when necessary as the product is drawn off. The integrator acts as a filter to average the output of the level sensor,

which is 'noisy' owing to the action of the stirrer, and to drive its output towards one or other switching point of the comparator circuit according to whether the mean liquid level is higher or lower than the desired level. A fraction of the comparator's output is fed back to its non-inverting terminal. This positive feedback determines the hysteresis or 'dead space' between the voltage levels at which the comparator switches, providing, in effect, a further filtering action to prevent the valves chattering on and off.

The integrator circuit also provides a means of modifying the shape of a waveform. Figure 7.3a shows a square wave centred about ground applied to an integrator. Since a constant input voltage results in a linear ramp output, the output wave-

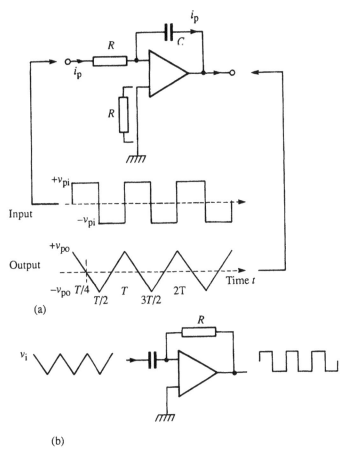

Figure 7.3 AC response of opamp integrator and differentiator.
(a) Opamp integrator with square wave input. Note that whilst the input is positive, the output becomes steadily more negative, as this circuit arrangement is an inverting integrator.

$$v_{po} = \frac{i_p(T/4)}{C} = \frac{(v_{pi}/R)(T/4)}{C} = \frac{Tv_{pi}}{4CR}$$

(b) The opamp differentiator.

form is a triangular wave as shown. However, don't forget that an integrator's gain at zero hertz is indefinitely large; so if there is a DC component at its input, be it ever so small, the mean level of the output waveform will wander up or down, according as the component is negative or positive respectively, until the opamp's output is stuck at a supply rail voltage. The DC component could arise from a number of causes, for example the input offset voltage of the opamp, or, if it is a bipolar input type, the drop caused by the input bias current flowing through the resistor R. (This could be partially compensated by adding R in series with the non-invert input as shown.) Even if these causes are absent, it might be that the square wave is not exactly centred about ground, or, if it is, that its mark/space ratio is not exactly 50/50. If the output of the integrator is connected to a comparator with hysteresis, the comparator will convert the triangular wave output of the integrator back into a square wave, which can be used to provide the input to the integrator in the first place. The circuit is a bang-bang servo which maintains the DC component at the input to the

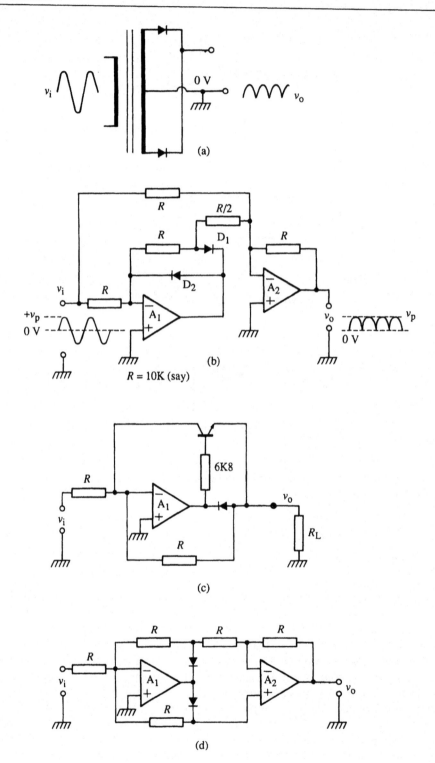

(a)

(b)

$R = 10K$ (say)

(c)

(d)

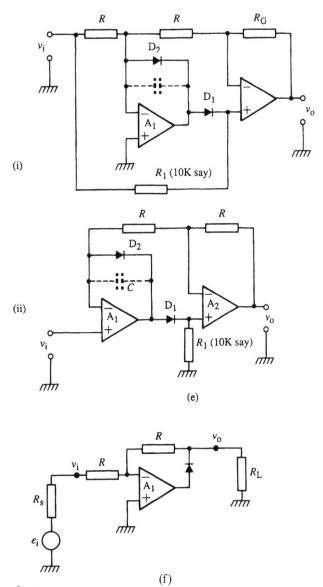

Figure 7.4 Full-wave rectification.

(a) Full-wave rectifier using diodes.

(b) Ideal rectifier circuit (accuracy depends on absolute values of all resistors). D_2 improves high-frequency performance by closing the NFB loop around A_1 on negative-going input half-cycles, thus preventing A_1 from being overdriven and saturating.

(c) Poor man's ideal rectifier. Only the two resistors R need be accurately matched; actual value is unimportant. Poor at high frequencies, since the opamp is saturated during negative half-cycles of the input and hence requires a recovery time. Make $R_L \gg R$.

(d) As (b) but less susceptable to errors due to opamp input voltage at low input levels.

(e) Two circuits requiring only one matched pair of resistors. In each case an extra stabilizing capacitor C may be needed as shown, as the loop includes both A_1 and A_2 (on input negative half-cycles, upper circuit: on positive half-cycles, lower circuit). (i) For unity gain, set R_g to zero: $v_{po} = v_{pi}$. Input resistance differs for +ve and −ve input half-cycles. (ii) Unity gain, very high input resistance.

(f) Another poor mans ideal rectifier. Requires $R_s \ll R, R \ll R_L$.

integrator at zero; it is also an example of a class of oscillator known as *relaxation oscillators*, which are described in Chapter 9.

Figure 7.3b shows another way of turning a triangular wave back into a square wave. A ramp of voltage applied to the capacitor will cause a current $I = C\,dv/dt$, where dv/dt is the rate of change of the ramp voltage in volts per second. The opamp's output voltage will cause an equal current I to flow through R and hence will remain at a constant positive or negative voltage on alternate half-cycles of the input triangular wave. The amplitude of the output square wave is proportional to the amplitude and frequency of the input triangular wave and to the *differentiator's CR* time constant.

If the input to a differentiator is a square wave, overload will usually result. This is because the slope dv/dt of the positive- and negative-going edges of a square wave is very large (ideally infinite), so that the current through the capacitor will exceed the maximum current that the opamp can force through R.

The absolute value circuit, also called an *ideal rectifier*, is an important and commonly met tool in signal processing. Figure 7.4a shows a full-wave rectifier of the *biphase* variety, such as is often used in power supply circuits. It can be used to provide the absolute value (i.e. the modulus) of a waveform, but suffers from two disadvantages: first, it requires a centre tapped transformer; and second, the smaller the peak-to-peak input voltage the poorer the performance, owing to the forward drop in the diodes. By enclosing a diode D_1 within the negative feedback loop of an opamp, we can effectively reduce its forward voltage to negligible proportions and thus simulate a perfect rectifier. Figure 7.4b shows one such circuit. A typical application might be as an AC/DC converter in an audio-frequency millivoltmeter, where the circuit's ability to provide a linear relation between the peak amplitude of the AC input and the DC output right down to small-signal levels is a big advantage over the passive circuit of Figure 7.4a. In this application, the DC output could drive a meter whose scale is calibrated in millivolts RMS. The circuit would actually respond to the average value of the modulus of the input, but the meter

sensitivity would typically be scaled to make the instrument read correctly on a sine wave. Figure 7.4c shows another absolute value circuit which is more remarkable for its ingenuity than anything else.[4] On positive-going inputs the opamp's feedback loop is closed via the diode and the lower 10 K resistor, whilst for negative-going inputs the opamp is open loop. Its output therefore flies up to the positive rail, saturating the NPN transistor. This connects the negative input directly to the output with no more than a few millivolts drop across the transistor, since it is being used in the inverted mode. Used in this way, a transistor's gain is low – of no consequence in this application – but the offset voltage between collector and base is much lower than the $V_{cc\,sat}$ of the device used in the normal mode. The circuit works well enough at low frequencies, but performance is poor at higher frequencies owing to the time required for the opamp to recover from overload each time the input goes positive. Figure 7.4d, e and f show other ideal rectifier circuits.

An ideal rectifier can be used to rectify a ground centred triangular voltage waveform, the result being a positive-going triangular wave of twice the frequency and half the amplitude. If this is translated back to a 0 V centred level and amplified by a factor of two, the same process can be repeated many times, giving a series of octave related triangular waves.

Another widely used function in analog signal processing is the *constant current generator*. Chapter 3 on active devices described how the collector slope resistance of a transistor is very high, so that to a first approximation the collector current is independent of collector voltage. The slope resistance is even higher when the transistor is used in the common base connection rather than the common emitter connection, and the same comment applies to FETs. In Figure 7.5a a variety of constant current sources is shown; the output current may be fixed or controlled by a signal voltage. In the latter case, the proportionality is upset by the V_{bc} of the transistor or the gate voltage in the case of an FET; this is a further imperfection on top of the output resistance which, though high, is not infinite.

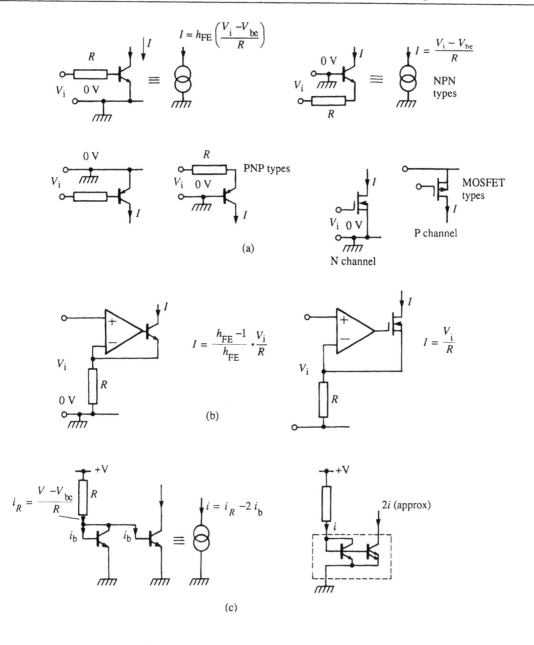

Figure 7.5   Constant currnt generators.

  (a) Unipolar (sink only or source only) constant currrent generators.

  (b) Opamp aided (improved) current sinks.

  (c) Current mirrors. By connecting two transisors as shown, the collector current of the right-hand device will 'mirror' that of the other, apart from the reduction due to base current. The circuit will only work if the transistors are perfectly matched, which means in practice that they must be a monolithic pair. Such devices are available, and, by using multiple emitter transistors, ratios other than $1:1$ can be obtained, e.g. $1:2$ (see right hand circuit) and $2:1$. Current mirrors are widely used in monolithic opamps.

A better constant current source can be arranged by enclosing the active device, be it FET or bipolar transistor, in the feedback loop of an opamp as in Figure 7.5b. The opamp output will take up whatever voltage is necessary to pass a current through the resistor R such that the voltage at the inverting input equals the input voltage. In the case of the bipolar type, if a change in collector voltage, due to some change in the load receiving the constant current, causes a change in the required V_{bc}, this will automatically be implemented by the opamp so as to keep the voltage across R unchanged. However, variations in the current gain of the transistor are not compensated for, as it is actually the emitter current of the transistor which is the controlled current. Thus the version using an FET as the controlled source provides improved performance.

The constant or controlled current sources in Figure 7.5a, b and c are strictly unipolar, that is to say that the output current can be set anywhere from zero to some large positive value when using the PNP version, or from zero to a negative value (a current sink rather than a source) with the NPN. Likewise, the voltage range that the output can take up – the *voltage compliance* – is limited to values negative with respect to the PNP's base voltage and conversely for the NPN type. Another type of constant current generator, called the *Howland current pump*,[5] circumvents both of these limitations. As you can see, the circuit has feed-back from the output to both the inverting and the non-inverting terminals in equal measure (Figure 7.6a). The feedback dividers therefore form a balanced bridge, with the result that variations in the output voltage produce only a common mode input to the opamp, and this has no effect. Thus with v_{i1}, and v_{i2} both at the same voltage, zero or otherwise, the output voltage is indeterminate. Now look what happens when an offset is introduced between v_{i1} and v_{i2} (Figure 7.6b). The only way the opamp's input terminals can be at the same voltage is if a current $I_L = (V_2 - V_1)/R_2$ flows through the load, as you can prove by substituting different values of V_1 and/or V_2 or, more satisfactorily, by analysing the circuit formally.

The arrangement is short-circuit stable and is also stable for all other finite values of Z_L, resistive or reactive, as under these conditions the negative feedback is stronger than the positive. With Z_L open-circuit the net feedback is ideally perfectly balanced, but in practice, owing to resistor tolerances, one or other will predominate. So the opamp's output will either fly off to one supply rail or the other, or hover in the region of 0 V at a magnified version of the opamp's input offset voltage. The output is truly bipolar, that is to say it can source or sink current at any voltage within the opamp's output voltage range. In this respect it differs from the *current mirror* of Figure 7.5c, which shares the limitations of the other constant current generators of Figure 7.5. If the load connected to a Howland current pump is a capacitor, then a ramp of output voltage is produced; the rate of change of voltage depends on the controlling voltage at the input and the size of the capacitor (Figure 7.6d). This application dates back many years;[6] the arrangement is a non-inverting integrator, with the further advantage that one end of the capacitor is grounded.

The base/emitter voltage of a bipolar transistor is logarithmically related to the collector current, and this is the basis of a class of circuits which can be used as elements of an analog computer. Analog computers were widely used at one time and are still employed in certain areas where limited accuracy is adequate but high processing speed and lower power consumption are essential, e.g. certain military applications. Figure 7.7a shows a *logarithmic amplifier*, that is to say one where the output voltage is proportional to the logarithm of the input voltage.[7] The logarithmic voltage/current relation of the transistor only holds over a wide range if the collector/base voltage is zero or very nearly so; this requirement is fulfilled by using the device in the feedback loop of an opamp as shown. Under these circumstances, the logarithmic relation can hold over up to seven or even eight decades of current, for some transistor types.

Figure 7.7b shows how two logamps, an opamp used as an adder, and an antilogamp can be used as a *multiplier* to obtain the product of two quantities, represented by the two input voltages. The arrangement shown is unipolar, so that both inputs must be constrained to be always positive, for which purpose

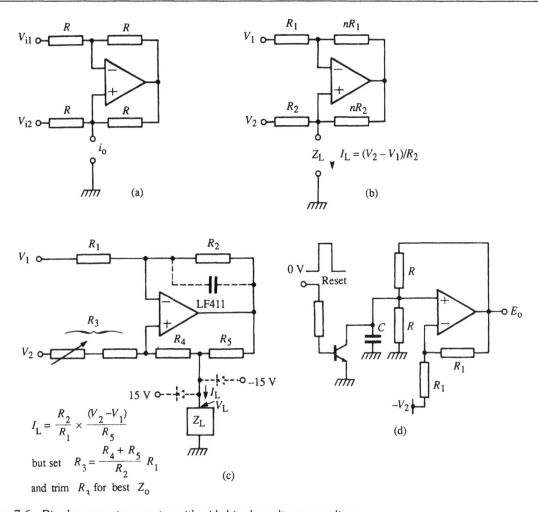

Figure 7.6   Bipolar current generator with wide bipolar voltage compliance.

(a) The basic Howland bipolar current generator. In this useful 'current pump' circuit, the output is taken not from the opamp's output terminal but from its non-inverting input terminal. It is a 'bipolar' voltage-controlled constant current generator, i.e. it can either source or sink current.

(b) Supplying load current. When supplying large load currents to Z_L (R_2 is low), opamp loading is minimized by making $R_1 \gg R_2$.

(c) Further modification for even higher load currents. The circuit has bipolar voltage compliance: it can source or sink current at both positive and negative voltages. Making $n < 1$ increases the voltage compliance range. (Reproduced by courtesy of *EDN* magazine.)

(d) Sawtooth generator using a non-inverting integrator with ground referenced capacitor. A circuit similar to this appeared as long ago as 1970:[6] did the author recognize it as an application of Howland's current pump?

an ideal rectifier circuit can be used. The correct polarity can be restored to the output by using comparators to determine the polarity of the input voltages and using these to control the polarity of a sign switched amplifier. One of these is shown in Figure 7.7c. If the FET (or bipolar transistor switch, as in Reference 8) is switched on, grounding the non-inverting input of the opamp, the circuit's gain is unity and inverting; whereas if the FET is off, the voltage at the non-inverting terminal of the opamp is the same as the input voltage, so the gain is unity and non-inverting.

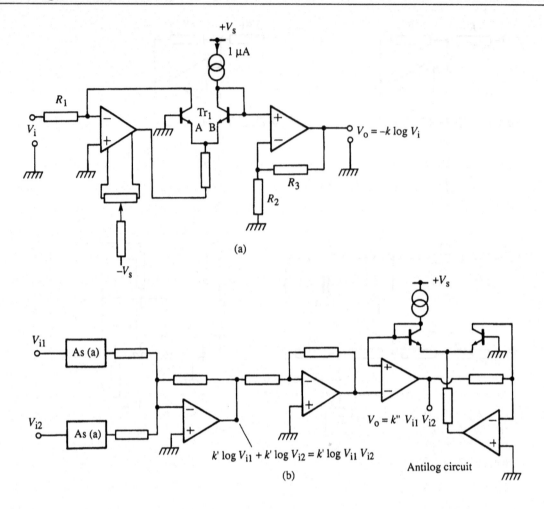

(a)

$V_o = k'' V_{i1} V_{i2}$

$k' \log V_{i1} + k' \log V_{i2} = k' \log V_{i1} V_{i2}$

Antilog circuit

(b)

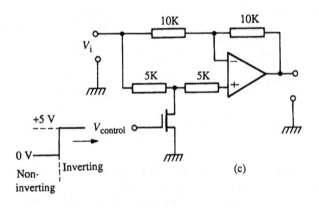

(c)

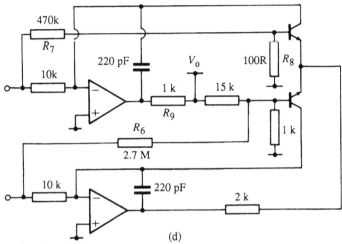

Figure 7.7   Log and antilog circuits.
(a) A logarithmic amplifier. k depends upon R_1, R_2, R_3 and Tr_1. Tr_1 is a monolithic dual transistor, e.g. 2N3860.
(b) Multiplier using logamps. Note that the usable range of V_o is no greater than in (a). Therefore the circuit cannot handle the whole combined range of $V_1 V_2$.
(c) Selectable sign of gain circuit: gives ×1 or ×(−1).
(d) Improved log ratio amplifier (reproduced by courtesy of *Electronic Product Design*).

If the same voltage is applied to both inputs of a multiplier circuit such as in Figure 7.7b, the output is the square of the input voltage. If this is smoothed (averaged) and then put through the inverse process, i.e. the square root taken, the result is the RMS value of the input voltage. The square root can be extracted by enclosing a squarer in the feedback loop of an opamp. Figure 7.7d shows a circuit which returns the logarithm of the ratio of two input voltages.[9]

Another type of low-speed analog multiplier, popular in the days before multiplier ICs were available, used the fact that the average value of a unipolar (e.g. positive-going from zero) pulse train of constant frequency is proportional to the product of the pulse width and the pulse height. A not dissimilar scheme employed pulses of constant width, but used one of the inputs to control the frequency of the pulse train, i.e. as a voltage-to-frequency converter.[10] This ingenious circuit produced an output $E_o = E_1 E_2 / E_3$, i.e. the product of two inputs divided by the third input. Even in the days of valves, applications such as analog multiplication and squaring were carried out, using multielectrode valves such as the nonode.[11]

Analog operations with a mixture of discrete devices and opamps are subject to errors due to imperfect matching of the logarithmic characteristics of the transistors and due to changes of temperature. These errors, together with a host of preset adjustment potentiometers for setting up, can be very largely avoided by using one of the integrated multiplier/divider or RMS linear integrated circuits which are available from a number of manufacturers. A typical example of a multiplier is the AD534 from Analog Devices, which is laser trimmed to an accuracy of 1% or better and which is a full 'four-quadrant' multiplier, i.e. the inputs are not restricted to positive voltages. The AD536 is a typical RMS-to-DC converter from the same manufacturer.

Converters

Frequency-to-voltage (F/V) and *voltage-to-frequency* (V/F) converters are important items in the analog engineer's toolbox of useful circuits, as they can be used for so many applications. Figure 7.8a shows a simple discrete V/F circuit[12] which produces an output in the range 0 to 10 kHz linearly related to a positive input control voltage,

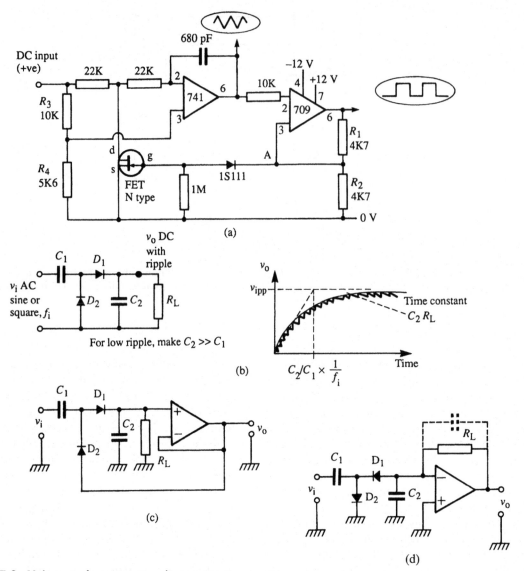

Figure 7.8 Voltage-to-frequency-to-voltage converter circuits.
   (a) V/F converter (voltage-controlled oscillator) (reproduced by courtesy of *Electronics and Wireless World*).
   (b) Simple pump and bucket F/V converter has exponential response when v_i peak-to-peak volts at f_i are first applied. $v_o/v_i = f_i C_1 R_L$, fairly linear up to 10%.
   (c) Linear pump and bucket F/V converter. Because of the bootstrap connection to the anode of D_2, C_1 is fully charged on *every* negative-going edge, regardless of v_o.
   (d) Another linear F/V converter.

with a scaling of 1 kHz per volt. Figure 7.8b shows a simple 'pump and bucket' *frequency counter* or F/V circuit. On each negative-going edge of the input square wave the pump capacitor C_1 is charged up to V volts (less one diode forward voltage drop), and on each positive-going edge a charge of $C_1 V$ coulombs is tipped into the bucket capacitor C_2. At least, that is what happens initially. However, as the voltage across C_2 rises to the point where charge is leaking away through

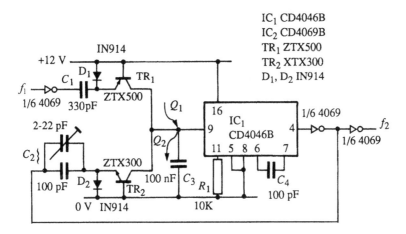

Figure 7.9   Circuit to provide an output at $f_2 = nf_1$ where n can be a non-integer, greater or less than unity, $f_2 = f_1(C_1/C_2)$. For correct operation, the peak-to-peak voltage of inputs at f_1 and f_2 to the two charge pumps must be equal, ensured here as they are derived from inverting buffers in the same CD4069 hex inverter. (Reproduced by courtesy of *Electronic Engineering*.)

R_L as fast as it is being added via C_1, the latter is no longer discharged completely on each cycle of the input, resulting in the non-linear (exponential) relation between input frequency and output voltage shown. Figure 7.8c and d show two solutions to this problem. In the first, the pump is charged up on negative-going edges not with respect to ground but with respect to the buffered output voltage. In the second, the charge flows into a short-circuit provided by the virtual earth at the inverting input of an opamp. The bucket should be large compared with the pump in order to keep down the size of the ripple on the DC output of the circuit, but not so large that the response to a change of input frequency is unduly slow $C_2 = 10C_1$ is generally a good compromise. In the case of Figure 7.8d, C_2 can alternatively be connected round the opamp as shown dashed, turning it into a *leaky integrator*.

Yet another variant is to replace the second diode by the base/emitter junction of a transistor, with the bucket capacitor in its collector circuit. This again provides a linear output as the pump is always charged and discharged by the full input voltage swing – except of course for the two diode drops. Figure 7.9 shows an ingenious application of this scheme.[13] Here, a single bucket receives an average rate of charge input from one pump balanced by removal via a complementary polarity pump. The voltage across the bucket is connected to the voltage-controlled oscillator forming part of $1C_1$ and the output frequency is connected to one of the complementary V/F inputs. The output frequency will therefore settle at value $F_2 = (C_1/C_2)F_1$, since any deviation from this ratio would result in inequality of the average current outputs of the two F/V converters in such a sense as to change F_2 so as to equalize them again.

As with multipliers and RMS circuits, V/F and F/V circuits are available in IC form from a number of manufacturers. The AD645JN from Analog Devices is an integrated V/F converter operating with a full-scale frequency up to 500 kHz and with a linearity of 0.03% for a full-scale frequency of 250 kHz. The AD651AQ from the same manufacturer is a synchronous V/F converter, that is to say that its full-scale frequency is the same as that of an externally provided clock waveform. It uses the charge balancing technique mentioned earlier and accepts a clock frequency of up to 2 MHz. At 100 kHz full-scale clock frequency the linearity is typically 0.002%. The 9400CJ from Teledyne Semiconductors is an IC which may be used as either a V/F or an F/V converter, with operation up to 100 kHz and 0.01% linearity at 10 kHz.

To an ever-increasing extent, analog signal processing is being carried out in numerical form, by first converting an analog waveform to a stream of numbers denoting the value of the voltage at successive instants. The number stream can be operated upon by a digital signal processor (DSP) IC in any number of ways (e.g. filtering, raising to a power, integrating, taking the modulus, autocorrelating, cross-correlating etc.) and then, if appropriate, reconverted to an analog voltage. From the point of view of the analog engineer, the important stages are the conversion to and from digital form, since any errors introduced, particularly in the analog-to-digital conversion, simply misrepresent the original signal and cannot afterwards be rectified.

There are many different forms of *analog-to-digital converter* (ADC, or A/D converter). Consider first the *flash converter*, in many ways conceptually the simplest though not historically the earliest type of ADC. The mode of operation, which should be clear from Figure 7.10, is as follows. The reference voltage, or plus and minus reference voltages if ground centred bipolar operation is required, is connected to a string of equal value resistors forming a precision voltage divider. The voltage difference between any two adjacent taps is equivalent to the least significant bit of the digital representation of the input analog voltage, which is thus quantized – in the case of the 7-bit MC10315L from Motorola, to a resolution of one part in 2^7 or about 0.8% of full scale. Each reference voltage tap is connected to one input of a string of 128 comparators, the analog input voltage being connected to the other input of each and every comparator. Every comparator whose reference input is at a lower voltage than

the analog input will produce a logic one at its output whilst all the others will indicate logic zero. On the rising edge of the clock input waveform the outputs of all the comparators are latched and encoded to 7-bit binary form, and the result is transferred to a set of output latches.

Although the device only provides 7-bit resolution, the divider chain and comparators are accurate to a resolution of one part in 2^8. It is possible to take advantage of this by cascading an MC10315L with an MC10317L to obtain 8-bit accuracy. Because all the comparators are identical and their outputs are latched at the same instant, they form a 'snapshot' of the input voltage at a time t_{ad} shortly after the positive clock edge, where t_{ad} is called the *aperture delay time*. Its value is generally not important except when using two similar ADCs to compare the voltages at two points in a circuit, as in a dual-channel digital storage oscilloscope (DSO). However, even in a single-channel application, sample-to-sample variations in t_{ad} are clearly undesirable, especially at the 15 mega-samples per second maximum sampling rate of the device when the input voltage is changing rapidly. The MC10315/7L devices can cope with input slew rates of up to 35 V/μs and exhibit an aperture uncertainty of only 80 ps. The successive approximation register is another type of ADC, capable of providing much greater resolution than the flash converter. However, to understand how it works, look first at the converse process of D/A conversion.

Figure 7.11a shows a current output high-speed 8-bit multiplying *digital-to-analog converter* (DAC, or D/A converter). Imagine the reference amplifier's inverting input connected to ground and its non-inverting input to, say, a +2.56 V reference

---

Figure 7.10   A high-speed analog-to-digital converter (ADC).

(a) Block diagram of Motorola MC10315L/7L 7-bit flash ADC. Features: 7-bit resolution and 8-bit accuracy plus overrange; direct interconnection for 8-bit conversion; 15 MHz sampling rate; input voltage ±2.0 volts: input capacitance ≤ 70 pF; power dissipation 1.2 W; no sample-and-hold required for video bandwidth signals; standard 24-pin package.

(b) Equivalent circuit of reference resistor ladder network, showing the input A_{in} applied to 128 comparators. $R \approx 0.5\,\Omega$. C_{eq} is the lumped equivalent value of capacitance, representing the distributed capacitance for each resistor R and the input capacitance for each comparator.

(Reproduced by courtesy of Motorola Inc.)

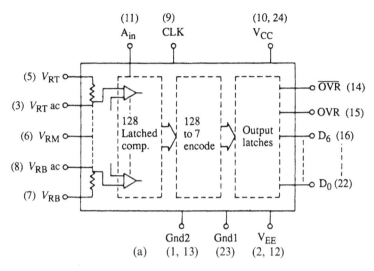

(a)

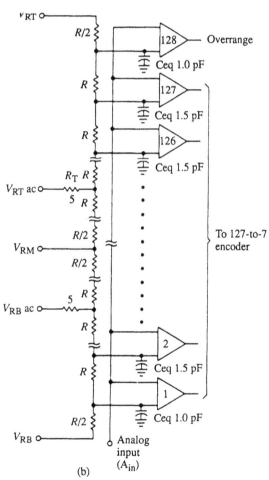

(b)

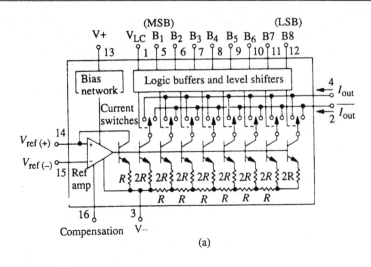

(a)

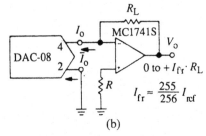

(b)

Figure 7.11
(a) An 8-bit digital-to-analog converter (DAC).
(b) An application producing a positive-going low-impedance output.
(Reproduction by courtesy of Motorola Inc.)

voltage source via a 1 K resistor. Then 2.56 mA will flow via the left-hand transistor and resistor R to the negative supply $V-$. The transistor acts as a constant current generator in a negative feedback loop; the feedback is taken to the opamp's non-inverting terminal as the transistor provides the necessary inversion. The transistor connected to the left-hand current switch will pass half as much current, as its emitter resistance is $2R$. You can easily verify that the next transistor current switch will receive half as much current as that, and so on for each successive stage to the right, except for the last. Theoretically the last isn't quite right as shown: a remainder current equal to the least significant bit is in fact shunted to ground, so that the maximum output current at I_{out} with all logic inputs B_1 to B_8 high (or at $\bar{I}_{out}$ with all low) is only $(255/256)I_{ref}$, or 2.55 mA in this case. Figure 7.11b shows the DAC arranged to give a positive-

going low-impedance voltage output of 0 to 10.20 V in 40 mV steps, according to the 8-bit binary code applied at the logic inputs. The device is called a *multiplying DAC* because its output is the product of the binary input and the reference current. For the device shown in Figure 7.11a the settling time of the output current to within 0.5 least significant bit (LSB) following an input code change from 0 to 255 or vice versa is typically less than 100 μs, whilst the reference current may be slewed at up to 4 mA/μs at least and typically more.

So to return to the *successive approximation register* (SAR) ADC, which works as shown in Figure 7.12a. The control logic sets the DAC output to half of full scale by setting the most significant bit (MSB) to 1 and all the other bits to 0. If the comparator indicates that the DAC output is greater than the analog input voltage, the

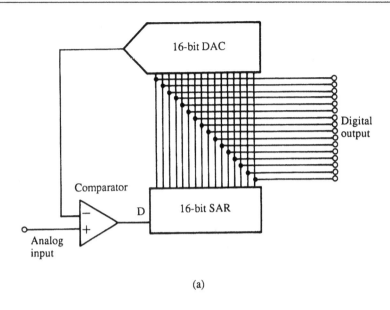

(a)

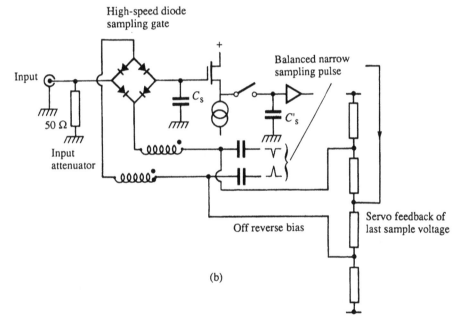

(b)

Figure 7.12
(a) A 16-bit successive approximation type of ADC (reproduced by courtesy of *Electronic Product Design*).
(b) Simplified schematic circuit of a sampling gate as employed in a high-speed analog sampling oscilloscope.

MSB is reset to 0; otherwise it is left set, and the control sets the next most significant bit (NMSB) to 1. The process is repeated so that after N comparisons the register controlling the DAC holds an n-bit digital representation of the input voltage. The resolution of the digital answer is one part in 256 or 0.39% (for an 8-bit device or 1 in 65 536 for a 16-bit device as in Figure 7.12), but the accuracy is another matter entirely. It depends upon the DAC's linearity and the accuracy of its reference voltage. It is obvious also that, with the sequential conversion algorithm used by the SAR DAC, the final answer will apply to the instant when the LSB comparison was made and can only be guaranteed correct if the input voltage changes negligibly during the course of a conversion.

Owing to the sequential nature of the conversion, SAR ADCs are inherently slower than flash types. They would effectively be very much slower indeed, because of the requirement that the input voltage must not change significantly during the course of a conversion, if special steps were not taken. The usual procedure is to precede the SAR DAC with a *sample-and-hold* (S/H) circuit. The LF398H from National Semiconductor is a monolithic BIFET sample-and-hold IC which will acquire the current value of the input voltage within 10 µs of being switched from the hold to the sample mode, however much the input has changed in the meantime, and will hold the current voltage with a low droop rate when switched back to the hold mode. Sample or hold mode is selected by means of a TTL and CMOS compatible control input, and the signal input characteristics do not change during hold mode. The two-stage AD585AQ from Analog Devices features a fast acquisition time of 3.0 µs to 0.01% of the steady state final voltage for a full-scale input change. Both devices incorporate offset null adjustment facilities. The devices described were state-of-the-art when introduced, but have long been outclassed by more recent introductions.

A discrete sample-and-hold circuit could of course be arranged using an *electronic switch* of some sort and a hold capacitor. The use of a bipolar transistor in the inverted mode as a switch has already been mentioned, but it is far from ideal. For one thing, it only works well for one polarity of the switched voltage. At one time symmetrical transistors were available. These were fabricated with identical emitter and collector junctions for use as cross-point switches in switching matrices, and had the advantage of working equally well (perhaps equally badly would be a better description) in either direction. Nowadays the FET is universally used as an electronic switch in applications up to a few megahertz. Whilst an individual FET can be employed, the usual arrangement is a monolithic switch using two enhancement mode FETs in parallel, one N channel and one P channel. The gates are driven by appropriate antiphase voltages derived from on-chip CMOS inverter stages; these require positive and negative power supply rails, usually ±15 V. By using two complementary FETs in parallel then, when the switch is on, one or other FET is fully enhanced regardless of the input voltage. The usable range of input voltage, with the switch off or on, extends to over 80% of the supply rail range, e.g. ±14 V or so with ±15 V supplies, depending upon what maximum on resistance is acceptable, because this rises as the input voltage approaches either rail. The DG series is a long-standing industry standard range of CMOS analog switches available from many manufacturers. For example, the DG300A from Maxim Integrated Products provides two separate single-pole single-throw (SPST) switches, each with a maximum on resistance $r_{ds\,on}$ of 50 Ω. In digital control applications, it is usually desirable to have the on/off control input for each switch latched. For example, the AD759xDI series are TTL, CMOS and microprocessor compatible di-electrically isolated CMOS switches featuring overvoltage protection up to +25 V above the power supplies. The AD7590 has four latch inputs, activated by a logic zero on the write (WR) input, each controlling a SPST switch, whilst the AD7592 has two logic inputs each controlling a single-pole double-throw (SPDT) switch. Single-pole multiway switches are known as *multiplexers*, a typical example being the DG508A. The Maxim version of this industry standard single-pole eight-way part has an $r_{ds\,on}$ of 300 Ω maximum, and its switching time is 1 µs maximum. Again, improved performance is offered by more recently introduced devices.

Whilst this is adequate for very many applications, there are cases where a very much more rapid switching time is essential. A good example is the *sampling gate* used in a sampling oscilloscope. The difference between a sampling oscilloscope and a digital storage oscilloscope (DSO) is that, unlike the latter, the sampling oscilloscope does not apply the samples it takes to an ADC. The sample is stored in a hold capacitor for one sampling period only, after which it is replaced by the next sample. The main requirement is for a very fast analog gate driven by a very narrow sampling pulse. In order to achieve the widest bandwidth and shortest rise time possible, the sampling pulse is made very narrow; in particular the trailing edge, where it switches from on to off, is particularly fast. In the Tektronix 7S11 plug-in with sampling head type S4 in a 7000 series mainframe, the rise time was only 25 ps, corresponding to a bandwidth of 14 GHz. To achieve this, a sampling pulse 200 ps wide but with a trailing edge aperture time of only 20 ps was used. Figure 7.12b shows a high-speed sampling gate using Schottky diodes as the switching element. A single diode acts as a switch, in the sense that it is on if the anode is positive to the cathode and off otherwise. By combining four diodes as shown in Figure 7.12b with a pulse transformer, they can be pulsed into conduction to provide a path between the input and the output, and switched off again to leave a sample of the input voltage at the instant of switch-off stored in the hold capacitor. The absence of stored charge in Schottky diodes permits very fast switching times, resulting in the performance described above. Whilst analog sampling oscilloscopes such as that mentioned are now part of history, the type of fast sampling gate described forms the critical signal capture stage in modern digital sampling oscilloscopes, the fastest of which offer bandwidths up to 50 GHz.

Further applications of analog processing

A number of applications of analog signal processing have been mentioned, in the course of reviewing the various circuits that are used for this purpose, but here are a few more, starting with one that is frankly frivolous.

Figure 7.13a includes an amplifier arranged to provide less gain to negative-going inputs than to positive. The result is to produce second-order (and other even-order) harmonic distortion. If this is applied only to the very lowest notes of an audio signal, say those below 100 Hz (separated out from the rest by a low-pass filter), correlated energy at harmonics of those low notes will be generated. This can then be added back into the

original signal as shown and will be heard by the ear (so the theory runs) as at the original low pitch, even through a loudspeaker system incapable of reproducing such low notes. Figure 7.13b is another gem from the armoury of the with-it audio engineer. It provides even-order distortion – or, with the alternative feedback network shown, both odd and even order. The result is both harmonic and intermodulation distortion of precisely the sort that in earlier years engineers spent great efforts trying to eliminate. It produces a sort of back-ground furriness to the sound, which can be heard on the soundtrack of any prewar black-and-white movie and which is now back in fashion under the label 'fuzz'. It's marvellous what you can sell to some people.

Here is another example of waveform processing: in this instance, a scheme to multiply the frequency of a waveform by three. The original application[14] was to obtain a clock frequency of around 100 kHz, for a microcontroller system incorporating an MM58174 real-time clock chip, from the latter's 32 768 Hz crystal derived clock output. The circuit, shown in Figure 7.14a, accepts either a sine or a square wave input and operates as follows. The first two inverters IC_{1a} and IC_{1b} are biased as linear amplifiers by virtue of being enclosed in a three-inverter loop with overall NFB at DC and low frequencies, via R_2. Their high gain ensures a square waveform at A, even if the input is a sine wave. The square wave is integrated by IC_{1c} to give a triangular wave at B, and again by IC_{1d} to give a parabolic waveform at C. This waveform is a good approximation to a sine wave, having only 3.5% total harmonic distortion. The 'sine wave' is subtracted from the square wave to give the waveform shown at D in Figure 7.14b: this is a square wave less its fundamental component and, as you can see by counting the zero crossings, it contains a substantial amount of third harmonic of the original frequency. C_4 is included to allow for excess phase shift through $1C_{1a}$ and IC_{1b}; it would not be necessary if the circuit were operating at a lower frequency. Finally, IC_{1e} and IC_{1f} slice waveform D through the zero crossing points to produce the wanted output at three times the input frequency. The circuit makes

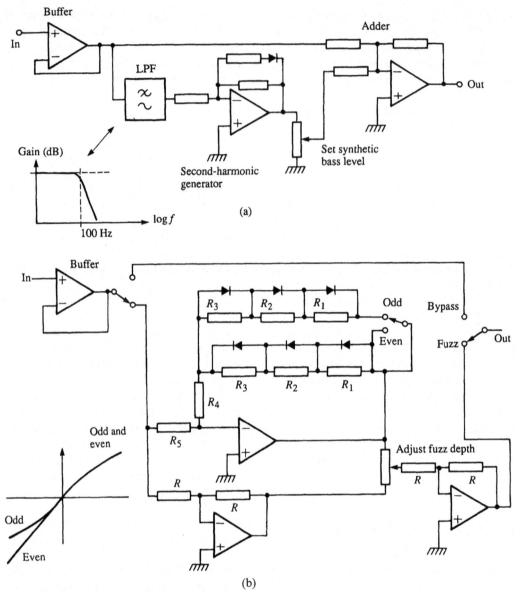

Figure 7.13 Audio processing circuits.

(a) Harmonic bass generator.

(b) Fuzz generator. R_1 to R_4 are graduated to produce the sequential break points shown. In practice the diodes exhibit a soft turn-on, providing a smoothly curved characteristic.

(approximate) use of the identity $\sin 3\theta = 3 \sin \theta - 4 \sin^3\theta$, where $\theta = \omega t$ and ω is the radian frequency. This identity can be used in reverse[15] to extract a sine wave at one-third of the frequency of a given sine wave: in the case of a frequency modulated (FM) wave this also has

the effect of reducing the modulation index, which in some cases may ease the measurement of the modulation index.

An application frequently arises for a linear circuit which will accept input signals over a wide range of levels and produce an output of

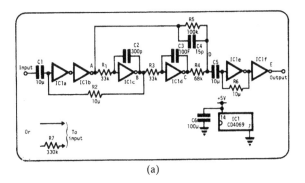

(a)

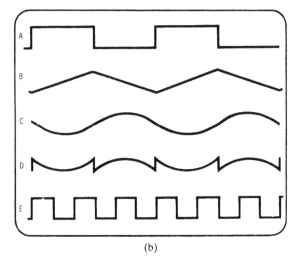

(b)

Figure 7.14
  (a) Clock frequency tripler circuit.
  (b) Waveforms at designated points in the circuit.
  (Reproduced by courtesy of *Electronic Engineering*.)

more or less constant level – that is, for an *automatic gain control* (AGC) circuit. At audio frequencies such circuits are readily implemented using opamps and FETs. One such published circuit[16] claims a 100 dB dynamic range and distortion-free operation over the frequency range 60 Hz to 30 kHz. However, any AGC circuit is only a linear amplifier in a limited sense. It will reproduce an input sine wave within the stated range without distortion and likewise will, in general, reproduce a complex waveform without distortion. But if there are variations in the amplitude of the signal occurring at a subaudio rate, then they will be suppressed by the AGC action, since it maintains a constant output level. For example, a 1 kHz sine wave whose amplitude

varies sinusoidally between 0 V and 2 V peak-to-peak at a rate of one cycle per second can be analysed (see Chapter 8) into a 1 kHz tone or 'carrier', and two other tones or 'sidebands' at 0.999 and 1.001 kHz. All three of these tones are within the amplifier's bandwidth and no other tones are present, yet the AGC action will result in an output of very nearly constant amplitude, i.e. the sidebands are suppressed.

Figure 7.15a shows the block diagram of an AGC circuit. The peak detector produces an output proportional to the peak level of the signal, and this is used to control the gain of the amplifier. The larger the input signal the larger is the output from the peak detector, and the more this reduces the gain of the amplifier. Figure 7.15b shows a typical sort of circuit where the gain-controlled amplifier is implemented by changing the effective input resistor R_1 of an opamp. The more positive the gate voltage of the FET, the lower its drain/source resistance. Performing the star/delta transformation (Figure 1.2b) on R_{1A}, R_{1B} and the FET resistance r_{ds}, this is equivalent to increasing R_1: the other two resistances of the delta simply shunt the source and the opamp's input, which is a virtual earth, and so are of no consequence. Since the amplifier's gain is determined by the ratio R_2/R_1, increasing the effective value of R_1 reduces the gain. Alternatively, the FET and R_{1A} may simply be regarded as a passive attenuator ahead of a fixed gain amplifier, although this is somewhat of an oversimplification, as it ignores the fact that the source resistance seen by R_{1B} changes.

At very small signal levels the positive-going peaks of the output may not be large enough to overcome the forward voltage of the diode, so up to that point the gain remains constant at the maximum level: this is described as an *AGC delay*. If an amplifier is fitted between the peak detector and the gate of the FET, a smaller change in output voltage will be sufficient to change the gain of the amplifier by a given amount, giving a 'tighter' AGC characteristic: this is described as *amplified AGC*. However, it will be clear on reflection that with a feedback AGC circuit such as Figure 7.15a the output must always rise slightly as the input is increased (Figure 7.15c),

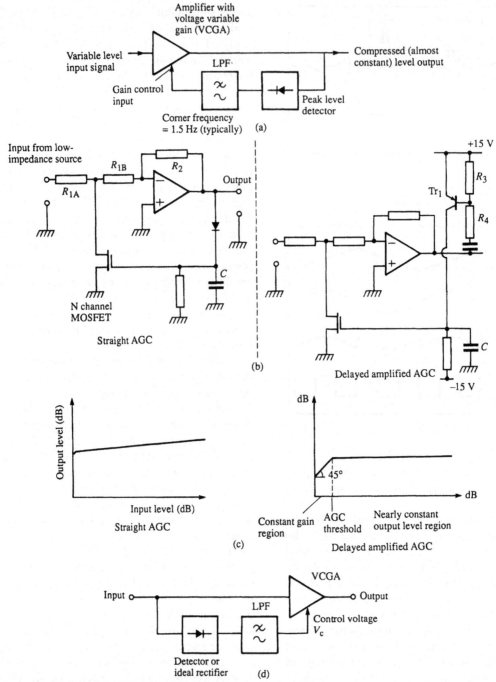

Figure 7.15   Automatic gain control schemes.
   (a) Block diagram of a feedback AGC system.
   (b) Audio-frequency AGC schemes. Tr_1 acts as both AGC diode and amplifier. Delay set by V_{be}, R_3 and R_4.
   (c) AGC characteristics.
   (d) Feedforward AGC. If gain of VCGA is inversely proportional to V_c, then above threshold the output level is constant.

since it is the rise in output level that causes the required reduction in the gain of the amplifier. The scheme is a form of negative feedback loop, and the usual stability requirements must be fulfilled. The higher the loop gain (the tighter the loop), the slower the loop must act or the higher the low-frequency limit of the audio band must be to enable the loop gain to be rolled off. In principle, loop stability can be predicted analytically as in the case of an NFB loop round any other amplifier, but often in the case of an AGC loop the law of gain reduction versus control voltage is unknown or incompletely specified.

Another point to note in connection with an AGC circuit such as Figure 7.15b is that the AGC *attack time* – the rate at which the gain is turned down following a sudden rise in input level – is determined by the capacitor C and the charging circuit resistance, i.e. the diode slope resistance and the source impedance of the circuit driving it. The rate of gain recovery following a sudden fall in input level, however, is determined by the time constant CR of the smoothing circuit following the detector. This makes the AGC loop more complicated to analyse theoretically. However, if the rate of gain recovery is not critical and can be made fairly slow, e.g. 20 dB per second or less, it is a powerful factor in ensuring loop stability. A typical example of an application where a fairly fast attack of around 5 ms and a much slower decay are acceptable, is the automatic record level circuitry in a cheap cassette recorder/dictation machine. Another example is the voice compression circuit used in single-sideband (SSB) transmitters to maintain the modulation at near the permitted maximum level, in order to increase the average output power or 'talk power'. Sophisticated ICs are available for this purpose, a typical example being the Plessey 5L6270 gain-controlled amplifier or voice operated gain adjusting device (VOGAD) circuit. It is designed to accept signals from a low-sensitivity microphone and to provide an essentially constant output signal for a 50 dB range of input. The dynamic range, attack and delay times are controlled by external components. Figure 7.15d shows a feedforward AGC circuit. With this arrangement it is possible to avoid even the slight rise in output level with increasing input

which occurs with amplified AGC. Indeed, depending on the circuit's characteristics, the output may even fall as the input increases. Where a very tight AGC loop with fast response times – requirements which conflict – is needed, the best approach is a looser feedback AGC loop followed by a narrow range feedforward AGC circuit.

Pulse modulation

Being concerned with analog circuitry, this book does not deal with digital signal processing (DSP) as such, only with the interfaces between digital and analog circuitry, such as A/D converters. However, there is a form of signal processing which is intermediate in nature between analog and digital, namely *time-discrete analog processing*. Here, the instantaneous voltage of a wave-form is sampled at regular intervals and represented by the value of some parameter of pulses occurring at the sampling instants. Typical examples are pulse amplitude modulation (PAM), pulse position modulation (PPM) and pulse width modulation (PWM). These are illustrated in Figure 7.16. PWM has already been discussed in Chapter 4 and so will not be further covered here. PPM is used as the modulation method in some forms of model radio-control systems.

Pulse amplitude modulation (PAM) is important since it is widely used in one particular form of audio signal processing, namely the *bucket brigade delay line* (BBD), also known as the *charge coupled device* (CCD), which provides a means of delaying an analog signal in time. The BBD is named from a not entirely fanciful similarity between its mode of action and a line of firefighters passing buckets from one to the next. It uses a series of capacitors to store charges proportional to the amplitude of the input waveform at successive instants corresponding to the rising edge of regular clock pulses. The capacitors are connected each to the next by an FET: the gates of alternate FETs are connected to one phase of the two-phase clock and those of the remaining FETs to the other. By this means the samples are effectively passed stage by stage along the line, arriving at the other end after the period occupied by $N/2$ clock pulses for an N-stage device; a pair of buckets is required to

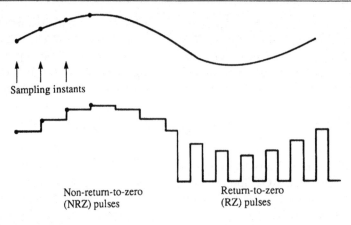

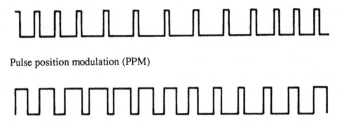

Figure 7.16 Pulse modulation.

provide one clock pulse period of delay. A straightforward delay is useful in sound reinforcement systems in large buildings such as a cathedral, to delay the signal from the speaker's microphone so that it is emitted from a loudspeaker column half-way down the hall at the same time as the direct sound waves arrive. This avoids a confusing post-echo from the direct sound. In other applications, an echo is deliberately produced to enrich the sound e.g. from an electronic organ. A discrete echo with a noticeable delay sounds very artificial and trying, as anyone who went to the Proms concerts in London's Royal Albert Hall fifty years ago can testify, so a shorter delay can be used and the delayed signal added back into the input of the BBD along with the original signal. This produces a multiple echo which, whilst an improvement, can also sound artificial if overdone. The MN3011 from National Panasonic is a 3328-stage BBD audio signal delay device with intermediate taps at stages 396, 662,

1194, 1726 and 2790 as well as the final output at stage 3328. The taps are chosen so that the ratios of the various delays are surds, i.e. irrational numbers. Thus when the outputs of the various taps, which may be deliberately attenuated by varying amounts, are combined and recirculated to the input along with the original signal, an approximate simulation is produced of the complex pattern of reverberation in a concert hall or cathedral.

An audio delay may also be used without recirculation to produce a *comb filter*, by simply combining the original signal with the delayed version. For example, if a short delay is used, say 10 ms, then a 100 Hz sine wave will be delayed by exactly one cycle, whilst 50 Hz will be delayed by half a cycle or 180°. If the original and delayed signals (both at the same level) are now combined, the net 50 Hz signal will be zero owing to cancellation, whilst the 100 Hz signal will double in amplitude. Furthermore, sine waves of 200, 300,

400 Hz etc. will be delayed by an even number of half-cycles and 150, 250, 350 Hz etc. by an odd number. So the response is the comb filter shown in Figure 7.17a. If the delayed signal is inverted, equivalent at any frequency to a 180° phase shift, then the peaks and troughs in Figure 7.17a change places and there is zero response at 0 Hz. If the delayed signal is attenuated relative to the original, or vice versa, before combining, then the peaks and troughs are less pronounced. If the clock frequency is altered, the delay changes and so do the positions and spacing of the peaks and troughs. By this means, the comb filter can be made to sweep up and down the audio band at will, giving rise to some novel and striking effects.

Recirculation of the delayed sound also results in a frequency response which is anything but level and, if the feedback level is too high, the output at one or more frequencies may build up indefinitely to give an unstable condition. The reason for this is not difficult to see, for if the delayed feedback signal at a frequency corresponding to a peak in Figure 7.17a is as large as the original signal, the latter may be removed and the output will never die away. If the feedback signal is a whisker larger than this, then it will build up indefinitely, as illustrated in Figure 7.17b. The effect of a regenerative comb filter on speech is intriguing, not to say weird. It sounds as though the speaker is in an echoey room with a set of tubular bells. Whenever the pitch of his voice coincides exactly with the frequency of one of these 'bells' it rings for a considerable period, dying away only slowly if the regeneration is only just short of oscillation. With a close comb spacing, several 'bells' may be sounding simultaneously, rendering the speech incomprehensible but not, strangely, preventing one from identifying the accent of the speaker!

Another application of CCDs is in *digital storage oscilloscopes* (DSOs). Usually the bandwidth of these is limited by the speed at which a flash converter type of ADC can operate. However, several manufacturers use CCDs to circumvent this problem. The input to the oscilloscope is fed into a very high-speed CCD delay line, operating basically in the same way as the audio BBDs described above, but with a clock frequency of up to 400 MHz. Thus at all times there is a record

of the latest segment of the input in the delay line, with older samples constantly 'falling off the end' and being lost as new samples are fed in at the front of the line. Following a selected triggering event, the high-speed clock can be stopped and the samples in the pipeline clocked out at a much slower rate to an ADC, which is thus relieved of the task of operating at up to 400 MHz. If the clock is stopped immediately on the occurrence of the trigger, the segment of signal passed to the ADC for digitization and subsequent storage in the random access memory (RAM) waveform store consists entirely of pretrigger information. If on the other hand the high-speed clock is allowed to continue for several or many pulses after the trigger event, the stored signal segment will bracket the trigger event, with part of the waveform referring to time before the trigger and part to time after, in any desired ratio up to 100% post-trigger. Even greater post-trigger delay can be employed to give a stored waveform segment referring to some later detail of interest in the waveform, similar to the A delayed by B function of a conventional (real-time) oscilloscope. What the conventional oscilloscope cannot do, of course, is to provide the pretrigger viewing capa- bility of the DSO.

Pulse amplitude modulation of audio signals is provided by the switched capacitor type of filter. These are now available in types implementing more and more complex filters; however, as they were mentioned in Chapter 6, they are not covered further here. *N-path band-pass filters* are also an application of the PAM principle in a way. They operate as in Figure 7.18, which illustrates the case where $N = 4$, i.e. a four-path filter. First, imagine that one of the four switches in Figure 7.18a is permanently closed and the other three open. Then clearly there is a low-pass response with a corner frequency of $f_c = 1/2\pi CR$. Now consider the case when the four switches are operated in sequence as indicated, so that each is on for exactly 25% of the time. Suppose the frequency of the four switch-control waveforms is f_s (where $f_s = f_{clock}/4$) and the input is a sine wave of frequency $f_i = f_s$. Then each switch is on during exactly the same quarter of the input sine wave every cycle, and so will in time charge up to the average value of the input

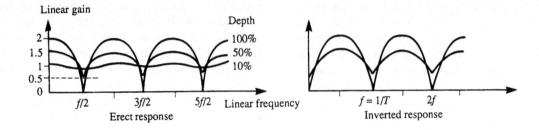

Erect response   Inverted response

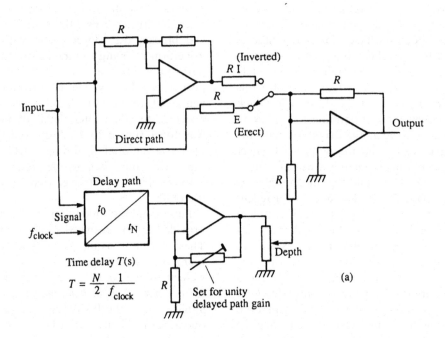

$$T = \frac{N}{2} \frac{1}{f_{clock}}$$

Set for unity
delayed path gain

(a)

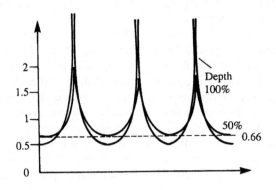

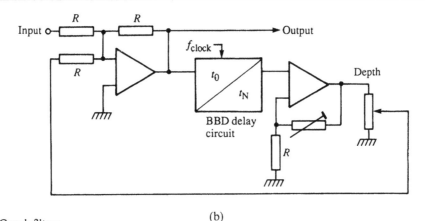

(b)

Figure 7.17   Comb filters.
  (a) Audio comb filter using BBD device.
  (b) Comb filter with regeneration.

voltage during that quarter-cycle. Thus the output will be a four-step approximation to a sine wave as shown. If the frequency of the input sine wave differs only very slightly from f_s, the phasing of the four steps relative to the sine wave will gradually drift through the patterns shown in Figure 7.18b, but otherwise the circuit operation is unchanged. However, if the input sine wave differs from f_s by a rather greater margin, either higher or lower in frequency, each switch will be closed during a rather different part of the input sine wave on successive closures, and so its associated capacitor won't have a chance to charge to an appropriate constant voltage. In fact, the circuit possesses a band-pass response similar to the low-pass response of the basic CR circuit, but mirrored symmetrically around the frequency f_s. Since each capacitor is only connected to the input via R for 25% of the time rather than continuously, the effective low-pass bandwidth is in fact $(1/4)(1/2\pi CR)$, and hence the pass bandwidth at f_s is $1/4\pi CR$ or more generally $1/N\pi CR$.

At audio frequencies, N-path filters can provide very high values of Q; for a two-pole band-pass (single-pole low-pass equivalent) filter such as in Figure 7.18a, the ratio of centre frequency to 3 dB bandwidth can be in the range 10 000 to 100 000. However, such performance is only achievable with care in both circuit design and layout, especially if using second-order sections as well to build up a higher-order filter. A second-order section may be realized using a Salen and Key (or, much better in this application, a Kundert) circuit as in Figure 7.18c. Another circuit arrangement has the advantage that both switched N-capacitor banks are connected to ground.[17] Each of the N capacitors must 'look' only at that $(1/N)$th of the input waveform occurring whilst its associated switch is closed, if the filter is to operate correctly. Any stray capacitance to ground at the junction of the N capacitors and the resistor R will 'smear' some charge of magnitude and polarity corresponding to the equilibrium voltage on one capacitor into the next capacitor, and from that into the following one, and so on. There is inevitably some stray capacitance associated with this node, for example the input capacitance of the opamp. In a second-order section such as that of Figure 7.18c, this results in one of the peaks being of greater amplitude than the other. This makes it difficult to build up a high-order filter with a Chebyshev pass band, since the complete filter exhibits a general attenuation slope across the pass band, superimposed upon the usual Chebyshev ripple.

As an example of a frequency selective filter, it can well be argued that the N-path filter should have been covered in the preceding chapter. However, it has been discussed here because of the time-discrete nature of its implementation, and it completes this review of analog signal processing in the time domain.

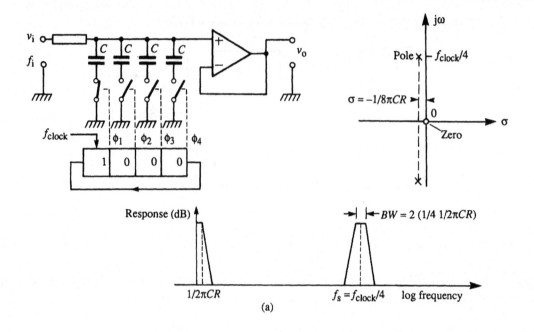

(a)

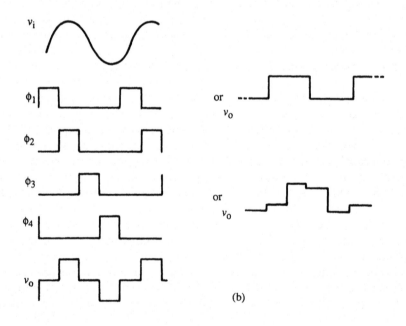

(b)

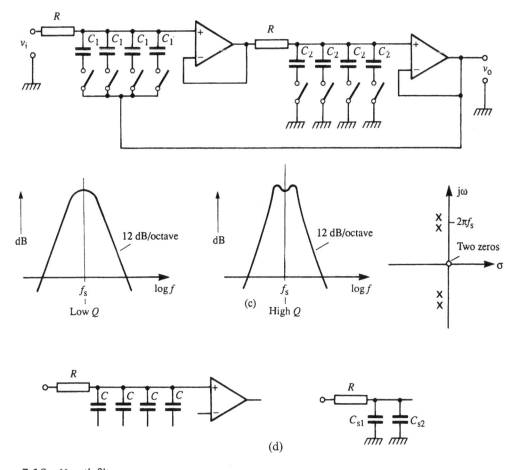

Figure 7.18 N-path filters.

(a) One-pole low-pass equivalent (LPE) N-path band-pass filter section. A single 1 circulating in a shift register is only one of many ways of producing the four-phase drive waveform shown in (b).

(b) Waveforms associated with (a). The exact shape of v_o, when $f_i = f_{clock}/4$ exactly, will depend upon the relative phasing of v_i and the clock waveform. For very small differences between f_i and $f_{clock}/4$, the output waveform will continuously cycle between the forms shown, and all intermediate shapes.

(c) Second-order N-path filter, showing circuit, frequency response and pole–zero plot. $Q = 1/2\sqrt{(C_1/C_2)}$, exactly as for the low-pass case.

(d) Stray capacitance. Showing the stray capacitance to ground, consisting of opamp input capacitance C_{s2} plus circuit and component capacitance to ground with all switches open C_{s1}.

References

1. Simple Digitally-Controlled Variable-Gain Linear DC Amplifier. A. Sedra, K. C. Smith, p. 362, *Electronic Engineering*, March 1969, Vol. 41, no. 493.
2. Controlled Gain Amp. Q. Rice, p. 18, *New Electronics*, 14 April, 1987.
3. High-Gain Amp Yields Low DC Output Offset. C. Paul, L. Burgner, p. 186, *EDN*, August 21, 1986, Vol. 31, no. 17. See also An Unusual Circuit Configuration Improves CMOS-MDAC Performance. N. Sevastopoulos, J. Cecil, T. Frederikson, *EDN*, March 1979.
4. Inexpensive Circuit Generates Unipolarity Output, p. 69, *Electronic Design*, November 22, 1969, Vol. 17, no. 24.

5. Improve Circuit Performance With A 1-opamp Current Pump. P. A. Pease, p. 85, *EDN*, January 20, 1983, Vol. 28, no. 2.

6. Improved Saw Tooth Generator Has Grounded Reference Point, p. 80, *Electronic Design*, February 15, 1970, Vol. 4, no. 18.

7. See e.g. *The BIFET Design Manual*. P. F. Nicholson, Texas Instruments Ltd, August 1980, and A Circuit With Logarithmic Transfer Response Over 9 Decades. J. F. Gibbons, *IEEE Trans on Circuit Theory*, Vol. CT-Il, September 1964, p. 378.

8. Simple Control For Sign Of Op-Amp Gain. *Electronic Design*, November 8, 1970, Vol. 18, no. 23.

9. Improved Log Ratio Amplifier. D. J. Faulkner, p. 23, *Electronic Product Design*, March 1986.

10. Integrated Analogue Divider/Multiplier. V. C. Roberts, *Electronic Engineering*, April 1969, Vol. 41, no. 494.

11. Multi-Electrode Valve Multipliers. V. A. Stephen and W. W. Forest, p. 185, *Electronic Engineering*, March 1964.

12. Voltage Controlled Triangle Generator. J. W. Howden, p. 540, *Wireless World*, November 1972.

13. Charge Balancing Frequency Multiplier. H. R. Goodwin, p. 25. *Electronic Engineering*, December 1983.

14. Clock Frequency Multiplier Using MM58174. I. March, p. 27, *Electronic Engineering*, July 1983.

15. Index Reduction of FM Waves By Feedback And Power-Law Non-Linearities. V. Beneš, p. 589, *BSTJ*, April 1965, Vol. XLIV, no. 4.

16. A 100 dB AGC Circuit Using The CA3130. Y. Gopola Rao, P. U. Mesh, p. 35, *Electronic Engineering*, November 1977.

17. A New Type of *N*-Path *N* Filters with Two Pairs of Complex Poles. E. Langer, Session 11, WAM 2.5, International Solid-State Circuits Conference, 1968.

Questions

1. Describe a three opamp instrumentation amplifier and explain how it works. What improvement in the ratio of normal mode to common mode signal is contributed by (i) the first stage, (ii) the output stage?

2. (i) Draw, and deduce the gain of; a two opamp instrumentation amplifier.
   (ii) How does an isolation amplifier differ from an instrumentation amplifier?

3. Derive from first principles the time response of an ideal inverting integrator with input and feedback components R and C respectively, to a unit step input, i.e. one which changes instantly from $0 V$ to, and remains at, $+1 V$.

4. Describe three different full-wave signal rectifier circuits, and analyse their mode of operation.

5. An adjustable unipolar current sink of 0–$100 mA$, controlled by an input voltage of exactly 0–$10 mV$ is required. Design such a circuit using an opamp, an FET and such other components as may be required.

6. Describe the operation of the Howland bipolar current pump. Show how it may be used to implement a non-inverting integrator, where one end of the integration capacitor is earthed.

7. The basic 'pump and bucket' F to V converter is only approximately linear over the lower portion of its output voltage range. Explain how an opamp may be added to the circuit to give a wide, linear output range.

8. Compare the operation of an SAR ADC with that of a Flash ADC. What are the main advantages and disadvantages of each?

9. Explain why a conventional (feedback) AGC system cannot produce an output whose amplitude (above threshold) is exactly constant. How can a feedforward AGC circuit circumvent this limitation?

10. Describe how a BBD may be used in a circuit to provide (i) an echo effect, (ii) a comb filter.

Chapter

8   Radio-frequency circuits

Radio-frequency equipment is used for a vast range of purposes, including heat treating special steels, medical diathermy treatment for cancer, heat sealing plastic bags, and experiments in atomic physics. Nevertheless, as the name implies, the original use was in connection with the transmission of information by radio waves. The earliest form of this was wireless telegraphy (WT) using Morse code. This was followed by wireless telephony and, much later, broadcasting – radio and television. So, before diving into RF circuits in detail, a word might be in order about the different forms of modulation employed to impress the information to be transmitted onto the radio wave. It is only a brief word, though, as this is a book particularly about analog electronic circuitry, not a general light-current electrical engineering textbook.

Modulation of radio waves

Figure 8.1a shows how information is transmitted by means of an interrupted continuous wave, often called simply *continuous wave* (CW). This type of modulation is frequently employed in the high-frequency (HF) band, i.e. from 1.6 to 30 MHz. In a simple transmitter either the oscillator would be 'keyed' on and off with a Morse key, or alternatively the drive signal or the power supply to the output stage would be likewise keyed. In the simplest possible transmitter there would be no separate output stage, only a keyed oscillator. Using CW, amateur radio enthusiasts can contact others in any country in the world using only a few watts, but only as and when propagation conditions are favourable.

Broadcasting on medium wave (MW) uses *amplitude modulation* which is illustrated in Figure 8.1b. Here, the frequency of the radio-frequency or *carrier* wave does not change, but its amplitude is modulated in sympathy with the programme material, usually speech or music. This gives rise to *sidebands*, which are limited to ±4.5 kHz about the carrier frequency by limiting the bandwidth of the *baseband* modulating signal to 4.5 kHz maximum. This helps to minimize interference between adjacent stations on the crowded MW band, where frequency allocations are only 9 kHz apart (10 kHz in the USA). With maximum modulation by a single sinusoidal tone, the transmitted power is 50% greater than with no modulation; this is the *100% modulation case*. Note that the power in the carrier is unchanged from the 0% or *unmodulated* case. Thus at best only one-third of the transmitted power actually conveys the programme information, and during average programme material the proportion is much lower even than this. For this reason, the *single-sideband* (SSB) mode of modulation has become very popular for voice communication at HF. With this type of modulation, illustrated in Figure 8.1c, only one of the two sidebands is transmitted, the other and the carrier being suppressed. As there is no carrier, all of the transmitted power represents wanted information, and as all of this is concentrated in one sideband, 'spectrum occupancy' is halved. At the receiver, the missing carrier must be supplied from a carrier reinsertion oscillator at exactly the appropriate frequency in order to demodulate the signal and recover the original. Although this is a trivial exercise with modern synthesized receivers, historically it was difficult. Amplitude modulation, with its uncritical tuning requirements, continues to be used by broadcasters for both local audiences on MW and international broadcasting on SW. There are a number of bands of frequencies allocated by international agreement to broadcasting in the short-wave band between 1.6 and 30 MHz.

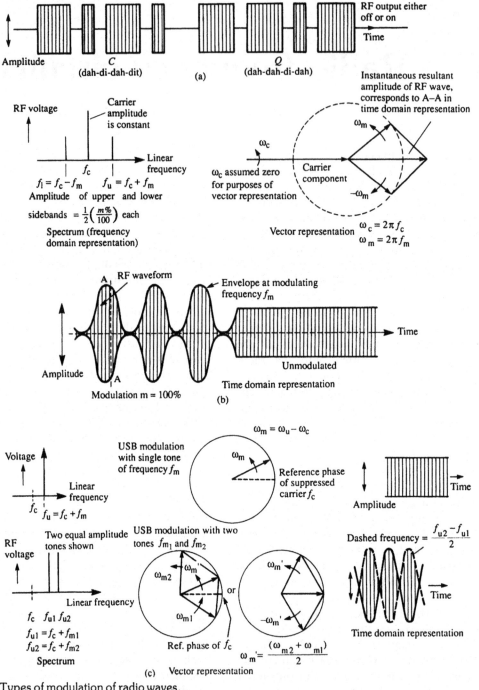

Figure 8.1   Types of modulation of radio waves.

(a) CW modulation. The letters CQ in Morse (seek you?) are used by amateurs to invite a response from any other amateur on the band, to set up a QSO (Morse conversation).

(b) AM : 100% modulation by a single sinusoidal tone shown.

(c) SSB(USB) modulation. Note that with two-tone modulation, the signal is indistinguishable from a double-sideband suppressed carrier signal with a suppressed carrier frequency of $(f_{u1} + f_{u2})/2$. This can be seen by subtracting the carrier component from the 100% AM signal in (b). The upper and lower halves of the envelope will then overlap as in (c), with the RF phase alternating between 0° and 180° in successive lobes.

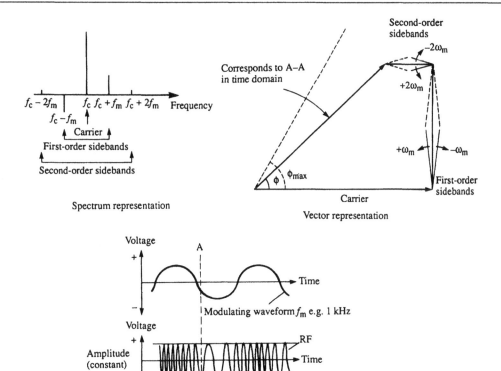

Spectrum representation

Vector representation

Frequency modulated RF carrier
(Frequency variation grossly exaggerated for clarity.
Actual RF carrier frequency would be much higher
than shown, e.g. 100 MHz)

Time domain representation

(d)

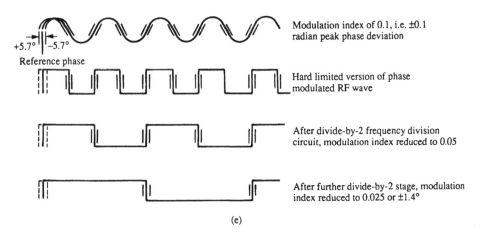

(e)

(d) FM. For maximum resultant phase deviation ϕ up to about 60° as shown, third- and higher-order sidebands are insignificant.

(e) Reduction of phase deviation when a phase modulated signal passes through a frequency divider chain, showing – for example – how a divide-by-4 (two-stage binary divider) reduced modulation index by a factor of 4.

Figure 8.ld illustrates *frequency modulation*. FM was proposed as a modulation method even before the establishment of AM broadcasting, but any enthusiasm for it waned as a result of an analysis which showed that it produced sidebands exceeding greatly the bandwidth of the baseband signal.[1] With the limited bandwidth available in the LW and MW bands, this was obviously an undesirable characteristic. However, following the Second World War the technology had advanced to the point where it was possible to use the considerable bandwidth available in the then largely unused very high-frequency (VHF) band. The lower part of the 30–300 MHz VHF band had already been used before the war for television, and now a high-quality sound broadcasting service was established using FM in the band 88-108 MHz. The standard adopted was a maximum deviation from the centre or carrier frequency of ±75 kHz, and a baseband frequency response extending from 50 Hz to 15 kHz. This represented real hi-fi compared with the 4.5 kHz limitation on MW, and the much lower level of interference from unwanted stations was a real blessing. The modulation index for an FM signal is defined in terms of a single sinusoidal modulating tone, as 'm', where $m = f_d/f_m$, the peak frequency deviation of the carrier, divided by the modulating frequency. It is shown below that m is also equal to the peak phase deviation of the carrier in radians. With the 75 kHz peak deviation being five times the highest modulating frequency, broadcast FM (also known as WBFM – wide band FM) is a type of spread spectrum signal. This confers a degree of immunity to adjacent- and co-channel interference due to the 'capture effect'. This is particularly effective on mono reception, the advantage being much less for stereo reception.

Figure 8.1 shows the characteristics of the various modulation methods in three ways: in the frequency domain, in the time domain, and as represented by vector diagrams. Each illustrates one aspect of the signal particularly well, and it is best to be familiar with all the representations. Choosing one and sticking to it is likely to be misleading since they each tell only a part of the story. Note that in Figure 8.ld a very low level of modulation is shown, corresponding to a low amplitude of the modulating sine wave (frequency f_m). Even so, it is clear that if only the sidebands at the modulating frequency are considered, the amplitude of the signal would be greatest at those instants when its phase deviation from the unmodulated position is greatest. It is the presence of the second-order sidebands at $2f_m$ which compensate for this, maintaining the amplitude constant. At wider deviations many more FM sidebands appear, all so related in amplitude and phase as to maintain the amplitude constant. They arise automatically as a result of frequency modulating an oscillator whose output amplitude is constant; their existence is predicted by the maths and confirmed by the spectrum analyser.

Note that the maximum phase deviation of the vector representing the FM signal will occur at the end of a half-cycle of the modulating frequency, since during the whole of this half-cycle the frequency will have been above (or below) the centre frequency. Thus the phase deviation is 90° out of phase with the frequency deviation. Note also that, for a given peak frequency deviation, the peak phase deviation is inversely proportional to the modulating frequency, as may be readily shown. Imagine the modulating signal is a 100 Hz square wave and the deviation is 1 kHz. Then during the 10 ms occupied by a single cycle of the modulation, the RF will be first 1000 Hz higher in frequency than the nominal carrier frequency and then, during the second 5 ms, 1000 Hz lower in frequency. So the phase of the RF will first advance steadily by five complete cycles (or 10π radians) and then crank back again by the same amount, i.e. the phase deviation is $\pm5\pi$ radians relative to the phase of the unmodulated carrier. Now the average value of a half-cycle of a sine wave is $2/\pi$ of that of a half-cycle of square wave of the same peak amplitude; so if the modulating signal had been a sine wave, the peak phase deviation would have been just ±10 radians. Note that the peak phase deviation in radians (for sine wave modulation) is just f_d/f_m, the peak frequency deviation divided by the modulating frequency: this is known as the *modulation index* of an FM signal. If the modulating sine wave had been 200 Hz, the deviation being 1 kHz as before, the shorter period of the modulating frequency would result in the peak-to-peak phase change being halved to ±5 radians; that is, for a given peak frequency deviation the peak phase deviation is inversely proportional to the modulating frequency.

For monophonic FM broadcasting the peak deviation at full modulation is 75 kHz, so the peak phase deviation corresponding to full sine wave modulation would be ±5 radians at 15 kHz and ±1500 radians at 50 Hz modulating frequency. If the modulation index of an FM signal is much less than unity, the second-order and higher-order FM sidebands are insignificant. If, on the other hand, the modulation index is very large compared with unity, there are a large number of significant sidebands and these occupy a bandwidth virtually identical to $2f_d$, i.e. the bandwidth over which the signal sweeps. The usual approximation for the bandwidth of an FM signal is $BW = 2(f_d + f_m)$.

You can see in Figure 8.1b that the vectors representing the two sidebands of an AM signal are always symmetrically disposed about the vector representing the carrier. As they rotate at the same rate but in opposite directions, their resultant is always directly adding to or reducing the length of the carrier vector. The second and higher even-order sidebands of an FM signal behave in the same way. But as Figure 8.1d shows, the first-order sidebands (at the modulating frequency) are symmetrical about a line at right angles to the carrier, and the same goes for higher odd-order sidebands. Note that if one of the first-order FM sidebands was reversed, they would look exactly like a pair of AM sidebands: this is why one of the first-order FM sideband signals in the frequency domain representation in Figure 8.1d has been shown as inverted. A spectrum analyser will show the carrier and sidebands of either an AM or a low-deviation FM signal as identical, as the analyser responds only to the amplitudes of the individual sidebands, not their phases. However, if the first-order sidebands displayed on the analyser are unequal in amplitude, this indicates that there is both AM and FM present on the modulated wave.

An important principle in connection with phase modulation is illustrated in Figure 8.1e. This shows how dividing the frequency of a phase or frequency modulated wave divides the modulation index in the same proportion. In the figure, a sinusoidal modulating waveform has been assumed; in this case the peak phase deviation in radians is numerically equal to the modulation index, i.e. to the peak frequency deviation divided by the modulating frequency, as noted above. However, whatever the modulating waveform – and even in the case of a non-repetitive signal such as noise – passing the modulated carrier through a divide-by-N circuit will reduce the peak deviation by a factor of N, as should be apparent from Figure 8.1e. For the time variations on the edges of the divider output remain unaffected but they now represent a smaller proportion of a complete cycle. Conversely, if a phase or frequency modulated signal is passed through a frequency multiplier (described later in this chapter), any phase noise on the signal is multiplied *pro rata*.

Low-power RF amplifiers

Having looked at some typical radio-frequency signals – there are many other sorts, for example frequency shift keying (FSK), numerous varieties of digital modulation, and of course television – it is time to look at some of the wide range of RF circuits, both passive and active, used to process them. These include amplifiers of all sorts, but only low-power RF amplifiers are discussed, for a very good reason. This is a very exciting time for the high-power RF engineer, with devices of ever higher power becoming available almost daily. There are regular improvements in high-power bipolar RF transistors. RF MOSFETs are improving in terms of both power handling and reduced capacitances, particularly the all-important drain/gate feedback capacitance C_{dg}; they are also available now as matched pairs in a single package, for push-pull applications. Meanwhile other exciting developments are on the horizon, including the static induction transistor (SIT). This device is half-way between a bipolar and an FET, and its notable feature is an unusually high voltage capability. This eases the difficulties associated with the design of high-power RF circuits due to the very low impedance levels at which lower-voltage

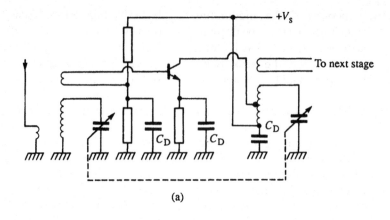

(a)

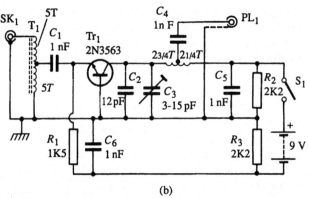

(b)

Figure 8.2   RF amplifier stages.
   (a) Common emitter RF amplifier stage with both input and output circuits tuned. C_D are decoupling capacitors.
   (b) Common base RF amplifier with aperiodic (broad band) input and tuned output stages (reproduced from 'VHF preamplifier for band II', Ian Hickman, *Practical Wireless*, June 1982, p. 68, by courtesy of *Practical Wireless*).

devices necessarily work. Even more exciting is the prospect of high-power devices using not silicon or gallium arsenide (GaAs), or even indium phosphide (InP), but diamond. The technology is currently being researched in the USA, Japan and the USSR, and already diodes (operating up to 700°C!) have been produced. With a carrier velocity three times that of silicon and a thermal conductivity twenty times that of silicon (four times that of copper, even) the possibilities are immense. So any detailed discussion of RF power devices is fated to be out of date by the time it appears in print. So only low-power amplifiers are discussed below.

Figure 8.2 shows two class A NPN *bipolar transistor* amplifier stages. In Figure 8.2a, both the input and output circuits are tuned. This is by no means the invariable practice but, for the input RF stage of a high-quality communications receiver, for example, it enables one to provide more selectivity than could be achieved with only one tuned circuit, whilst avoiding some of the complications of coupled tuned circuits. The latter can provide a better band-pass shape – in particular a flatter pass band – but, for a communications receiver covering say 2 to 30 MHz, two single tuned circuits such as in Figure 8.2a provide an adequate pass band width in any case. With the continuing heavy usage of the 2 to 30 MHz HF band, which seems to become even more congested yearly rather than dying as the pundits were once predicting, RF stages are

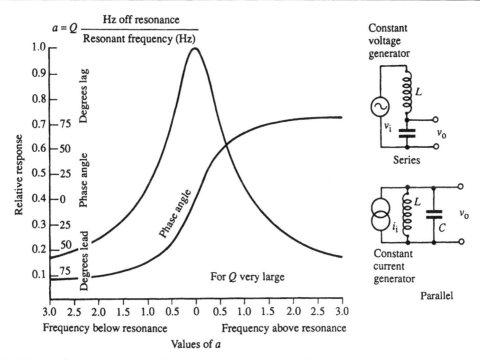

Figure 8.3 Universal resonance curve for series resonant circuit. For a Q of greater, the phase and amplitude curves depart by only a very small amount from the above. Also applies to the response of a parallel tuned circuit, for $Q > 20$. In both cases, curves give v_0/v_{max} in magnitude and phase.

coming back into favour again. However, an RF amplifier with both input and output circuits tuned needs very careful design to ensure stability, especially when using the *common emitter* configuration. The potential source of trouble is the collector/base capacitance, which provides a path by which energy from the output tuned circuit can be fed back to the base input circuit. The common emitter amplifier provides inverting gain, so that the output is effectively 180° out of phase with the input. The current fed back through the collector base capacitance will of course lead the collector voltage by 90°. At a frequency somewhat below *resonance* (Figure 8.3) the collector voltage will lead the collector current, and the feedback current via the collector/base capacitance will produce a leading voltage across the input tuned circuit. At the frequency where the lead in each tuned circuit is 45°, there is thus a total of 180° of lead, cancelling out the inherent phase reversal of the stage and the feedback becomes positive. The higher the stage

gain and the higher the Q of the tuned circuits, the more likely is the feedback to be sufficient to cause oscillation, since when the phase shift in each tuned circuit is 45°, its amplitude response is only 3 dB down (see Figure 8.3). Even if oscillation does not result, the stage is likely to show a much steeper rate of fall of gain with detuning on one side of the tuned frequency than on the other – a sure sign of significant internal feedback. The *grounded base* stage of Figure 8.2b may prove a better choice, since some bipolar transistors exhibit a significantly smaller feedback capacitance in the grounded base connection, i.e. C_{ce} is smaller than their C_{cb}. The N channel *junction depletion FET* (JFET) is also a useful RF amplifier, and can be used in either the grounded source or grounded gate configuration, corresponding to the circuits of Figure 8.2. It is particularly useful in the grounded gate circuit as a VHF amplifier.

For ease of reference, Figure 8.3 is repeated as Appendix 2.

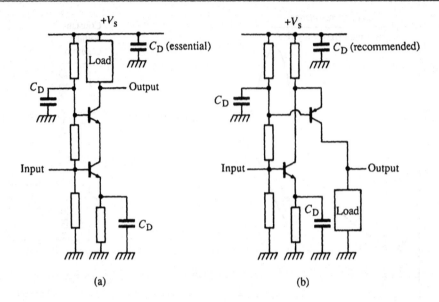

Figure 8.4
  (a) Cascode amplifier.
  (b) Complementary cascode. The load may be a resistor, an *RL* combination (peaking circuit), a tuned circuit or a wide band RF transformer. C_D are decoupling capacitors.

Stability

There are a number of circuit arrangements which are used to ensure the *stability* of an RF amplifier stage. One of these, the cascode, is shown in Figure 8.4a. The *cascode* stage consists of two active devices; bipolar transistors are shown in the figure, but JFETs or RF MOSFETs are equally applicable. The input transistor is used in the grounded emitter configuration, which provides much more current gain than the grounded base configuration. However, there is no significant feedback from the collector circuit to the base tuned circuit since the collector load of the input transistor consists of the very low emitter input impedance of the second transistor. This is used in the grounded base configuration, which again results in very low feedback from its output to its input. With a suitable type of transistor the cascode circuit can provide well over 20 dB of gain at 100 MHz together with a reverse isolation of 70 dB. *Reverse isolation* is an important parameter of any RF amplifier, and is simply determined by measuring the 'gain' of the circuit when connected back to front, i.e. with the signal input applied to

the output port and the 'output' taken from the input port. This is easily done in the case of a stand-alone amplifier module, but is not so easy when the amplifier is embedded in a string of circuitry in equipment. In the days of valves, one could easily derive a stage's reverse isolation (knowing its forward gain beforehand) by simply disconnecting one of the heater leads and seeing how much the gain fell! When a valve is cold it provides no amplification, so signals can only pass via the interelectrode capacitances, and these are virtually the same whether the valve is hot or cold. With no gain provided by the valve, the forward and reverse isolation are identical. Much the same dodge can be used with transistors by open-circuiting the emitter to DC but leaving it connected as before at AC. However, the results are not nearly so reliable as in the valve case, as many of the transistor's parasitic reactances will change substantially when the collector current is reduced to zero. For an RF amplifier stage to be stable, clearly its reverse isolation should exceed its forward gain by a reasonable margin, which need not be anything like the 40 to 80 dB obtainable with the cascode mentioned above. A difference of 20 dB is fine and of 10 dB adequate, whilst some commercially available broad band RF amplifier modules exhibit a reverse isolation which falls to as little as 3 dB in excess of the forward gain at the top end of their frequency range.

An interesting feature of the cascode stage of Figure 8.4a arises from the grounded base connection of the output transistor. In this connection its collector/base breakdown voltage is higher than in the common emitter connection, often by a considerable margin, as transistor data sheets will show. This fact makes the cascode circuit a favourite choice for amplifiers which have to handle a very wide range of frequencies whilst producing a very large peak-to-peak output voltage swing. Examples include the range from DC to RF in the Y deflection amplifier of an oscilloscope, and that from 50 Hz to RF in the video output amplifiers in a TV set. Figure 8.4b shows a *complementary cascode* stage. This has the advantage of not drawing any appreciable RF current from the positive supply rail, easing decoupling requirements.

Figure 8.5a shows what is in effect a cascode circuit, but in the *dual-gate RF MOSFET* the two devices are integrated into one, the drain region of the input device acting as the source of the output section. Thus the dual-gate MOSFET is a 'semiconductor tetrode' and, as in the thermionic tetrode and pentode, the feedback capacitance internal to the device is reduced to a very low level (for the Motorola MFE140 the drain/gate$_1$ capacitance amounts to little more than 0.02 pF). The dual-gate MOSFET exhibits a very high output slope resistance, again like its thermionic counterpart, and also an AGC capability. The circuit of Figure 8.5b provides up to 27 dB gain at 60 MHz when the AGC voltage V_{gg} is +8 V and up to 60 dB of gain reduction as V_{gg} is reduced to below 0 V.

A common technique to increase the stability margin of transistor RF amplifiers is *mismatching*. This simply means accepting a stage gain less than the maximum that could be achieved in the absence of feedback. In particular, if the collector (or drain) load impedance is reduced, the stage will have a lower voltage gain, so the voltage available

to drive current through the feedback capacitance (C_{cb} in a bipolar transistor, C_{dg} or C_{rss} in an FET) is reduced *pro rata*. Likewise, if the source impedance seen by the base (or gate) is reduced, the current fed back will produce less voltage drop across the input circuit. Both measures reduce gain and increase stability: it may well be cheaper to recover the gain thus sacrificed by simply adding another amplifier stage than to add circuit complexity to obtain the extra gain from fewer stages by *unilateralization*. This cumbersome term is used to indicate any type of scheme to reduce the effective internal feedback in an amplifier stage, i.e. to make the signal flow in only one direction – forward. Data sheets for RF transistors often quote a figure for the maximum available gain (MAG) and a higher figure for maximum unilateralized gain (MUG).

The traditional term for unilateralization is *neutralization*, and I shall use this term hereafter as it is just a little shorter, even though they are not quite the same thing. Figure 8.6a shows one popular neutralization scheme, sometimes known as *bridge* neutralization. The output tuned circuit is centre tapped so that the voltage at one end of the inductor is equal in amplitude to, and in antiphase with, the collector voltage. The neutralizing capacitor C_n has the same value as the typical value of the transistor's C_{cb}, or C_n can be a trimmer capacitance, set to the same value as the C_{cb} of the particular transistor. The criterion for setting the capacitor is that the response of the stage should be symmetrical. This occurs when there is no net feedback, either positive or negative. The series capacitance of C_{cb} and C_n appears across the output tuned circuit and is absorbed into its tuning capacitance, whilst the parallel capacitance of C_{cb} and C_n appears across the input tuned circuit and is absorbed into its tuning capacitance. Neutralization can be very effective for small-signal amplifiers, but is less so for stages handling a large voltage swing. This is because the feedback capacitance C_{cb}, owing to the capacitance of the reverse biased collector/base junction, is not constant but varies (approximately) inversely as the square of the collector/base voltage.

Neutralization can be applied to a push-pull stage as in Figure 8.6b, but great care is necessary when so doing. The scheme works fine just so long

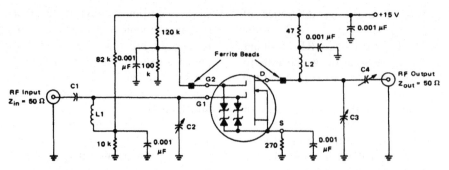

The following component values are for a stern stability factor = 2.0.

L1,L2 126 nH PAUL SMITH CO. SK-138-1
       4-½ Turns (yellow)
C1 Nominal 7.0 pF Adjusted for source impedance of
    approximately 1000 Ω, JOHANSON JMC2951

C2 Nominal 4.0 pF ARCO 402
C3 Nominal 13.73 pF ARCO 403
C4 Nominal 4.36 pF JOHANSON JMC2951

All Decoupling Capacitors are Ceramic Discs.

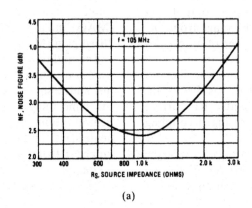

(a)

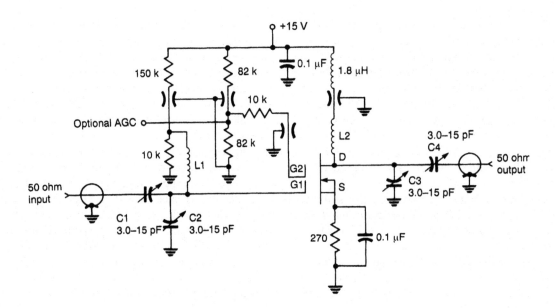

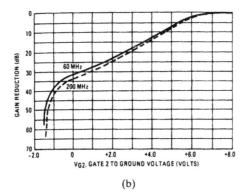

(b)

Figure 8.5 Dual-gate MOSFET RF amplifiers.
(a) Low-noise dual-gate MOSFET VHF amplifier stage and noise figure curve. The Motorola MFE140 shown incorporates gate protection Zener diodes, to guard against static electricity discharge damage.
(b) Dual-gate MOSFET VHF amplifier with AGC, with gain reduction curve. Maximum gain 27 (20) db at 60 (200) MHz with no gain reduction (V_{g2} at +7.5 V). The Motorola MPF131 provides an AGC range featuring up to 60 dB of gain reduction.
(Reproduced by courtesy of Motorola Inc.)

as the voltage at the collectors can be guaranteed to be in antiphase. This will indeed be the case at the resonant frequency where the collector load is a tuned circuit, or over the desired band of output frequencies where the collector load is a wide band RF transformer. However, at some other (usually higher) frequency this may no longer apply, owing to leakage inductance between the two halves of the collector circuit's inductor or transformer. The two collector voltages may then be able to vary in phase with each other, and the circuit simply becomes two identical amplifiers in parallel, each with a total feedback capacitance equal to twice its internal feedback capacitance. If the amplifier devices still have substantial power gain left at the frequency at which this condition exists, then the circuit can oscillate in a parallel single-ended mode.

Linearity

All of the amplifier circuits discussed so far have operated in class A, that is to say the peak current swing is less than the standing current, so that at no time is the transistor cut off. Where the collector circuit of an amplifier is a tuned circuit, this will have a 'flywheel' effect so that the collector voltage is approximately sinusoidal even

though the collector current is not. Thus a transistor can amplify the signal even though it conducts for more than 180° but less than 360° – i.e. operates in class AB. Likewise for class B (180° conduction angle) and class C (conduction angle less than 180°). These modes offer higher efficiency than class A, but whether one or other of them is appropriate in any given situation depends upon the particular application. Consider the earlier low and intermediate power amplifier stages of an FM transmitter, for example. Here, the amplitude of the signal to be transmitted is constant and it is the only signal present; there are no unwanted signals such as one inevitably finds in the earlier stages of a receiver. Consequently a class B or C stage is entirely appropriate in this application. However, an AM or SSB transmitter requires a linear amplifier, i.e. one that faithfully reproduces the variations in signal amplitude which constitute the *envelope* of the signal.

In the receiver the requirement for *linearity* is even more pressing, at least in the earlier stages where many unwanted signals, some probably very much larger than the wanted signal, are present. Chapter 4 covered the mechanism by which second-order non-linearity – second-harmonic distortion – results in sum and difference products when more than one signal is present,

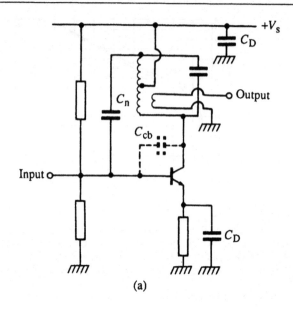

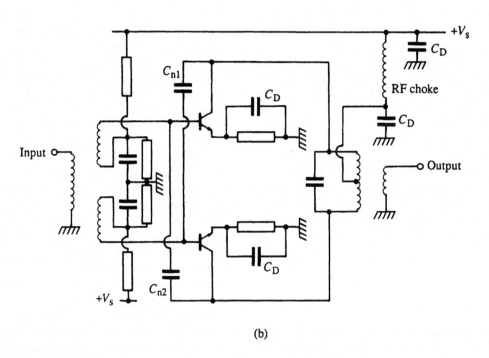

Figure 8.6   Neutralization.

(a) Bridge neutralization. The internal feedback path is not an ideal capacitor C_{cb} as shown, but will have an in-phase component also. If the phase angle of the neutralization via C_n is adjusted, e.g. by means of an appropriate series resistance, the neutralization is more exact – at that particular frequency. The stage is then described as 'unilateralized' at that frequency.

(b) Cross-neutralization, push-pull amplifier.

and third-order non-linearity in products of the form $2f_1 \pm f_2$. These latter *intermodulation products*, resulting from two unwanted frequencies f_1 and f_2, are particularly embarrassing in radio reception. Imagine that f_1 is, say, 20 kHz higher than the wanted signal at f_0, and that f_2 is 20 kHz higher still. Then $2f_1 - f_2$ turns out to be exactly at f_0. If the two unwanted frequencies were on the low-frequency side, f_2 being 20 kHz lower than f_0 and f_1 20 kHz lower still, then it would be the intermodulation product $2f_2 - f_1$ that falls on the wanted frequency. The intermediate-frequency (IF) amplifier section of a superheterodyne receiver – or *superhet* for short, shown in block diagram form in Figure 8.7a – is preceded by a highly selective filter which, in a good quality communications receiver, will attenuate frequencies 20 kHz or more off tune by at least 80 dB. However, it is not possible to provide that sort of selectivity in a tunable filter; the comparative ease of obtaining high selectivity at a fixed frequency is the whole *raison d'être* of the superhet. So the RF amplifier stage (if any) and the mixer must be exceedingly linear to avoid interference caused by third-order intermodulation products.

In a *double-conversion superhet* such as shown in Figure 8.7b, this requirement applies also to the first IF amplifier and second mixer, although the probability of interference from odd-order intermodulation products is reduced by the *roofing filter* preceding the first IF amplifier. This is always a crystal filter offering 30 or 40 dB of attenuation at frequencies 30 kHz or more off tune. Indeed, recent developments in crystal filter design and manufacture permit the roofing crystal filter to be replaced by a crystal filter, operating at 70 MHz, with the same selectivity as previously obtained in the second IF filter at 1.4 MHz, enabling the design of an 'up-converting single superhet'. An up-converting superhet removes the *image* problem encountered with a down-converting single superhet such as in Figure 8.7a. With a 1.4 MHz IF and a local oscillator tuning from 3 to 31.4 MHz, it is difficult to provide enough selectivity at the top end of the HF band. For example, when the receiver is tuned to 25 MHz the local oscillator frequency

will be 26.4 MHz, and an unwanted signal at 27.8 MHz will also produce an IF output from the mixer at 1.4 MHz. This represents a fractional detuning of only 11.2%, and reference to the universal tuned circuit curves of Figure 8.3 will verify that even with high-Q tuned signal frequency circuits, it is difficult adequately to suppress the image response; hence the popularity of the up-converting double superhet of Figure 8.7b. Here, signal frequency tuned circuits can be replaced by *suboctave filters* (band-pass filters each covering a frequency range of about 1.5 to 1), or simply omitted entirely – although this sacrifices the protection against second-order intermodulation products afforded by suboctave filters.

The linearity of amplifiers, both discrete components and multistage amplifier stages, and of mixers is often quoted in terms of *intercept points*. You may not realize it, but the theory behind these has already been covered in Chapter 4. You may recall that if the input to an amplifier with some second-order curvature in its transfer characteristic is increased by 1 dB, the second-harmonic distortion rises by 2 dB, and the sum and difference terms due to two different input frequencies applied simultaneously behave likewise. Also, with third-order (S-shaped) curvature, both third-harmonic and third-order intermodulation products rise three times as fast as the input, at least for small inputs.

Of course, for very large inputs an amplifier will be driven into limiting and the output will eventually cease to rise: the output is said to be *compressed*. Figure 8.8a illustrates this: the point where the gain is 1 dB less than it would have been if overload did not occur is called the compression level. It is found that for levels up to about 10 dB below compression, it is a good rule that Nth-order intermodulation products rise by N dB for every 1 dB by which the two inputs rise. Figure 8.8b shows the behaviour of an imaginary but not untypical amplifier. The level of second- and third-order intermodulation products, as well as of the wanted output, have been plotted against input level. All three characteristics have then been produced on past the region of linearity, and it can be seen that eventually they cross. The higher

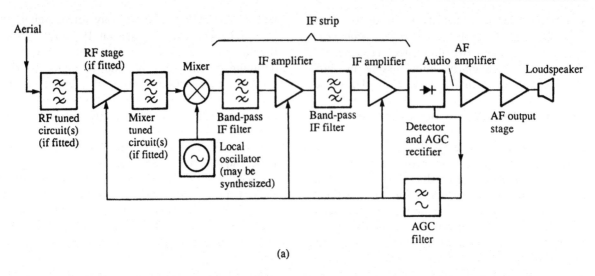

(a)

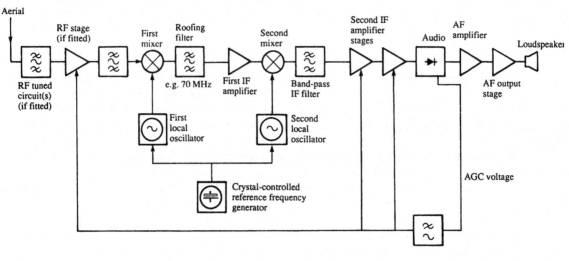

(b)

Figure 8.7 Supersonic heterodyne (superhet) receivers.
(a) Single-conversion superhet. Several filters may be used throughout the IF strip.
(b) Double-conversion superhet, with synthesized first local oscillator and second local oscillator both crystal reference controlled.

the level at which an amplifier's second- and third-order intercept points occur, the less problem there will be with unwanted responses due to intermodulation products, provided always that it also has a high enough compression point to cope linearly with the largest signals. A mixer (frequency changer) can be characterized in a similar way, except of course that the intermodulation products (colloquially called 'intermods') now appear translated to the intermediate frequency.

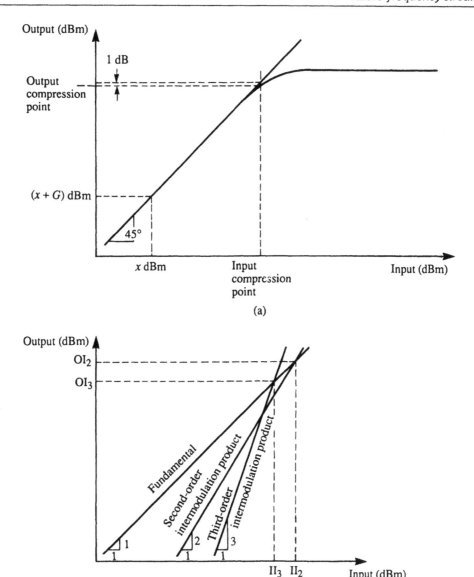

Figure 8.8 Compression and intermodulation.
(a) Compression point of an amplifier, mixer or other device with gain G dB. (Single tone input).
(b) Second- and third-order input and output intercept points (II and OI); see text. (Two tones of equal amplitude)

Noise and dynamic range

For an amplifier forming part of a receiver, high linearity is only one of several very desirable qualities. The input stage must exhibit a low noise figure, as indeed must all the stages preceding the IF filter defining the final bandwidth. For it makes sense to supply most of the gain after this filter; this way, large unwanted signals are amplified as little as possible before being rejected by the filter. Remember that unwanted signals may be 60, 80 or even 100 dB larger than the wanted signal!

The *noise figure* of an amplifier is related to the amount of noise at its output, in the absence of any intentional input, and its gain. Noise is an unavoidable nuisance, and not only in amplifiers. Chapter 1 showed how a current in a metallic conductor consists of a flow of electrons jostling their way through a more or less orderly jungle of atoms, of copper maybe or some other metal; and Chapter 3 how current is produced by carriers – electrons or holes – flowing in a semiconductor. Since at room temperature – indeed at any temperature above absolute zero – the atoms are in a state of thermal agitation, the flow of current will not be smooth and orderly but noisy, like the boisterous rushing of a mountain stream. Like the noise of a stream, no one frequency predominates. Electrical noise of this sort is called thermal agitation noise or just *thermal noise*, and its intensity is independent of frequency (or 'white') for most practical purposes. The available noise power associated with a resistor is independent of its resistance and is equal to -174 dBm/Hz e.g. -139 dB relative to a level of 1 milliwatt in a 3 kHz bandwidth.

This means that the wider the bandwidth of a filter, the more noise it lets through. It would seem that if we have no filter at all to limit the bandwidth, there would be an infinite amount of noise power available from a resistor – free heating for evermore! This anomaly had theoretical physicists in the late nineteenth century worrying about an ultraviolet catastrophy, but all is well; at room temperature thermal noise begins to tail off beyond 1000 GHz (10% down), the noise density falling to 50% at 7500 GHz. At very low temperatures such as are used with maser amplifiers, say 1 kelvin ($-272°C$), the noise density is already 10% down by 5 GHz (see Figure 8.9b).

Returning to RF amplifiers then, if one is driven from a 50 Ω source there will be noise power fed into its input therefrom (see Figure 8.9a). If the amplifier is matched to the source, i.e. its input impedance is 50 Ω resistive, the RMS noise voltage at the amplifier's input v_n is equal to half the source resistor's open-circuit noise voltage, i.e. to

$\sqrt{(kTRB)}$, where R is 50 Ω, k is Boltzmann's constant $= 1.3803 \times 10^{-23}$ joules per kelvin, T is the absolute temperature in kelvin (i.e. degrees centigrade plus 273) and B is the bandwidth of interest. At a temperature of 290 K (17°C or roughly room temperature) this works out at 24.6 nV in 50 Ω in a 3 kHz bandwidth. If the amplifier were perfectly noise free and had a gain of 20 dB (i.e. a voltage gain of $\times 10$, assuming its output impedance is also 50 Ω), we would expect 0.246 µV RMS noise at its output: if the output noise voltage were twice this, 0.492 µV RMS, we would describe the amplifier as having a noise figure of 6 dB. Thus the noise figure simply expresses the ratio of the actual noise output of an amplifier to the noise output of an ideal noise-free amplifier of the same gain. The amplifier's equivalent input noise is its actual output noise divided by its gain.

The *dynamic range* of an amplifier means the ratio between the smallest input signal which is larger than the equivalent input noise, and the largest input signal which produces an output below the compression level, expressed in decibels.

Impedances and gain

The catalogue of desirable features of an amplifier is still not complete; in addition to low noise, high linearity and wide dynamic range, the input and output impedances need to be well defined, and the gain also. Further, steps to define these three parameters should not result in deterioration of any of the others. Figure 8.10a shows a broadband RF amplifier with its gain, input impedance and output impedance determined by negative feedback.[2] The resistors used in the feedback network necessarily contribute some noise to the circuit. This can be avoided by the scheme known as *lossless feedback*,[3] shown in Figure 8.10b. Here, the gain and the input and output impedances are determined by the ampere-turn ratios of the windings of the transformer.

Whilst in a high-quality receiver the stages preceding the final bandwidth crystal filter need to be exceedingly linear, this requirement is relaxed in the stages following the filter; a little distortion in these will merely degrade the wanted signal marginally,

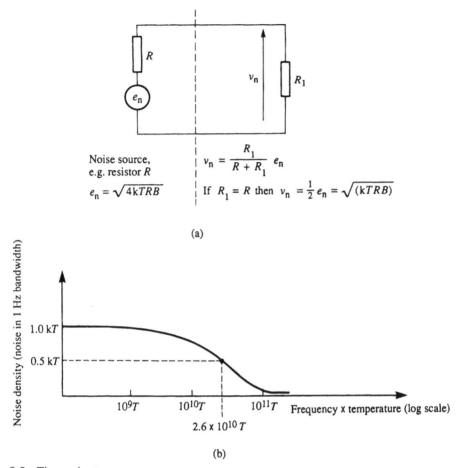

(a)

(b)

Figure 8.9   Thermal noise.

(a) A noisy source such as a resistor can be represented by a noise-free resistor R of the same resistance, in series with a noise voltage generator of EMF $e_n = \sqrt{(4kTRB)}$ volts. Available noise power $= v_n^2/R = (e_n/2)^2/R = P_n$ say. At room temperature (290 K) $P_n = -204$ dBW in a 1 Hz bandwidth $= -174$ dBm in a 1 Hz bandwidth. If $B = 3000$ Hz then $P_n = -139$ dBm, and if $R = R_1 = 50\,\Omega$ then $v_n = 0.246\,\mu V$ in 3 kHz bandwidth.

(b) Thermal noise is 'white' for all practical purposes. The available noise power density falls to 50% at a frequency of $2.6 \times 10^{10}T$, i.e. at about 8000 GHz at room temperature, or 26 GHz at $T = 1$ K.

since by that stage in the circuit all the unwanted signals have been rejected by the filter. It is usual to apply *automatic gain control* so that the level of the wanted signal at the receiver's output does not vary by more than a few decibels for an input level change of 100 dB or more. This is achieved by measuring the level of the signal, for example the level of the carrier in the case of an AM signal, and using this to control the gain of the receiver. Since most of the gain is in the IF amplifier, this is where most of the gain reduction occurs, starting with the penultimate

stage and progressing towards the earlier stages the larger the gain reduction required. The final IF stage may also be gain controlled, but this must be done in such a way that it can still handle the largest received signals. Finally, in the presence of a very large wanted signal it may be necessary to reduce the gain of the RF amplifier. The application of AGC is usually 'scheduled' to reduce the gain of successive stages in the order described, as this ensures that the overall noise figure of the receiver is not compromised.

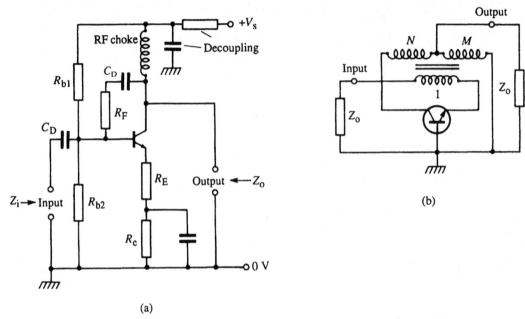

(a)

(b)

Figure 8.10   Input and output impedance determining arrangements.

(a) Gain, input and output impedances determined by resistive feedback. R_{b1}, R_{b2} and R_e determine the stage DC conditions. Assuming the current gain of the transistor is 10 at the required operating frequency, then for input and output impedances in the region of 50 Ω, $R_F = 50^2/R_E$. For example, if $R_E = 10\,\Omega$, $R_F = 250\,\Omega$, then $Z_i \approx 35\,\Omega$, $Z_o \approx 65\,\Omega$ and stage gain ≈ 10 dB, while if $R_E = 4.7\,\Omega$, $R_F = 470\,\Omega$, then $Z_i \approx 25\,\Omega$, $Z_o \approx 95\,\Omega$ and gain ≈ 15 dB. C_D are blocking capacitors, e.g. 0.1 μF.

(b) Gain, input and output impedances determined by lossless (transformer) feedback. The absence of resistive feedback components results in a lower noise figure and higher compression and third-order intercept points. Under certain simplifying assumptions, a two-way match to Z_o results if $N = M^2 - M - 1$. Then power gain = M^2, impedance seen by emitter = $2Z_o$ and by the collector = $(N + M)Z_o$. This circuit arrangement is used in various broadband RF amplifier modules produced by Anzac Electronics Division of Adams Russel and is protected by US Patent 3 891 934: 1975 (DC biasing arrangements not shown). (Reprinted by permission of *Microwave Journal*.)

A number of different schemes are used to vary the gain of radio-frequency amplifier stages, one of which, the dual-gate FET, has already been mentioned. The gain of a bipolar transistor can also be reduced, by reducing its collector current, but this also reduces its signal handling capability, so that only a few tens of millivolts of RF signal may be applied to the base. The available output is also reduced when AGC is applied. At one time, bipolar transistors designed specifically for gain-controlled IF amplifier stages were available. These used forward rather than reverse control, i.e. the collector current was increased to reduce gain. This had the advantage that the signal handling capability of the stage was actually increased rather than reduced with large signals. The change of gain was brought about by a spectacular fall in the f_T of the transistor as the collector current increased. At the constant intermediate frequency at which the device was designed to operate, this resulted in a fall in stage gain.

Discrete transistor IF stages are giving way to integrated circuits purpose designed to provide stable gain and wide range AGC capability. A typical example is the Plessey SL600/6000 series of devices, the SL610C and 611C being RF amplifiers and the 612C an IF amplifier. The devices provide 20 to 34 dB gain according to type, and a 50 dB AGC range. The range also contains the SL621 AGC generator. When receiving an AM signal, the automatic gain control voltage can be derived from the strength of the carrier component at the detector. With an SSB signal there is no carrier; the signal effectively disappears in pauses between words or sentences. So audio derived AGC is used, with a fast attack capable of reducing the gain to maintain constant output in just a few milliseconds, and at a rate of decay or recovery of gain of typically 20 dB per second. The disadvantage of this scheme is that a stray plop of interference can wind the receiver's gain right down, blanking the wanted signal for several seconds. The SL621 avoids this problem. It provides a 'hold' period to maintain the AGC level during pauses in speech, but will nevertheless smoothly follow the fading signals characteristic of HF communication. In addition, interaction between two detector time constants, a level detector and a charge/discharge pulse generator, prevent stray plops and crashes from inappropriately winding the receiver gain down.

In critical applications such as the RF stage of a professional communications receiver, a different approach to gain variation is often employed. As noted above, with an increasing input signal level the AGC scheduling would reduce the RF stage gain last. But if it is difficult to achieve sufficient linearity in the RF stage in the first place, it is virtually impossible to maintain adequate linearity if the gain is reduced. So instead the gain is left constant and an electronically controlled *attenuator* is introduced ahead of the RF stage. The attenuator uses PIN diodes, whose mode of operation was described in Chapter 3. PIN diodes can only operate as current-controlled linear variable resistors at frequencies at which the minority carrier lifetime in the intrinsic region is long compared with the period of one cycle of the RF. Even so, PIN diodes are available capable of operation down to 1 MHz or so, and can exhibit an on resistance, when carrying a current of several tens of millamperes, of an ohm or less. When off, the diode looks like a capacitance of 1 pF or less, depending on type. Whilst a single PIN diode can provide control of attenuation when used as a current-controlled variable resistor in series with the signal path, the source and load

circuits will be mismatched when attenuation is introduced. Two or more diodes can therefore be used, and the current through each controlled in such a manner as to implement an L pad,[4] which is matched in one direction (see Figure 8.11a), or a T or π pad, which is matched from both sides. In principle, an attenuator matched both ways can be implemented with only two diodes if the bridged T circuit is used (Figure 8.11b).

It is only when receiving signals where the modulation results in variations of signal amplitude, such as AM and SSB, that AGC is required. With FM, PM and certain other signal types, no information is contained in the signal amplitude – other than an indication as to how strong the signal is. Any variations in amplitude are therefore entirely adventitious and are due to fading or noise or interference. The effect of fading can be suppressed, and that of noise or interference reduced by using a *limiting IF strip*, i.e. one in which there is sufficient gain to overload the last IF stage even with the smallest usable signal. With larger signals, more and more of the IF stages operate in overload; all the stages are designed to overload 'cleanly', that is to accept an input as large as their output. Thus stages in limiting provide a gain of unity; in this way the effective gain of the IF strip is always just sufficient to produce a limited output, however small or large the input, without the need for any form of AGC. Here again, ICs have taken over from discrete devices in limiting IF strips, and other stages as well. For example, the Plessey SL6652 is a complete single-chip mixer/oscillator, IF amplifier and detector for FM cellular radio, cordless telephones and low-power radio applications. Its limiting IF strip has a maximum gain to small signals, before limiting sets in, of 90 dB, whilst the whole chip typically draws a mere 1.5 mA from a supply in the range 2.5 to 7.5 V.

In contrast to FM and PM signals, for some signals the amplitude is the only useful information. For example, in a low-cost radar receiver a successive detection *log IF strip* is used to detect the returns from targets. As the strength of a return varies enormously depending upon the range and size of the target, an IF strip with a wide dynamic range is needed. The Plessey

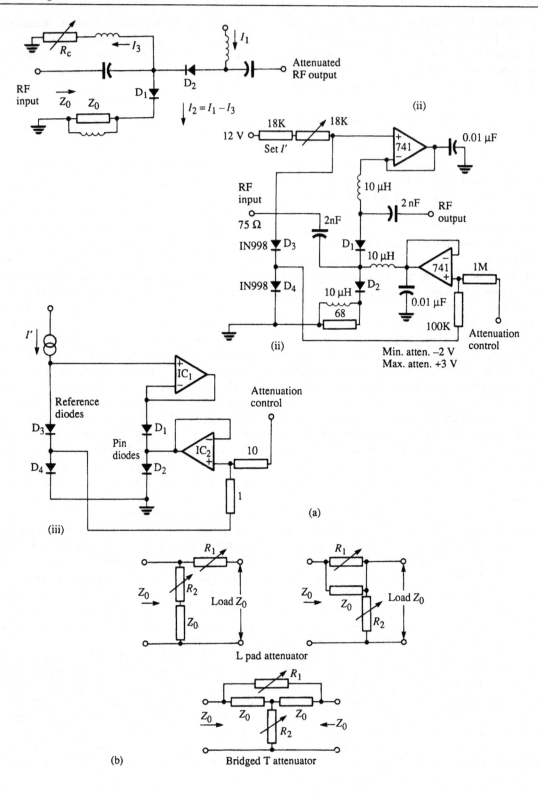

(ii)

$I_2 = I_1 - I_3$

(ii)

Min. atten. –2 V
Max. atten. +3 V

(iii)

(a)

L pad attenuator

Bridged T attenuator

(b)

SL1613C is an IC wideband log IF stage with 12 dB gain RF input to RF output and a rectified output providing 1 mA video current for a 500 mV RMS signal input. The video output currents of successive stages are summed to provide an output whose amplitude is proportional to the logarithm of the signal amplitude, with a video rise time of only 70 ns. Six or more stages may be cascaded to provide 60 MHz IF strips with up to 108 dB gain with better than 2 dB log linearity.

Mixers

Most modern receivers are of the superheterodyne type, with most of the amplification provided by the IF stages. This applies to broadcast receivers of all sorts, both sound and television; to professional communications, both civil and military, whether at HF (up to 30 MHz), VHF or UHF; and to receivers of other sorts, such as radar and navigation beacons. A *frequency changer, converter or mixer* – all names for the same thing – is used to translate the incoming signal from whatever frequency it was transmitted at to a fixed frequency, which is more convenient for providing high selectivity. In a single superhet such as Figure 8.7a the RF signal is applied, following amplification by one or more RF stages if fitted, to a mixer. This stage has two input ports and one output port. To the second input port is applied an RF signal generated locally in the receiver; this is called the local oscillator (LO). The mixer is a non-linear device and thus produces sum and difference frequency components. For example, if the receiver of Figure 8.7a were tuned to receive a signal at 10 MHz (it might be the WWV standard time and frequency transmission, broadcast from

the USA on that frequency) the local oscillator frequency could be either 8.6 MHz or 11.4 MHz, since in either case the difference frequency is equal to 1.4 MHz, the intermediate frequency. The sum frequency will also appear at the ouput of the mixer, but the IF filter rejects not only the sum frequency but the original RF and local oscillator signals as well, accepting only the wanted 1.4 MHz IF. In many cases the local oscillator frequency will be higher than the signal frequency ('high side injection', 'LO runs high'); for example, the first LO in the 100 kHz to 30 MHz double superhet of Figure 8.7b would run from 70.1 MHz to 100 MHz.

It has already been noted that any device with second-order curvature of its transfer characteristic will produce not only second-harmonic distortion but also second-order intermodulation products, i.e. sum and difference tones. The mixer in an early valve superhet, also called the 'first detector', worked in exactly this manner: a half-wave rectifier circuit would do just as well. However, this type of mixer exhibits a large number of spurious responses. At its broadest, a receiver *spurious response* is any frequency at which a receiver produces an output other than the wanted frequency to which it is tuned. One example, the *image frequency* (formerly called the 'second channel'), has already been noted: this is really a special case. Given sufficient front end selectivity, there will be no image response since no energy at that frequency can reach the mixer. In the up-converting superhet of Figure 8.7b, the image frequency will always be higher than 70.2 MHz, so a low-pass filter at the front end can suppress the image response entirely. This same filter will also prevent a response at the IF

Figure 8.11 Voltage-controlled RF attenuators using PIN diodes.
(a) (i) Pair of PIN diodes in L pad configuration, used to attenuate RF signals controlled by DC. Both I_1 and I_2 must be varied appropriately to control attenuation and keep Z_o constant. (ii) Working PIN diode attenuator must provide separation of the DC control current and RF signal paths. (iii) Constant attenuator impedance and temperature compensation are attained when the PIN diodes are matched against reference diodes in this arrangement. Opamp IC_1 keeps the voltage drive to both sets of diodes equal, and IC_2 acts as a current sink control for the PIN diodes and as a temperature compensator. Control of attenuation is logarithmic (dB law).
(b) L pad attenuators can provide a constant characteristic impedance Z_o as the attenuation is varied, but only at the input terminals. A bridged T configuration can keep Z_o constant at both input and output terminals.

frequency by preventing any signals at 70 MHz reaching the mixer. However, the image and IF rejection are usually quoted separately in a receiver's specification, the term 'spurious response' being reserved for unwanted responses due to much subtler and more insidious causes.

A mixer necessarily works by being non-linear. It would be nice if the mixer produced only the wanted IF output, usually the difference frequency between the RF signal and local oscillator inputs. In practice the mixer may also produce an output at the intermediate frequency due to signals not at the wanted RF at all. A mixer, being a non-linear device, will produce harmonics of the frequencies present at its inputs, and these harmonics themselves are in effect inputs to the mixer. So imagine the single superhet of Figure 8.7a tuned to receive a signal at 25 MHz. The LO will be at 26.4 MHz, and the second harmonic of this, at 52.8 MHz, will be lurking in the mixer just waiting to cause trouble. Imagine an unwanted signal at 25.7 MHz, too close to the wanted frequency to be much attenuated by the RF tuned circuits. The second harmonic of this, at 51.4 MHz, is exactly 1.4 MHz away from the second harmonic of the local oscillator and will therefore be translated to IF. This is variously called the *2–2 response* or the 'half IF away' response, being removed from the wanted frequency by half the IF frequency. Similarly, the *3–3 response* will occur at a frequency removed from the wanted frequency by 1.4/3 MHz. Clearly these responses will not be a problem in the up-converting superhet of Figure 8.7b, which is one reason for the popularity of this design. It is not, however, entirely immune from spurious responses. Imagine that it is tuned to 23 MHz, so that its first LO is at 93 MHz, and that there is a strong unwanted signal at 23.2 MHz. The fifth harmonic of the latter, at 116 MHz, is removed from the second harmonic (186 MHz) of the LO by 70 MHz. Admittedly this is a seventh-order response, and fortunately the magnitude of spurious responses falls off fairly rapidly as the order increases. But it does indicate that the ideal mixer is a very peculiar device: it must be very linear to two or more unwanted signals applied at the RF port (to avoid unwanted responses due to intermodulation), and should

ideally only produce an output due to the RF and LO signals themselves, not their harmonics.

The spurious responses just described are termed *external* spurious responses, in that they appear in response to an externally applied signal which bears a particular relation to the LO frequency, and thus to the wanted frequency. *Internal* spurious responses, on the other hand, are totally self-generated in the receiver. Most professional communications receivers nowadays contain a microprocessor to service the front panel, to accept frequency setting data from a remote control input, to display the tuned frequency, and so on. Harmonics of the microprocessor's clock frequency can beat with either the first or the second local oscillator, to produce the same effect as an externally applied CW interfering signal. Needless to say, in a well-designed receiver such responses are usually at, or below, the receiver's noise level. However, there is also the possibility of the odd spurious response due to interaction of the first and the second LO, which makes the up-converting single superhet an attractive proposition now that advances in crystal filter technology make it possible. Most modern communications receivers have the odd internal 'spur' in addition to the inevitable external spurious responses or 'spurii'.

A dual-gate FET can be used as a multiplicative mixer by applying the RF and LO voltages to gate 1 and 2 respectively. If the RF and LO voltages are represented by pure sinusoidal waveforms $\sin r$ and $\sin L$, where $\sin r$ stands for $\sin(2\pi f_{RF}t)$ and $\sin (L)$ for $\sin (2\pi f_{LO}t)$, then, ignoring a few constants, the mutual conductance can be represented by $\sin r \sin L$. So the drain output current can be represented by $[\cos(r - L) - \cos(r + L)]/2$, courtesy of your friendly neighbourhood maths textbook, i.e. it contains the sum and difference frequencies. The constants ignored in such a cavalier fashion are responsible for the presence in the drain current of components at the RF and LO frequencies, so the dual-gate FET mixer is described as *unbalanced*. However, if its operation were ideally multiplicative then these would be the only unwanted outputs, i.e. it would be free of spurious responses.

The presence in a mixer's output of components

at the RF and LO frequencies can be a serious embarrassment. Consider the communications receiver of Figure 8.7b, for example. Such a receiver is typically specified to operate right down to an input frequency of 10 kHz. At this tuned frequency the LO will be running at 70.001 MHz, which is uncomfortably close to the IF at 70 MHz, bearing in mind that the LO signal is very large compared with a weak RF signal. So a balanced mixer is used. A *single-balanced* mixer is arranged so that the signal at one of the input ports (usually the LO port) does not appear at the output port; thus it can effectively 'reject' the LO. In a *double-balanced* mixer (DBM) neither of the inputs appears at the output, at least in the ideal case – and in practice this condition is nearly met, with RF and LO rejection figures typically greater than 20 dB.

Figure 8.12 shows three DBMs. The first is the basic *diode ring* mixer, so called because if you follow round the four diodes you will find they are connected head to tail (anode to cathode) like four dogs chasing each other in a circle. On positive-going half-cycles of the LO drive two of the diodes conduct, connecting one phase of the RF input to the IF port. On the other half-cycle the other two diodes conduct, reversing the phase fed to the IF port. A very large LO drive is used, so that for virtually all the time either one pair of diodes or the other is conducting heavily: the diodes (which are selected for close matching, or are monolithic) are in fact used simply as switches. The ring DBM is double balanced, produces the sum and difference frequencies, and exhibits about half as many spurious responses as an unbalanced mixer. The *conversion loss* (ratio of IF output power to RF input power) is about 7 dB; this is attributable to several different causes. Half of the input RF energy will contribute to the sum output and half to the difference: as only one of these is required there is an inherent 3 dB conversion loss, the other 3 or 4 dB being due to resistive losses in the on resistance of the Schottky diodes, and to transformer losses. The IF port is 'DC. coupled', and thus operates down to 0 Hz. This is the mode of operation when the diode DBM is used as a phase sensitive detector, the RF and LO frequencies then being identical. Where an IF response down to DC is not required, the inputs can be applied differently. For example, the LO can be applied to the DC coupled port and the IF output taken from one of the transformer coupled ports. Whilst this has certain advantages in special cases, it is not usually used in a receiver, since LO radiation via the receiver's input port is then likely to be worse.

Another well-known scheme (not illustrated here) uses MOSFETs instead of diodes as the switches.[5] It is thus, like the Schottky diode ring DBM, a passive mixer, since the active devices are used solely as voltage-controlled switches and not as amplifiers. Reference 6 describes a single-balanced active MOSFET mixer providing 16 dB conversion gain and an output third-order intercept point of +45 dBm.

Figure 8.12b shows a double-balanced active mixer of the *seven-transistor tree* variety; the interconnection arrangement of the upper four transistors is often referred to as a *Gilbert cell*. The emitter-to-emitter resistance R sets the conversion gain of the stage; the lower it is made the higher the gain but the worse the linearty, i.e. the lower the third-order intercept point. This circuit is available in IC form (see Figure 8.12c) from a number of manufacturers under type numbers such as LM1496/1596 (National Semiconductor), MC1496/1596 (Motorola, Mullard/Signetics) and SG1496/1596 (Silicon General), whilst derivatives with higher dynamic range are also available.

Finally, in this whistle-stop tour of mixers, Figure 8.12d shows one of the simplest of the many ingenious ways in which the performance of the basic Schottky diode ring DBM has been improved – almost invariably, as here, at the expense of a requirement for greater LO power (up to +27 dBm is not uncommon). The resistors in series with the diodes waste LO power and increase the insertion loss, but they have beneficial effects as well. They permit a larger LO drive to be applied, which reduces the fraction of the LO cycle which is taken up by commutation, that is changing from one pair of diodes conducting to the other pair. They stabilize the effective on resistance of the diodes, which would otherwise vary throughout each half-cycle owing to the sinusoidal current waveform. Finally, they cause an additional voltage drop across the on diodes; this

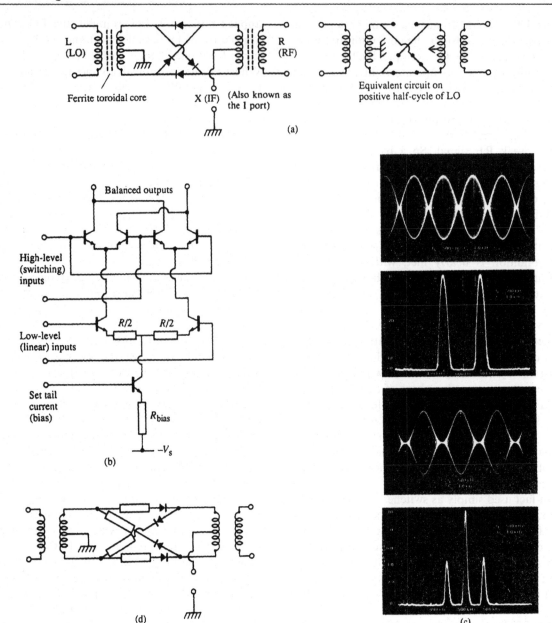

Figure 8.12   Double-balanced mixers (DBMs).

(a) The ring modulator. The frequency range at the R and L ports is limited by the transformers, as also is the upper frequency at the X port. However, the low-frequency response of the X port extends down to 0 Hz (DC).

(b) Basic seven-transistor tree active double-balanced mixer. Emitter-to-emitter resistance R, in conjunction with the load impedances at the outputs, sets the conversion gain.

(c) The transistor tree circuit can be used as a demodulator (see text). It can also, as here, be used as a modulator, producing a double-sideband suppressed carrier output if the carrier is nulled, or AM if the null control is offset. The MC1496 includes twin constant current tails for the linear stage, so that the gain setting resistor does not need to be split as in (b). (Reproduced by courtesy of Motorola Inc.)

(d) High dynamic range DBM (see text).

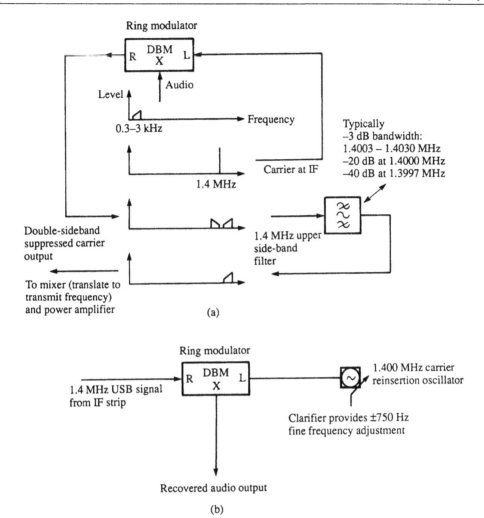

Figure 8.13 DBM used as modulator and demodulator.
(a) DBM used as a modulator in an HF SSB transmitter. The carrier rejection of the mixer plus the 20 dB selectivity of the USB filter at 1.4 MHz ensure that the residual carrier level is more than 40 dB down on the peak transmitter power.
(b) DBM used as an SSB demodulator in an HF SSB receiver.

increases the reverse bias of the off diodes, thus reducing their reverse capacitance.

Demodulators

The DBM is also popular as both a modulator and a demodulator. A modern transmitter works rather like a superhet receiver in reverse, that is to say that the signal to be transmitted is modulated onto a carrier at a fixed IF and then translated to the final transmit frequency by a mixer, for amplification in the power output stages. In an SSB transmitter, the voice signal to be transmitted can be applied to the DC coupled port (also known as the X or I port) of a double-balanced mixer, whilst the LO signal is applied to one of the transformer coupled ports as in Figure 8.13. The output from the other transformer coupled port is a double-sideband suppressed carrier signal as shown, which can then be filtered to leave the SSB signal, either upper side-

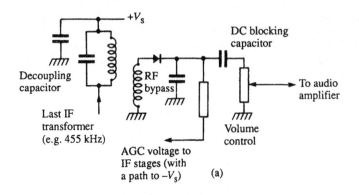

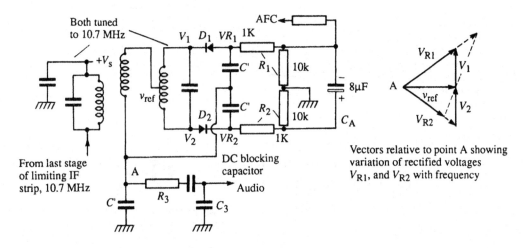

Figure 8.14   AM and FM demodulators (detectors).

(a) Diode AM detector. In the 'infinite impedance detector', e.g. Tr₃ in Figure 8.21, a transistor base/emitter junction is used in place of the diode. The emitter is bypassed to RF but not to audio, the audio signal being taken from the emitter. Since only a small RF base current is drawn, the arrangement imposes much less damping on the previous stage, e.g. the last IF transformer, whilst the transistor, acting as an emitter follower, provides a low-impedance audio output.

(b) Ratio detector for FM, with de-emphasis. C' =RF bypass capacitor, 330 pF.

band (USB) or lower sideband (LSB) as required. (Amateur radio practice is to use LSB below 10 MHz and USB above, but in commercial and military applications USB is the norm regardless of frequency.) In the receiver, the reverse process can be applied to demodulate an SSB signal, i.e. the output of the IF strip is applied to one of the transformer coupled ports of a diode ring mixer, and a carrier wave at the frequency of the missing suppressed carrier at the other. The beat frequency between the two is simply the original modulating voice signal, but offset by a few cycles if the

reinserted carrier is not at exactly the appropriate frequency. This results in reduced intelligibility and has been likened to the sound of Donald Duck talking through a drainpipe. A control called a *clarifier* is usually provided on an SSB receiver to permit adjustment of the frequency of the reinserted carrier for maximum intelligibility. In practice an IC such as 1496 DBM is often used for the demodulator: linearity is not of paramount importance in this application, since any signal in the pass band of the IF is either the wanted signal or unavoidable cochannel interference.

Having touched on the subject of SSB demodulation, it is appropriate to cover here demodulators – often called *detectors* – for other types of signals as well. Figure 8.14a shows a diode detector as used in an AM broadcast receiver. It recovers the audio modulation riding on a DC level proportional to the strength of the carrier component the signal. This DC level is used as an AGC voltage, being fed back to control the gain of the IF stages, so as to produce an effectively constant signal even though the actual level may change due to fading. The result is usually acceptable, but AGC can give rise to unfortunate effects. For example, on medium wave after dark, signals from distant stations can be received but the nature of the propagation (via reflections from the ionosphere) can give rise to frequency selective fading, resulting in quite sharp notches in the received RF spectrum. If one of these coincides with the carrier component of an AM signal, the AGC will increase the IF gain to compensate. At the same time, as the sidebands have not faded in sympathy, the result is that the signal is effectively modulated by greater than 100%, resulting in gross distortion in the detected audio. It is unfortunate that this coincides with the increased output due to AGC action, resulting in a very loud and unpleasant noise!

Figure 8.14b shows one of the types of demodulator used for FM signals. It depends for its action upon the change of phase of the voltage across a parallel tuned circuit relative to the current as the signal frequency deviates first higher then lower in frequency than the resonant frequency. The reference voltage v_{ref} in the small closely coupled winding at the earthy end of the collector tuned circuit is in phase with the voltage across the latter. The centre tapped tuned circuit is very lightly coupled to the collector tuned circuit, so the reference voltage is in quadrature with the voltage across the centre tapped tuned circuit. The resulting voltages applied to the detector diodes are as indicated by the vector diagrams. Capacitors C' have a value of around 330 pF, so that they present a very low impedance at the usual FM IF of 10.7 MHz but a very high impedance at audio frequencies. As the detected output voltage from one diode rises, that from the other falls, so that

the recovered audio appears in antiphase across R_1 and R_2, whilst the voltage across C_A is constant. $R_3 C_3$ provides de-emphasis to remove the treble boost applied at the transmitter for the purpose of improving the signal/noise ratio at high frequencies: the time constant is 50 μs (75 μs is used in the USA). This type of frequency discriminator, known as the *ratio detector*, was popular in the early days of FM broadcasting, since it provided a measure of AM rejection to back up the limiting action of the IF strip. Any rapid increase or decrease in the peak-to-peak IF voltage applied to the diodes would result in an increase or decrease of the damping on the centre tapped tuned circuit by the detectors, as C_4 was charged up or discharged again. This tended to stabilize the detected output level, whilst slow variations in level, due to fading for example, were unaffected. Modern FM receivers use IC IF strips with more than enough gain to provide hard limiting on the smallest usable signal, so an on-chip discriminator based upon quadrature detection by the Gilbert cell is normally used.

Oscillators

The next major category of circuit considered in this chapter is the RF oscillator. Every transmitter needs (at least) one, and receivers of the superhet variety also need one in the shape of the local oscillator. The frequency of oscillation is determined by a tuned circuit of some description. The hallmarks of a good oscillator are stability (of both output frequency and output level), good waveform (low harmonic content) and low noise. An oscillator can be considered either as an amplifier whose output is applied via a band-pass filter back to its input, so as to provide positive feedback with a loop gain of just unity at one frequency; or as a circuit in which an active device is arranged to reflect a negative resistance in parallel with a tuned circuit, of value just sufficient to cancel out the losses and raise its Q to infinity. In practice, there is seldom any real difference between these apparently divergent views: Figure 8.15 illustrates the two approaches.

In Figure 8.15a a single tuned circuit with no coupled windings is employed. For the circuit to

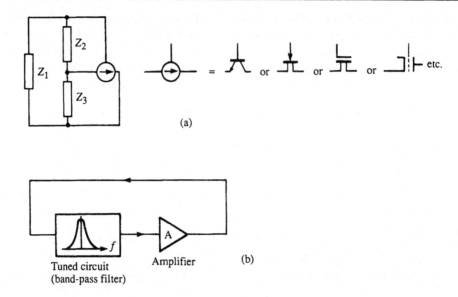

(a)

(b)

Tuned circuit
(band-pass filter)

Amplifier

Figure 8.15   Oscillator types.
 (a) Negative resistance oscillator: see text.
 (b) Filter/amplifier oscillator.

oscillate, Z_2 and Z_3 must be impedances of the same sign (both positive, i.e. inductances, or both negative, i.e. capacitances) whilst Z_1 must be of the opposite sign. The funny symbol is a shorthand sign for any three-terminal active device, be it valve, bipolar transistor or FET. Figure 8.16 shows a number of Figure 8.15a type oscillators, with their usual names. Of these, the *Clapp* (or *Gouriet*) is a circuit where the value of the two capacitors of the corresponding Colpitts oscillator has been increased and the original operating frequency restored by connecting another capacitor in series with them. To understand how this improves the stability of the oscillator, remember that any excess phase shift through the active maintaining device, resulting in its phase shift departing from exactly 180°, must be compensated for by a shift of the frequency of oscillation away from the resonant frequency of the tuned circuit, so that the voltage applied to the 'grid' lags or leads the 'anode' current by the opposite amount. This restores zero net loop phase shift, one of the necessary conditions for oscillation. By just how much the frequency of operation has to change to allow for any non-ideal phase shift in the active

device depends upon the Q of the tuned circuit. True, the Q is infinity, in the sense that the amplitude of the oscillation is not dying away, but that is only because the active device is making up the losses as they occur. As far as rate of change of phase with frequency in the tuned circuit is concerned, the Q is determined by the dynamic resistance R_d of the tuned circuit itself, in parallel with the loading reflected across it by the presence of the active device. In the case of a valve or FET, the anode or drain slope resistance is often the main factor: in the case of a bipolar transistor, the low base input impedance is equally important.

The additional capacitor C_1 in the Clapp circuit effectively acts in the same way as a step-down transformer, reducing the resistive loading on the tuned circuit, so that its loaded Q approaches more nearly to its unloaded Q. This improves the frequency stability by increasing the isolation of the tuned circuit from the vagaries of the maintaining circuit, but of course does nothing to reduce frequency drift due to variation of the value of the inductance and of the capacitors with time and temperature variations. The improved isolation of the tuned circuit

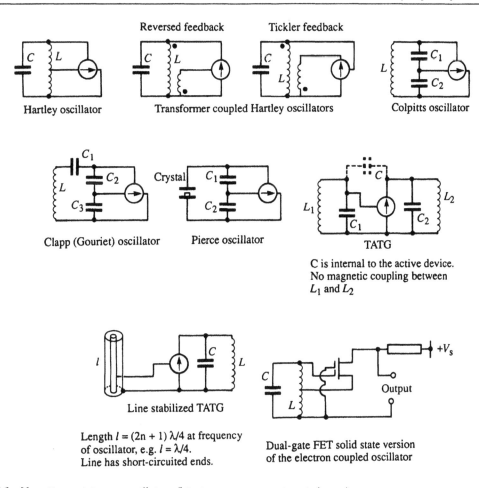

Figure 8.16 Negative resistance oscillators (biasing arrangements not shown).

from the active device cuts both ways. There is less drive voltage available at the active device's input and, at the same time, the load resistance reflected into its output circuit is reduced: both of these factors reduce the stage gain. Thus the Clapp circuit needs a device, be it valve, transistor or FET, with a high power gain. Clearly the higher is the unloaded Q of the tuned circuit, the lower are the losses to be made up and hence the less gain is demanded of the maintaining circuit. Assuming a high output slope resistance in the active device, the losses will nearly all be in the inductor. If this is replaced by a crystal, which at a frequency slightly below its parallel resonant frequency will look inductive, a very high-Q resonant circuit results, and indeed the Clapp

oscillator is a deservedly popular configuration for a high-stability crystal oscillator.

Figure 8.17 shows oscillator circuits of the Figure 8.15b variety. The TATG circuit in Figure 8.16 (named from its valve origins: tuned anode, tuned grid) is like the Meissner oscillator in Figure 8.17, except that the feedback occurs internally in the device. The *line stabilized* oscillator is an interesting circuit, sometimes used at UHF where a coaxial line one wavelength long becomes a manageable proposition. By increasing the rate of change of loop phase shift with frequency, the line increases the stability of the frequency of oscillation. A *surface acoustic wave* (SAW) device can provide at UHF a delay equal to many wavelengths. If the SAW device provides N complete

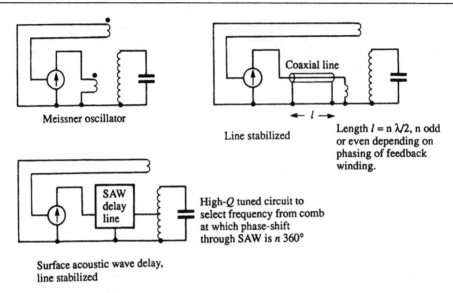

Figure 8.17   Filter/amplifier oscillators.

cycles of delay, the rate of change of phase shift with frequency will be N times as great as for a single-wavelength delay. The SAW stabilized oscillator can thus oscillate at any one of a 'comb' of closely spaced frequencies, a conventional tuned circuit being used to force operation at the desired frequency.

Figure 8.18 shows two oscillator circuits which use two active devices. In principle, two devices can provide a higher gain in the maintaining amplifier and thus permit it to be more lightly coupled to the tuned circuit, improving stability. But on the other hand the tuned circuit has to cope with the vagaries of two active devices instead of just one. The maintaining amplifier need not use discrete devices at all. The maintaining device can

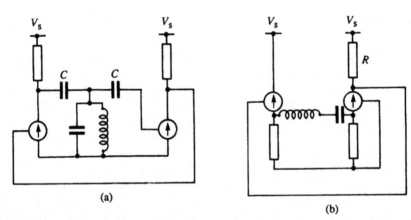

Figure 8.18   Two-device oscillators (see also Figure 9.14a).
(a) Franklin oscillator. The two stages provide a very high non-inverting gain. Consequently the two capacitors C can be very small and the tuned circuit operates at close to its unloaded value of Q.
(b) Emitter coupled oscillator. This circuit is unusual in employing a series tuned resonant circuit. Alternatively it is suitable for a crystal operating at or near series resonance, in which case R can be replaced by a tuned circuit to ensure operation at the fundamental or desired harmonic, as appropriate.

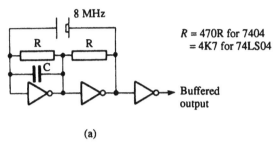

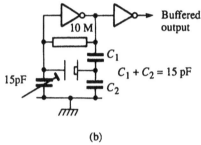

Figure 8.19 Crystal-controlled computer clock oscillators.
(a) TTL type with crystal operating at series resonance.
(b) CMOS type wih crystal operating at parallel resonance.

be an integrated circuit amplifier, or even an inverting logic gate used as an amplifier, as shown in Figure 8.19.

Where high stability of frequency is required, a crystal oscillator is the usual choice. For the most critical applications, an ovened crystal oscillator can be used. Here, the crystal itself and the maintaining amplifier are housed within a container, the interior of which is maintained at a constant temperature such as +75°C. Oven-controlled crystal oscillators (OCXOs) can provide a temperature coefficient of output frequency in the range 10^{-7} to 10^{-9} per °C, but stabilities of much better than one part in 10^6 per annum are difficult to achieve. The best stability is provided by the glass encapsulated crystal, the worst by the solder seal metal can crystal, with cold weld metal cans providing intermediate performance. Where the time taken for oven warm-up is unacceptable and the heater cannot be left permanently switched on, a temperature-compensated crystal oscillator

(TCXO) is used. In this, the ambient temperature is sensed by one or more thermistors and a voltage with an appropriatc law is derived for application to a voltage-controlled variable capacitor (varicap). Both OCXOs and TCXOs are provided with adjustment means – a trimmer capacitor or varicap diode controlled by a potentiometer – with sufficient range to cover several years drift, allowing periodic readjustment to the nominal frequency.

In any oscillator circuit, some mechanism is needed to maintain the loop gain at unity at the desired amplitude of oscillation. Thus the gain must fall if the amplitude rises and vice versa. In principle, one could have a detector circuit which measures the amplitude of oscillation, compares it with a reference voltage and adjusts the amplifier's gain accordingly, just like an AGC loop. In this scheme, called an *automatic level control* (ALC) loop, the amplifier operates in a linear manner, for example in class A. However, it requires a detector circuit with a very rapid response, otherwise the level will 'hunt' or, worse, the oscillator will 'squegg' (operate only in short bursts). Most oscillator circuits therefore forsake class A and allow the collector current to be non-sinusoidal. This does not of itself ensure a stable amplitude of oscillation, but the circuit is arranged so that as the amplitude of oscillation increases, the device biases itself further back into class C. Thus the energy delivered to the tuned circuit at the fundamental frequency decreases, or at least increases less rapidly than the losses, leading to an equilibrium amplitude. In a transistor oscillator, stability is often brought about by the collector voltage bottoming, thus imposing heavy additional damping upon the tuned circuit. This is most undesirable from the point of view of frequency stability, and the current switching circuit of Figure 8.20a is much to be preferred.

Figure 8.20 also shows various ways in which the net loop gain of an oscillator can vary with amplitude. The characteristic of Figure 8.20b is often met though not particularly desirable. That of Figure 8.20c will not commence to oscillate unless kicked into oscillation by a transient such as at switch-on, a most undesirable characteristic. That of Figure 8.20d is representative of the

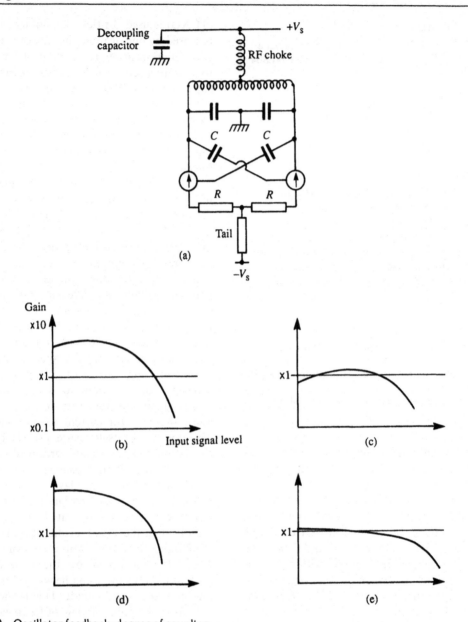

Figure 8.20 Oscillator feedback: degree of coupling.
    (a) Class D or current switching oscillator, also known as the Vakar oscillator. With R zero, the active devices act as switches, passing push-pull square waves of current. Capacitors C may be replaced by a feedback winding. R may be zero, or raised until circuit only just oscillates. 'Tail' resistor approximates a constant current sink.
    (b–e) Characteristics (see text).

current switching and Vakar oscillators and is very suitable for a high-stability oscillator. That of Figure 8.20e results in an amplitude of oscillation which is very prone to amplitude variations due to outside influences. It is therefore excellent for a simple radio receiver designed to achieve most of its sensitivity by means of *reaction*, also known as regeneration, such as that shown in Figure 8.21.[7] With this circuit, as the reaction is turned up, the effective circuit Q rises towards infinity, providing

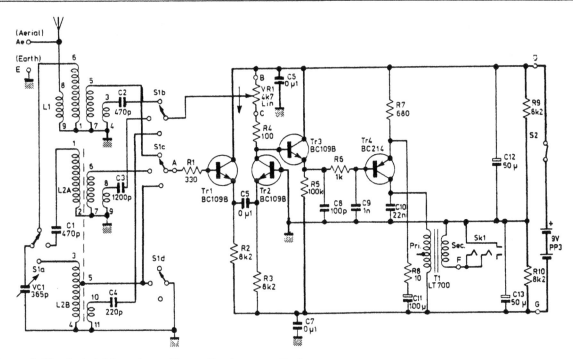

Figure 8.21 A straight receiver with reaction (regeneration).
(Reproduced by courtesy of *Practical Wireless*.)

a surprising degree of sensitivity. The greatest sensitivity occurs when the RF amplifier is actually oscillating very weakly; it is thus able to receive both CW and SSB signals. With AM signals its frequency becomes locked to that of the incoming signal and its amplitude varies in sympathy; anyone who has never played with a 'straight' set (i.e. not a superhet) with reaction has missed an experience.

It is not always convenient to generate an RF signal using an oscillator running at that frequency: an example is when a crystal-controlled VHF or UHF frequency is required, as crystals are only readily available for frequencies up to around 70 MHz. A common procedure in these cases is to generate the signal at a frequency of a few tens of megahertz and then multiply it in a series of doubler and/or tripler stages. A multiplier stage is simply a class C amplifier with frequency f MHz applied to its input and with a tuned circuit resonant at Nf MHz as its collector load. The collector current contains harmonics of the input frequency, since for a single-ended amplifier stage

only class A operation provides distortion-free amplification. The output tuned circuit selects the desired harmonic. In principle, the bias and drive level can be adjusted to optimize the proportion of the desired harmonic in the collector current; however, whilst this is worth doing in a one-off circuit, it is difficult to achieve in production.

It is important at any frequency, and particularly in RF circuits, to ensure that the signals to be amplified, multiplied, converted to another frequency or whatever, only proceed by the intended paths and do not sneak into places where they are not wanted, there to cause spurious responses, oscillations or worse. The main means of achieving this are *decoupling*, to prevent RF signals travelling along the DC supply rails, and *screening*, to avoid unintended capacitive or inductive coupling between circuits. At radio frequencies, screens of non-magnetic metal are equally effective at suppressing unwanted magnetic coupling as well as electrostatic coupling. Supply rail decoupling is achieved by bypassing RF currents to ground with decoupling capacitors whose reactance is very low at the frequency involved, whilst placing a high series impedance in the supply rail, in the form of an inductance so as not to incur any voltage drop at DC. For a more detailed coverage of radio-frequency technology, see Ref. 8.

References

1. Notes on the Theory of Modulation. J. R. Carson. *Proc. I.R.E.* Vol. 10, p. 57. February 1922.
2. *Solid State Design for the Radio Amateur*. Hayward and DeMaw. 2nd printing 1986. p. 189, American Radio Relay League Inc.
3. High Dynamic Range Transistor Amplifiers Using Lossless Feedback, D. E. Norton, p. 53. *Microwave Journal.* May 1976.
4. Need a PIN-Diode Attenuator? R. S. Viles, *Electronic Design* 7, p. 100. March 29, 1977.
5. Symmetric MOSFET Mixers of High Dynamic Range, R. P. Rafuse, p. 122. Session XI, 1968 International Solid State Conference.
6. Single Balanced Active Mixer Using MOS-FETs. E. S. Oxner, p. 292, *Power FETs and Their Applications*, 1982, Prentice-Hall.
7. The PW Imp 3-Waveband Receiver. I. Hickman, p. 41. *Practical Wireless*, May 1979. NOTE: 'Plessey' devices are now manufactured by GEC Plessey Semiconductors Ltd.
8. *Practical RF Handbook*, 2nd edition, 1997, Ian Hickman, Butterworth-Heinemann.

Questions

1. (i) Describe and contrast the distinguishing features of amplitude modulation and frequency modulation, with particular reference to the phase of the sidebands relative to the carrier.

   (ii) A VHF broadcast FM transmitter radiates a signal modulated with a 1 kHz sine wave, with 63 kHz deviation. What is the peak phase deviation?

2. In question 1(ii), approximately how many significant sidebands are there?

3. In a small-signal common emitter RF amplifier, what is the mechanism contributing to potential instability? How does the cascode circuit circumvent the problem?

4. Describe two different methods used to ensure the stability of a single transistor common emitter small-signal RF amplifier stage.

5. A small-signal class A RF amplifier produces an output of -10 dBm for an input of -20 dBm. When two such signals are applied simultaneously, the third-order intermodulation products at the output are at -50 dBm. What is the amplifier stage's third-order intercept point?

6. Draw the circuit diagram of a continuously variable bridged Tee attenuator. Why, in practice, is the attenuation range limited at both the minimum and maximum extremes?

7. What measures are incorporated in some Schottky Quad diode double balanced mixers, to increase linearity and dynamic range?

8. Draw a block diagram of an HF SSB transmitter, showing clearly the stages in the production of the single sideband signal.

9. Describe the operation of the ratio detector for broadcast FM signals. How is the necessary quadrature reference voltage obtained?

10. Compare and contrast the Hartley and Colpitts oscillators. What feature of the Clapp variant of the Colpitts oscillator contributes to its superior frequency stability?

9   Signal sources

Signal sources play an important role in electronic test and measurements, but their use is far from limited to that. They form an essential part of many common types of equipment. For example, a stabilized power supply needs an accurate DC voltage source as a reference against which to compare its output voltage. Many pieces of electronic equipment incorporate an audio-frequency signal source as an essential part of their operation, from the mellifluous warble of a modern push-button telephone to the ear-shattering squeal of a domestic smoke detector. And RF sources – oscillators – form an essential part of every radio transmitter and of virtually every receiver. So let's start with the DC signal source or voltage reference circuit.

Voltage references

The traditional voltage reference was the Weston standard cell, and these are still used in calibration laboratories. However, in most electronic instruments nowadays, from power supplies to digital voltmeters (DVMs), an electronic reference is used instead.

A Zener diode exhibits a voltage drop, when conducting in the reverse direction, which is to a first approximation independent of the current flowing through it, i.e. it has a low slope resistance. Thus if a Zener diode is supplied with current via a resistor from say the raw supply of a power supply (Figure 9.1a), the voltage variations across the Zener – both AC due to supply frequency ripple and DC due to fluctuations of the mains voltage – will be substantially less than on the raw supply, provided that the value of the resistor is much greater than the diode's slope resistance. In practice, this means that about as many volts must be 'thrown away' across the resistor as appear across

the diode. Even so, the improvement is inadequate for any purposes other than the cheapest and simplest stabilized power supply. Figure 9.1b shows how the performance of the regulator can be notably improved by using the high drain slope resistance of a junction FET in place of the resistor. Unfortunately an FET is a lot dearer than a resistor. Two-lead FETs with the gate and source internally connected as shown are available as 'constant current diodes' and work very well; unfortunately they are even more expensive than FETs, which themselves have always commanded a price ratio relative to small-signal bipolar transistors of about five to one. If an FET is used, the problem of the usual $5:1$ spread in I_{dss} can be alleviated by including a source bias resistor, as in Figure 9.1c, or even by adjusting it for a given drain current as in Figure 9.1d.

Zener diodes have been much improved over the years. Earlier types left one with the difficult choice of going for lowest slope resistance – which was found in devices with a rating of about 8.2 V – or for lowest temperature coefficient (TC or 'tempco'), then found in 5.1 V devices. With modern devices such as the Philips BZX79 series, lowest TC and lowest slope resistance occur for the same voltage rating device, i.e. $+0.4\%/°C$ and $10\,\Omega$ at 5 mA respectively in the BZX79 C6V2 with its 6.2 V $\pm5\%$ voltage rating. A point to watch out for is that the measurement of a Zener diode's slope resistance is usually an adiabatic measurement. This means that a small alternating current is superimposed upon the steady DC and the resulting alternating potential is measured. The frequency of the AC is such that the diode's temperature does not have time to change in sympathy with each cycle of the current. If now there is a change in the value of the steady DC component of current through the diode, there will

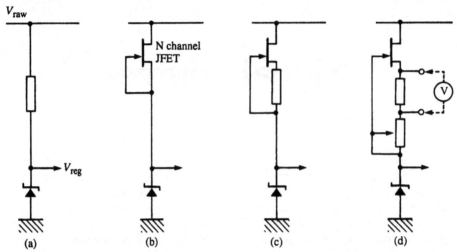

Figure 9.1   Zener DC voltage references, simple and improved (reproduced by courtesy of *New Electronics*).

be an accompanying instantaneous small change in voltage δV due to δI, the change in current flowing through the slope resistance R_s, followed by a slower change of voltage due to the TC as the operating temperature of the diode changes. This clearly highlights the benefit of a range of diodes where the minimum slope resistance and TC can be had in one and the same device.

Returning to Figure 9.1a, this arrangement can provide a stabilization ratio V_{raw}/V_{reg} of about 100:1 or 1%, whereas the FET aided version improves on this by a factor of about 30, depending on the FET's slope resistance. However, a useful if not quite so great improvement can be provided by the arrangement of Figure 9.2.[1] Here the diode current is stabilized at a value of approximately $0.6/R_2$, since the PNP transistor's V_{be} changes little with change of emitter current. Consequently, if V_{raw} increases, most of the resultant increase in current through R_1 is shunted via the collector to ground rather than through the Zener diode. Where a modest performance, about 10 times better than Figure 9.1a, is adequate, the circuit of Figure 9.2 offers a very cheap solution. Where substantially better performance is required, a voltage reference IC is nowadays the obvious choice. These are available from most manufacturers of linear ICs and operate upon the *bandgap* principle. A typical example is the micropower two-lead LM385-1.2 from National

Semiconductor, which is used in series with a resistor or constant current circuit, just like a Zener diode. This 1.2 V reference device is available in 1% or 2% selection tolerance, operates over a current range of 10 μA to 20 mA, and features a dynamic impedance of 1 Ω; the suffix X version features a TC at 100 μA of less than 30 PPM/°C. A 2.5 V device, the LM385-2.5, is also available. Other commonly available reference voltage ICs come in various output voltages, including 5.0 V, 10.0 V and 10.24 V.

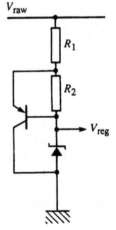

Figure 9.2   Inexpensive improved Zener voltage references (reproduced by courtesy of *New Electronics*).

Non-sinusoidal waveform generators

Sources of AC signals can be divided into two main categories: sine wave generators, and generators of non-sinusoidal waveforms. The latter can be subdivided again into pulse generators and other types. *Pulse generators* provide pulses of positive- or negative-going polarity with respect to earth or to a presettable DC offset voltage. The pulse repetition frequency, pulse width, amplitude and polarity are all adjustable; on some pulse generators, so too are the rise and fall times. Commonly also the output may be set to provide 'double pulses', that is pulse pairs with variable separation, and a pulse delay with respect to a prepulse, which is available at a separate output for test and synchronization purposes. Pulse generators of this type are used mainly for test purposes in digital systems, so they are not considered further here. So let's press straight on and look at those 'other types'.

Non-sinusoidal or *astable waveform generators* may be categorized as operating in one of two modes, both of which are varieties of *relaxation oscillator*. As the name implies, the oscillation frequency is determined by the time taken by the circuit to relax or recover from a positive extreme of voltage excursion, towards a switching level at which a transient occurs. The transient carries the output voltage to a negative extreme and the circuit then proceeds to relax towards the switching level again, but from the opposite polarity. On reaching it, the circuit switches rapidly again, finishing up back at the positive extreme.

The two modes are those in which *differentiated* (phase advanced) *positive* feedback is combined with broad band negative feedback on the one hand, and types in which broad band positive feedback is combined with *integrated* (phase retarded) *negative* feedback on the other. Figures 9.3 and 9.4 show both discrete component and IC versions of these two types respectively. The circuit operation should be clear from the circuit diagrams and waveforms given.

There is no reason why such an oscillator should not use differentiated positive feedback and integrated negative feedback, as in Figure 9.5a; indeed, there is a definite advantage in so doing.

It results in a greater angle between the two changing voltage levels at the point at which regeneration occurs, and this makes that instant less susceptible to influence by external or internal circuit noise. Thus the frequency of oscillation is more stable, a worthwhile improvement since the frequency purity of astable oscillators generally is very much poorer than that of sinusoidal oscillators using an *LC* resonant circuit. In the latter the stored energy is much greater than any circuit noise, which consequently has less effect. However, a circuit such as Figure 9.5a contains two time constants, both of which play a part in determining the frequency. The circuit of Figure 9.3b will provide a 10:1 variation of frequency for a 10:1 variation of the resistance *R* forming part of the frequency determining time constant *CR*. The same applies to the circuit of Figure 9.5a only if more than one resistor is varied in sympathy. Thus the circuit of Figure 9.5a is more attractive in fixed frequency applications or where a tuning range of less than an octave is required. For wide frequency applications, as in a function generator providing sine, triangular and square output waveforms, it is not uncommon to opt for the economy of single resistor control.

Figure 9.5b shows a popular and simple astable oscillator circuit. There is only a single path around the circuit, for both the positive and the negative feedback. At any time (except during the switching transients) only one of the two transistors conducts, both tail currents being supplied via the 1K resistor or from the +15 V rail.

The circuit of Figure 9.6 works on a slightly more sophisticated principle than the circuits of Figures 9.3 and 9.4, where the feedback voltage relaxes exponentially.[2] It uses the *Howland current pump*, a circuit discussed earlier (Figure 7.6), to charge a capacitor, providing a linearly rising ramp. When this reaches the trigger level of half the supply rail voltage (at the non-inverting input of the comparator), the trigger level, and the voltage drive to the current pump, both reverse their polarity, setting the voltage on the capacitor charging linearly in the opposite direction. The frequency is directly proportional to the output of the current pump and hence to the setting of the 10 K potentiometer, which can be a multiturn type

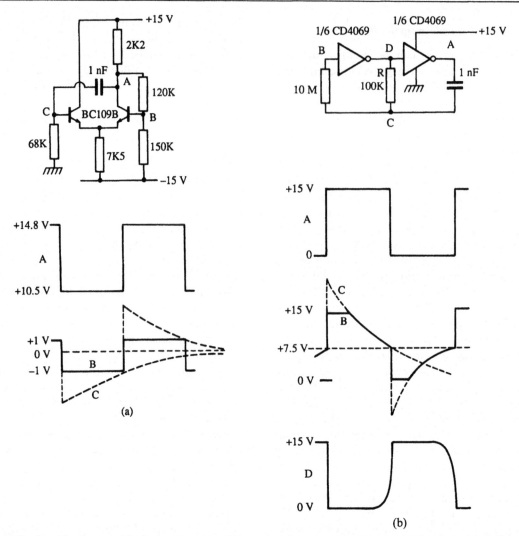

Figure 9.3 Astable (free-running) circuits using differentiated positive feedback and flat (broad band) negative feedback.

(a) Cross-coupled astable circuit. The dashed line shows the 0 V level at which the discharge at point C is aiming when it reaches the switching level.

(b) Astable circuit using CMOS inverters. The waveform at B is similar to that at C except that the excursions outside the 0 V and + 15 V supply rails have been clipped off by the device's internal gate protection diodes.

with a ten-turn digital dial. With the values shown the circuit provides five frequency ranges from 0 to 1 Hz up to 0 to 10 kHz, with direct read-out of frequency. Each range determining capacitor has an associated 4K7 preset resistor associated with it, enabling the full-scale frequency to be set up for each range, even though ordinary 10% tolerance capacitors are used. The circuit provides buffered

low-impedance triangular and square wave outputs.

Most function generators provide a sine wave output of sorts. The popular 8038 function generator IC includes an on-chip shaping stage to produce a sine wave output by shaping the triangle waveform. This operates purely on a waveform shaping basis and thus works equally well at any

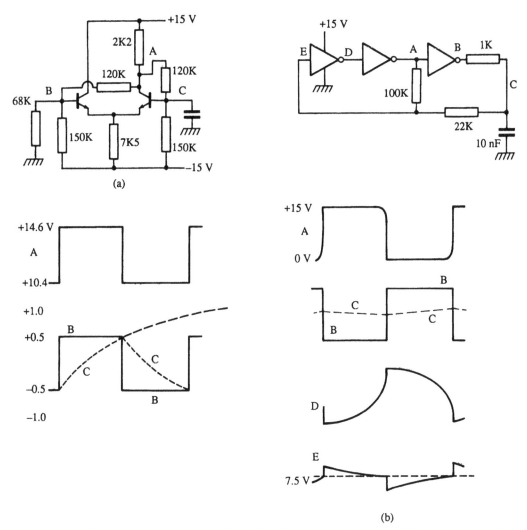

Figure 9.4 Astable (free-running) circuits using broad band positive and integrated (delayed) negative feedback.
(a) Cross-coupled astable circuit.
(b) Astable circuit using CMOS inverters.

frequency. An alternative scheme is to use an integrator: a triangular (linear) waveform is integrated to a parabolic (square law) waveform which forms a passable imitation of a sine wave,the total harmonic distortion being about 3.5%. However, the disadvantage of the integrator approach is that the output amplitude varies inversely with frequency, unless the value of the integrator's input resistor is varied to compensate for this.

An aperiodic (non-frequency dependent) method of shaping a triangular wave into an approximation to a sine wave is to use an amplifier which runs gently into saturation on each peak of the triangular waveform. Unlike the integrator method, where the sharp point at the peak of the triangle wave becomes a slope discontinuity at the zero crossing point of the pseudo-sine wave, it is difficult with the aperiodic shaping method to avoid some residual trace of the point at the peak of the sine-shaped waveform. A scheme which has been used to avoid this is to slice off the peaks of the triangular wave before feeding it

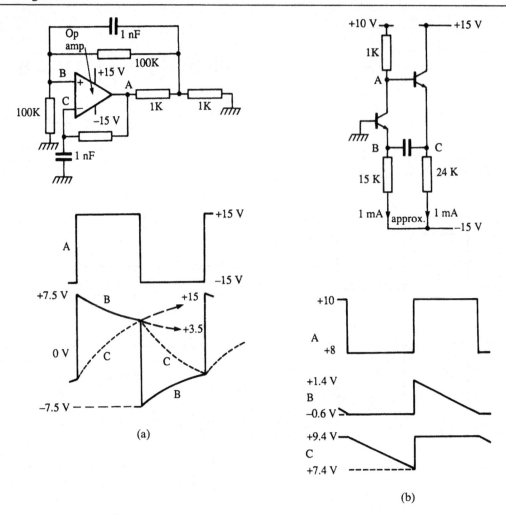

Figure 9.5 Other types of astable circuit.

(a) Astable circuit using both differentiated positive and integrated negative feedback. Aiming potentials of points B and C prior to switching shown dashed.

(b) The Bowes, White or emitter coupled astable does not have separate positive and negative feedback paths, so differing from the oscillators of Figures 9.3, 9.4 and 9.5a.

to the shaping circuit.[3] In the reference cited, by choosing the optimum degree of preclipping and of non-linearity of the shaping amplifier gain, distortion as low as 0.2% is achieved at low frequencies (a times ten improvement on the results usually achieved by this method). The shaping amplifier is implemented in an IC using a 1 GHz device process, resulting in good conversion of triangular waveforms to sine waves at frequencies up to 100 MHz.

Some function generators are capable of producing other waveforms besides the usual square/triangle/sine waves. A popular waveform is the sawtooth and its close cousin the asymmetrical triangle (see Figure 9.7). This figure also indicates how a stepwise approximation to any arbitrary waveform can be produced by storing the data values corresponding to say 256 successive samples of the waveform over one whole cycle in a read only memory (ROM), and then reading

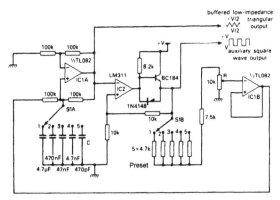

Figure 9.6 Function generator using a Howland current pump. The five 4K7 preset potentiometers enable the maximum frequency of the ranges to be set to 1 Hz to 10 kHz exactly; range capacitors *C* can thus be inexpensive 10% or even 20% tolerance types. If 10K resistor *R* is a ten-turn digital dial potentiometer, it will indicate the output frequency directly. +V and −V supplies must be equal, but frequency is independent of the value of *V*. (Reproduced by courtesy of *New Electronics*.)

them out sequentially to a digital-to-analog converter (DAC). In this way it is possible to reproduce natural sounds which have been recorded and digitized, for example the sound of a diapason or reed pipe from a real pipe organ, as is done in some electronic organs. The step nature of the output will correspond to very high-frequency harmonics of the fundamental, which in the organ application may well be beyond the range of hearing, but where necessary the steps can be smoothed off with a low-pass filter. This can still have a high enough cut-off frequency to pass all the harmonics of interest in the output waveform.

Another way of achieving a smooth, step-free output waveform is to make use of the multiplying capability of a DAC. The output current from a DAC is equal to the input bit code times the reference voltage input. Figure 9.8 shows two multiplying DACs with reverse sawtooth waveforms applied to their reference inputs so that as the output of the *P* DAC decreases, that of the Q DAC increases. Sample values are fed to the DACs at the same rate as the sawtooth frequency. When the output of the *P* DAC reaches zero, its input code is changed to that currently present at the input of the *Q* DAC. Immediately after this the sawtooth waveforms fly back to their initial values, so that the output from the *Q* DAC is now zero, and its input bit code is promptly changed to that

of the next waveform sample. The output currents of the two DACs are summed to give a smoothly changing voltage output from the opamp. The generation of a sine wave by this means is illustrated in Figure 9.9, but any arbitrary waveform can be produced once the appropriate values are stored in ROM. In practice, both the new DAC values are simply applied at the same instant that the sawtooth waveforms fly back to their starting values: any 'glitch' in the output voltage, if appreciable, can be smoothed out with a little integrating capacitor across the summing opamp's feedback resistor, which in Figure 9.8 is internal to the DAC.

An interesting application of this is for writing data on the screen of a real-time oscilloscope. Such an oscilloscope uses the electron beam to write the traces under control of the *X* and *Y* deflection plates, but it does not produce a raster scan like a TV display, so some other means is needed if information such as control settings is to be displayed on the screen. Two completely separate but complementary voltage waveform generators such as Figure 9.9 can be used to produce the appropriate *X* and *Y* deflection voltages to write alphanumeric data on the screen, the appropriate DAC data being stored in ROM. This scheme is used on many makes of oscilloscope. When the display read-out is on, it is possible under certain conditions to observe short breaks in the trace

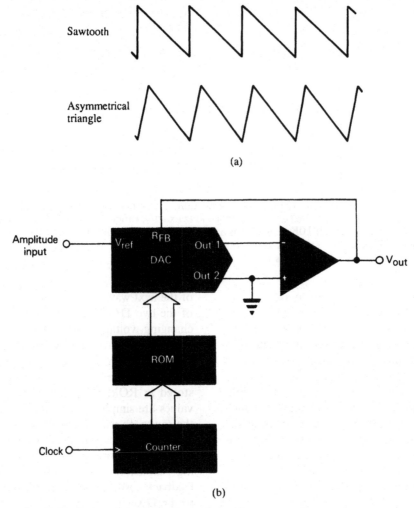

Sawtooth

Asymmetrical
triangle

(a)

Amplitude
input

V_{ref}

R_{FB}

DAC

Out 1

Out 2

V_{out}

ROM

Clock

Counter

(b)

Figure 9.7   Generalized triangle waveform and universal waveform generator.
(a) Sawtooth and asymmetrical triangle waveforms; both are generally provided by the more versatile type of function generator. The sawtooth and the triangular wave (Figure 9.6) can both be considered as limiting cases of the asymmetrical triangular wave.
(b) Simple ROM waveform generator (reproduced by courtesy of *Electronic Engineering*).

where the beam goes away temporarily to write the read-out data.

Sine wave generators

Turning now to sine wave generators, let's look first at *audio-frequency generators*. These generally do not use *LC* tuned circuits to determine the frequency, and therefore have a degree of frequency stability intermediate between that of tuned circuit oscillators and relaxation oscillators, and in some cases not much better than the latter. To measure the distortion of a high-fidelity audio power amplifier, one needs, in addition to a distortion meter, a sine wave source of exceptional purity. Not only must the source's distortion be exceedingly low, but its frequency stability must be of a very high order. This is because the usual sort of distortion meter works by rejecting the fundamental component of the amplifier's

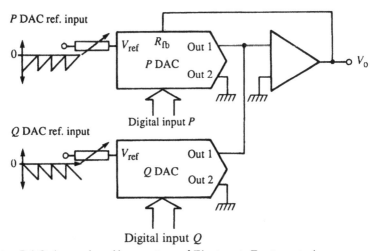

Figure 9.8 Interpolating DACs (reproduced by courtesy of *Electronic Engineering*).

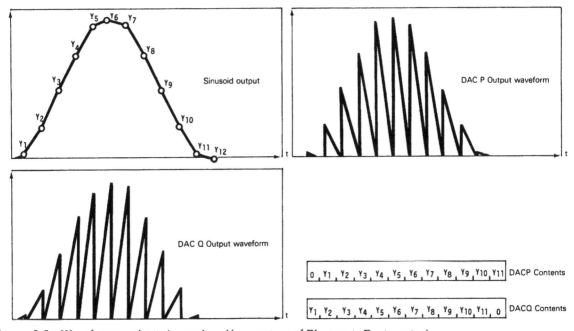

Figure 9.9 Waveform synthesis (reproduced by courtesy of *Electronic Engineering*).

output with a narrow notch filter, so that the harmonics, residual noise and hum can be measured. Their level relative to the total output signal, expressed as a percentage, is the *total harmonic distortion* (THD) or, more strictly, the total residual signal if noise and hum are significant. Clearly, if the frequency of the sine wave generator drifts it will be difficult to set and keep

it in the notch long enough to take a measurement. However, even if its drift is negligible, it may exhibit very short-term frequency fluctuations. Thus it will 'shuffle about' in the notch, resulting in a higher residual output than if its frequency were perfectly steady, as it tends to peep out first one side of the notch and then the other.

Vector representing pure sine wave of frequency *f* Hz

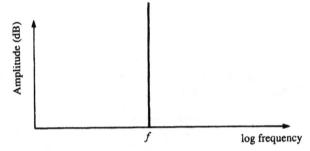

Pure sine wave of frequency *f* Hz represented in the frequency domain

(a)

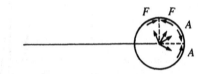

Sine wave with AM and FM noise sidebands (*A*, *F*), grossly exaggerated

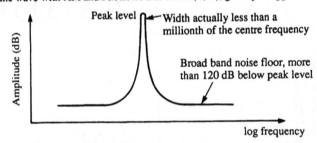

Corresponding frequency domain representation

Figure 9.10   Sine waves.
  (a) Ideal pure sine wave.
  (b) Real-life sine wave.

Now this is simply an explanation in the time domain of something which can equally well be explained in the frequency domain. Figure 9.10a shows an ideal sinusoidal signal, whilst Figure 9.10b shows, much exaggerated for clarity, a practical sine wave, warts and all. In addition to the ideal sine wave there are *close-in noise side-bands* of two sorts, AM and FM. These represent energy at frequencies very close to that of the sine wave, falling rapidly in amplitude as the frequency difference increases. The FM noise sidebands are the manifestation in the frequency domain of slight phase variations which were noted as frequency shuffle in the time domain and which are shown as FM sidebands in Figure 9.10b. There are also AM sidebands corresponding to slight amplitude variations in the sine wave, and these also will contribute to the residual. The residual may be considered as being responsible for it being impossible to say exactly where the tip of the vector in Figure 9.10b is at any time; it will be somewhere in the much exaggerated 'circle of uncertainty' shown. (Note that noise sidebands, both AM and FM, are also found either side of the output frequency of an *LC* oscillator and even of a crystal oscillator; it is just that in those cases they are restricted by the high *Q* of the frequency determining components to a very much narrower fractional bandwidth about the centre frequency.) In a well-designed audio oscillator, the energy in the noise sidebands which is not rejected by the notch of the distortion meter is always lower in level than the energy of the harmonics.

Figure 9.11 shows an audio oscillator using the popular *Wien bridge* configuration. In Figure 9.11a you can see the principle of the thing. By using the idea of extremums – replacing a capacitor by an open-circuit at 0 Hz and by a short-circuit at infinite frequency – there will clearly be no signal at B at these frequencies. It turns out (the sums are not difficult, have a go) that at the frequency $f = 1/2\pi RC$ the amplitude at B is one-third of that at A and the two waveforms are in phase. At other frequencies the attenuation is greater and the waveforms are out of phase. If the bridge is just out of balance sufficiently to provide the necessary input to the maintaining amplifier, then the latter will drive the bridge at an amplitude adequate to produce the said input. If this sounds like a specious circular argument, it is: in the practical circuit of Figure 9.11b the necessary degree of bridge imbalance is provided by a thermistor. The usual type is an R53, which has a cold resistance of 5K (or $5 \times 10^3 \, \Omega$, hence the type number). At switch-on, the bridge is unbalanced by much more than is necessary, so that the positive feedback via the *CR* network exceeds the negative feedback via the thermistor/resistor combination. Therefore the circuit commences to oscillate at the frequency at which the phase shift and attenuation of the *CR* network is least. As the amplitude of the oscillation builds up, the current through the thermistor heats it up. Now the thermistor consists of a pellet of amor-

phous semiconductor whose resistance falls rapidly with increasing temperature; the negative feedback via the thermistor/resistor arm therefore increases, and the bridge approaches balance. At an output voltage of about 3 V peak to peak, the dissipation in the thermistor, with the circuit values shown, is approaching the rated maximum, corresponding to a temperature of the pellet inside its evacuated glass envelope of 125°C, and the output amplitude is stabilized. Oscillators operating on this principle are commercially available from many manufacturers, such are their popularity. The oscillator can even be made to cover the frequency range 10 Hz to 10 MHz, although it is not then possible to optimize the circuit for the lowest possible distortion in the audio-frequency range.

The main problems with a thermistor stabilized Wien bridge oscillator are amplitude bounce and poorish distortion. The former is due to the thermistor: it is found that on changing frequency, the amplitude of the output oscillates up and down several times before settling to a steady value. Running the thermistor near its maximum permitted dissipation helps to minimize this. The other problem is due to the limited selectivity of the Wien bridge, which does little to reduce any distortion in the maintaining amplifier, and (at frequencies below 100 Hz) to the finite thermal time constant of the thermistor.

The Wien bridge oscillator shown in Figure 9.11b uses a two-gang variable resistor to vary both resistors of the frequency determining network simultaneously. This keeps constant the attenuation through the network at the zero phase shift frequency. It can also provide a 10 to 1 frequency tuning range for a 10 to 1 resistance variation, as can be deduced from the formula for the frequency quoted above. There are numerous sine wave oscillator circuits which provide frequency variation using only a single variable resistor, but in these the frequency ratio obtained is only equal to the square root of the resistance variation.[4,5]

An improved audio oscillator can be based on the *state variable filter*. Oscillation is ensured by the addition of fixed positive feedback and variable negative feedback applied to the inverting and

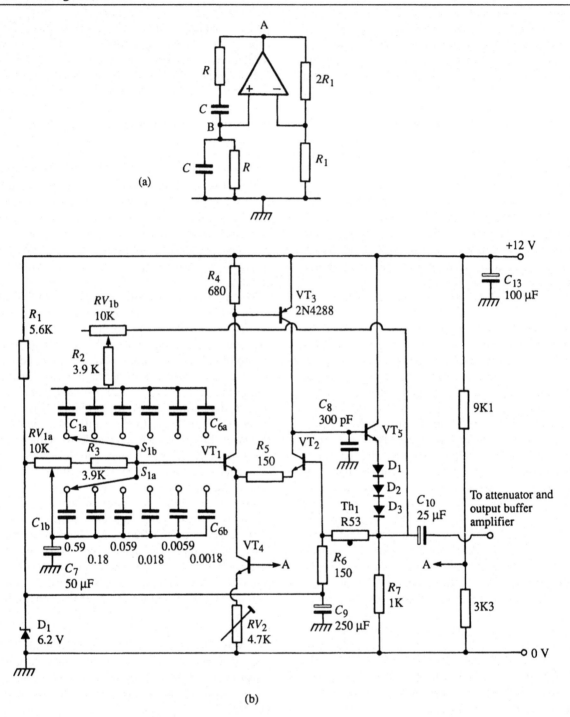

Figure 9.11   Audio-frequency Wien bridge sine wave oscillator.
(a) Principle of oscillator using Wien bridge.
(b) Low-distortion sine wave oscillator: 20–66 Hz, 66–200 Hz etc. up to 6.6–20 kHz. RV_1 is semilog; S_1 frequency range; all transistors BC109 except VT_3: D_1–D_3 IN4148.

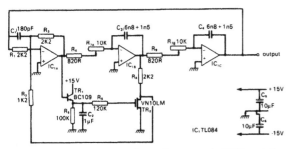

Figure 9.12 SVF-based oscillator with FET stabilizing: the FET costs one-tenth of an R53 thermistor (reproduced by courtesy of *New Electronics*).

non-inverting inputs respectively. The degree of NFB can be controlled by an FET, used as a variable resistor. In turn, the FET's resistance is controlled by a DC voltage proportional to the oscillator's peak-to-peak output voltage, so as to make the positive and negative feedback balance at the desired output amplitude. A variation on this scheme is shown in the SVF-based oscillator of Figure 9.12, which covers the frequency range 200–2000 Hz: here the PFB is variable, whilst the NFB or damping is provided by the phase advance due to C_1.[6] Interestingly, Tr_1 can be replaced by a IN4148 diode with virtually no increase in the harmonic distortion, which is about 0.02%. This is because the resultant slight dent in the positive peak of the sine wave at IC_{1a}'s output, as C_2 is topped up, is composed of high-order even harmonics. These are heavily attenuated in the two following integrators IC_{1b} and IC_{1c}.

The circuit of Figure 9.12 exhibits two undesirable features, which a little lateral circuit design can circumvent. First, if other values of C_3 and C_4 are switched in, to provide 20–200 Hz and 2–20 kHz ranges, the smoothing time constant R_5C_2 is inadequate on the lower range, leading to increased distortion owing to the FET's resistance varying in sympathy with the ripple. Worse still, the time constant is excessive on the top frequency range, leading to amplitude bounce just like a thermistor stabilized oscillator and even complete instability of the level control loop. Second, the frequency is inversely proportional to the integrator time constants, leading to a very non-linear frequency scale if the two-gang fine frequency control potentiometer (pot) R_7 has a

linear resistance law. The scale is excessively open at the low-frequency end and terribly cramped at the top end. A somewhat better, more linear, scale results if a reverse taper log pot is used, but this is a rather specialized component. If the frequency scale is marked on the skirt of the knob rather than on the panel of the instrument, a normal log pot can be used, but there is still a problem due to the wide selection tolerance and poor law repeatability of log pots.

Both of these drawbacks are avoided by the circuit of Figure 9.13. Here, the degree of NFB applied to the non-inverting terminal is fixed whilst the PFB is applied to the inverting terminal via a diode clipping network. Thus the oscillator works as a high-Q filter with a small square wave input of approximately fixed amplitude at its corner frequency. Amplitude control does not involve any control loop time constant and there is no rise in distortion below 100 Hz, such as is always found with thermistor-controlled circuits. The NFB or damping is set by the two resistors feeding back to the non-inverting terminal of IC_1. The ratio is one part in 83 and, as noted in Chapter 8, for feedback organized in this way the Q is one-third of the ratio, that is just over 27 in this case. If you assume that the input to the filter is a perfect square wave at the corner frequency, one would expect the third-harmonic component, which in a square wave amounts to one-third of the fundamental amplitude, to be attenuated by a factor of nine at the low-pass output from IC_3. For that is the theoretical attenuation of a frequency three times higher than the corner frequency at 12 dB/octave, relative to the flat response at low frequencies. Meanwhile, the fundamental at the corner frequency is actually accentuated by the factor Q, in this case 27. So for a per unit input, the output from IC_3 should consist of 27 per unit at the fundamental (approximately) and one-ninth of one-third at the third harmonic. So the third-harmonic component in the output should be $1/27^2$ per unit, which works out at about 0.15%. Despite a few approximations (the fundamental of a per unit square wave is itself slightly larger than unity, and the fifth and higher odd harmonics also contribute marginally to the distortion at the low-pass output), that is just about the distortion level

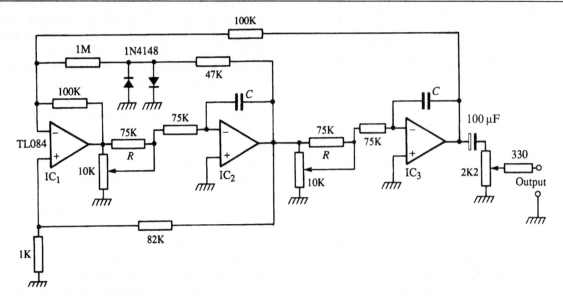

Figure 9.13 Low-distortion audio-frequency oscillator. Additional 75K resistors *R* result in an almost linear frequency scale. *C* is 10 nF for 0–200 Hz; 1 nF for 0–2 kHz; 100 pF for 0–20 kHz.

measured in the output of the circuit of Figure 9.13 at all frequencies from 20 Hz to 20 kHz. Furthermore, if the integrators' frequency determining resistors are high in value compared with the value of the two-gang fine frequency control pot, a linear law pot will provide a substantially linear scale.

Having covered audio-frequency oscillators in some detail, let's turn our attention next to *radio-frequency oscillators*. The basic oscillator circuits have already been covered in the previous chapter, so here I will just look at a couple of interesting variations before moving on to see how oscillators can be integrated with other circuits to increase their flexibility and accuracy. Figure 9.14a shows a two-transistor RF oscillator designed to be free of all time constants other than that of the tuned circuit itself, so that it cannot 'squegg', i.e. oscillate in bursts instead of continuously, as sometimes happens.[7] *Squegging* is a form of relaxation oscillation usually involving a *CR* time constant forming part of the stage's biasing circuit. The two transistors form a DC coupled pair with 100% NFB. They can only oscillate at the frequency at which the tuned circuit provides phase inversion or 180° phase shift. Further, if the total resistance of

$R_2 + R_3$ is greater than R_1, then the tuned circuit must also provide a voltage step-up. C_1 and C_2 are in series as far as determining the frequency goes, and, by making $C_2 \gg C_1$, a wide tuning range can be achieved with the variable capacitor. If waveform is unimportant, R_2 and R_3 can be replaced by a single 10K resistor.

The question of suppressing unwanted modes of oscillation is particularly important in crystal oscillators, since in most cases the quartz crystal (which is simply a high-Q electromechanical resonator) can vibrate in several different modes, rather like the harmonics of a violin string or the overtones of a bell. Indeed many crystals are designed specifically to operate at a harmonic (often the third) rather than at their fundamental resonance, since for a given frequency the crystal is then larger and has a higher Q. In the highest-quality crystal oscillators, the *strain compensated* (SC) cut is used. This is a 'doubly rotated' crystal, where the angle of the cut is offset from two of the three cystallographic axes. It has the advantage of a much lower temperature coefficient than the commoner cuts such as AT or BT, together with less susceptibility to shock and improved ageing characteristics (Figure 9.14b). These ad-

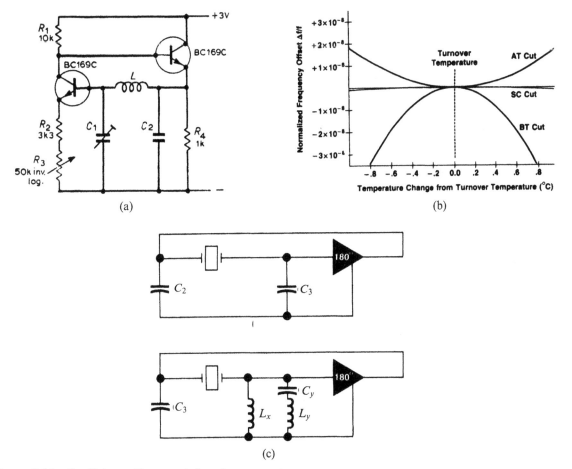

Figure 9.14 Oscillators with unwanted mode suppression.
(a) This *LC* oscillator can only oscillate at the frequency at which the tuned circuit supplies 180° phase shift and voltage step-up (reproduced by courtesy of *Electronics and Wireless World*).
(b) Temperature performance of SC, AT and BT crystal cuts.
(c) Standard Colpitts oscillator (top) and the same oscillator with SC mode suppression (10811A/B oscillator). (Parts (b) and (c) reproduced with permission of Hewlett-Packard Co.)

vantages do not come without an appropriate price tag, in that the SC crystal is more difficult and expensive to produce and has more spurious modes than other crystal types. For example, the 10 MHz SC crystal used in the Hewlett-Packard 10811A/B ovened reference oscillator is designed to run in the third-overtone C mode resonance. The third-overtone B mode resonance is at 10.9 MHz, the fundamental A mode is at 7 MHz, and below that are the strong fundamental B and C modes. The circuit of Figure 9.14c

shows the SC cut crystal in the 10811A/B oscillator connected in what is basically a Colpitts oscillator, so as to provide the 180° phase inversion at the input of the inverting maintaining amplifier.[8] With the correct choice of values for L_x, L_y and C_y, they will appear as a capacitive reactance over a narrow band of frequencies centred on the desired mode at 10.0 MHz, but as an inductive reactance at all other frequencies. Thus all of the unwanted modes are suppressed.

Voltage-controlled oscillators and phase detectors

Increasingly today receivers, and more particularly transmitters, have digital read-out and control of frequency. Consequently, where the old-time RF designer strove to design a variable frequency oscillator tuned with a mechanical variable capacitor, whose frequency (at any given capacitor setting) was very stable regardless of temperature and supply voltage, the design problem nowadays is usually slightly different. The oscillator is likely to be a voltage-controlled oscillator (VCO) forming part of a *phase lock loop* synthesizer, where the oscillator's output frequency is locked to a stable reference such as a crystal oscillator. Not that synthesizers are the only application for VCOs. Figure 9.15 shows a VCO covering the frequency range 55–105 MHz, which is suitable for the generation of very wide deviation FM.[9] Alternatively, if its output is mixed with a fixed 55 MHz oscillator, a very economical 0–50 MHz sweeper results.

Probably the earliest description of a phase lock loop as such is to be found in the article 'La Reception synchrone' by H. de Bellescize, published in *L'Onde Electronique*, vol. 11, pp. 230–40, June 1932. This described the synchronous reception of radio signals, in which a local oscillator operates at the same frequency as the incoming signal. If the latter is an amplitude modulated wave, such as in MW broadcasting, the audio can be recovered by mixing the received signal with the local oscillator, provided the local oscillator has exactly the same frequency as the carrier and approximately the same phase. This can be achieved by locking the frequency and phase of the local oscillator to that of the signal's carrier – in other words, a phase lock loop. When the local oscillator and the mixer are one and the same stage, as in Figure 8.21, the result is a simple synchrodyne receiver, in which the carrier of the received signal takes control of the frequency and phase of the (just oscillating) detector circuit. In a synthesizer employing a phase lock loop, however, the oscillator, the phase detector (mixer) and the all-important loop filter are all separate, distinct stages.

The operating principle of a synthesizer incor-

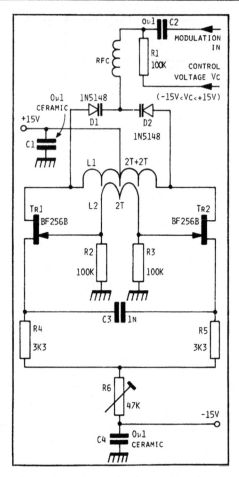

Figure 9.15   A wide range VCO, suitable for generating very wide deviation FM signals (reproduced by courtesy of *New Electronics*).

porating a phase lock loop is indicated in the block diagram of Figure 9.16a. A sample of the output of the oscillator is fed by a buffer amplifier to a variable ratio divider; let's call the division ratio N. The divider output is compared with a comparison frequency, derived from a stable reference frequency source such as a crystal oscillator. An error voltage is derived which, after smoothing, is fed to the VCO in such a sense as to reduce the frequency difference between the variable ratio divider's a output and the comparison frequency. If the comparison is performed by a frequency discriminator, there will be a standing frequency error in the synthesizer's output, albeit small if the loop

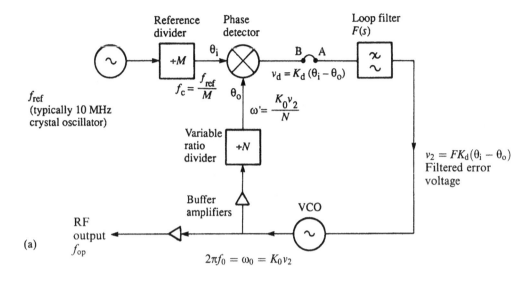

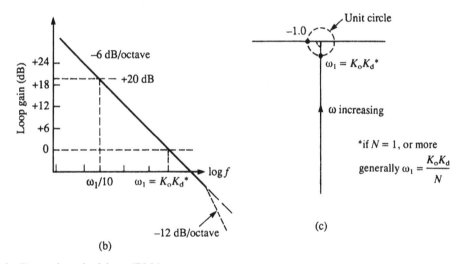

Figure 9.16   Basic phase lock loop (PLL).
(a) Phase lock loop synthesizer.
(b) Bode plot, first-order loop.
(c) Nyquist diagram, first-order loop.

gain is high. Such an arrangement is called a *frequency lock loop* (FLL); these are used in some specialized applications. However, the typical modern synthesizer operates as a phase lock loop, where there is only a standing phase difference between the ratio N divider's output and the comparison frequency The oscillator's output frequency is simply Nf_c, where f_c is the comparison frequency. Thus f_c equals 12.5 kHz gives a simple

means of generating any of the transmit frequencies used in the VHF private mobile radio (PMR) band, used by taxi operators, delivery drivers etc., which is channelized in steps of 12.5 kHz in Europe (15 kHz in North America).

In fact, there is a practical difficulty in that variable ratio divide-by-N counters which work at frequencies up to 150 MHz or more are not available, but this problem is circumvented by the use of a prescaler. If a fixed prescaler ratio, say divide by 10, is used, then the comparison frequency must be reduced to 1.25 kHz to compensate. However, the lower the comparison frequency, the more difficult it is to avoid comparison frequency ripple at the output of the phase comparator passing through the loop filter and reaching the VCO, causing comparison frequency FM sidebands. Of course one could just use a lower cut-off frequency in the loop filter, but this makes the synthesizer slower to settle to the new channel frequency following a change in N. The solution is a 'variable modulus prescaler' such as a divide by 10/11 type. Such prescalers are available in many ratios, through divide by 64/65 up to divide by 512/514, thus a high comparison frequency can still be used: their detailed mode of operation is not covered in this book, as it is a purely digital topic.

A phase lock loop synthesizer is an NFB loop and, as with any NFB loop, care must be taken to roll off all the loop gain safely before the phase shift exceeds 180°. This is easier if the loop gain itself does not vary wildly over the frequency range covered by the synthesizer. In this respect, a VCO whose output frequency is a linear function of the control voltage is a big help (see Figure 9.17).[10] The other elements of the loop equally need to be correctly proportioned to achieve satisfactory operation, so let's analyse the loop in a little more

detail. Returning to Figure 9.16a, the parameters of the various blocks forming the circuit have been marked in, following for the most part the terminology used in what is probably the most widely known treatise on phase lock loops.[11] Assuming that the loop is in lock, then both inputs to the phase detector are at the comparison frequency f_c, but with a standing phase difference $\theta_i - \theta_o$. This results in a voltage v_d out of the phase detector equal to $K_d(\theta_i - \theta_o)$.

In fact, as Figure 9.20 shows, the phase detector output will usually include ripple, i.e. alternating frequency components at the comparison frequency or at $2f_c$, although there are types of phase detector which produce very little (ideally zero) ripple. The ripple is suppressed by the low-pass loop filter, which passes v_2, the DC component of v_d, to the VCO. Assuming that the VCO's output radian frequency ω_o is linearly related to v_2 then $\omega_o = K_o v_2 = K_o F K_d(\theta_i - \theta_o)$, where F is the response of the low-pass filter. Because the loop is in lock, ω' (i.e. ω_o/N) is the same radian frequency as ω_c, the reference. If the loop gain $K_o F K_d/N$ is high, then for any frequency in the synthesizer's operating range, $\theta_i - \theta_o$ will be small. The loop gain must at least be high enough to tune the VCO over the frequency range without $\theta_i - \theta_o$ exceeding $\pm 90°$ or $\pm 180°$ whichever is the maximum range of operation of the particular phase detector being used.

Let's check up on the dimensions of the various parameters. K_d is measured in volts per radian phase difference between the two phase detector inputs. F has units simply of volts per volt at any given frequency. K_o is in hertz per volt, i.e. radians per second per volt. Thus whilst the filtered error voltage v_2 is proportional to the difference in phase between the two phase detector inputs, v_2 directly controls not the VCO's phase but its frequency. Any change in frequency of ω_o/N, however small, away from exact equality with ω_c will result in the phase difference $\theta_i - \theta_o$ increasing indefinitely with time. Thus the phase detector acts as a perfect integrator, whose gain falls at 6 dB per octave from an infinitely large value at DC. It is this infinite gain of the phase detector, considered as a frequency comparator, which is responsible for there being zero net average frequency error

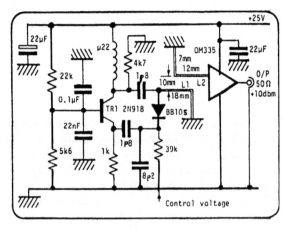

Figure 9.17   A linear VCO (reproduced by courtesy of *Electronic Engineering*).

between the comparison frequency and f_o/N. (Some other writers alternatively consider the VCO as the integrator, producing an output phase which is proportional to the integral of the error voltage. It comes to the same thing either way; it's just that the explanation I have given seems clearer to me. You pays your money and you takes your choice.)

Consider a first-order loop, that is, one in which the filter F is omitted, or where $F = 1$ at all frequencies, which comes to the same thing. At some frequency ω_1 the loop gain, which is falling at 6 dB/octave due to the phase detector, will be unity (0 dB). This is illustrated in Figure 9.16b and c, which show the critical loop unity-gain frequency ω_1 on both an amplitude (Bode) plot and a vector (Nyquist) diagram. To find ω_1 in terms of the loop parameters K_o and K_d without resort to the higher mathematics, break the loop at B, the output of the phase detector, and insert at A a DC voltage exactly equal to what was there previously. Now superimpose upon this DC level a sinusoidal signal, say 1 V peak. The resultant peak FM deviation of ω_o will be K_o radians per second. If the frequency of the superimposed sinusoidal signal were itself K_o radians/second, then the modulation index would be unity, corresponding to a peak VCO phase deviation of ± 1 radian. This would result in a phase detector output of K_1 volts (assuming for the moment that $N = 1$). If we increase the frequency of the input at A from K_o to $K_o K_d$, the peak deviation will now be $1/K_d$ and so the voltage at B will be unity. So the loop unity-gain frequency ω_1 is $K_o K_d$ radians/second, or, more generally, $K_o K_2/N$, as shown in Figure 9.16b and c. With a first-order loop there is no independent choice of gain and bandwidth; quite simply $\omega_1 = K_o K_d/N_1$. We could reintroduce the filter F as a simple passive CR cutting off at a corner frequency well above ω_1, as indicated by the dashed line in Figure 9.16b and by the teacup handle at the origin in Figure 9.16c, to help suppress any comparison frequency ripple. This technically makes it a low-gain second-order loop, but it still behaves basically as a first-order loop provided that the corner frequency of the filter is well clear of ω_1 as shown.

Synthesizers usually make use of a high-gain

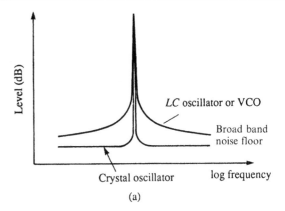

(a)

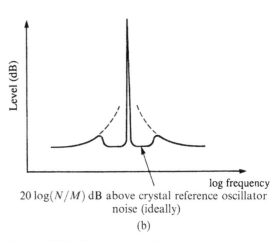

(b)

Figure 9.18 Purity of radio-frequency signal sources.
(a) Comparison of spectral purity of a crystal and an *LC* oscillator.
(b) At low-frequency offsets, where the loop gain is still high, the purity of the VCO (a buffered version of which forms the synthesizer's output) can approach that of the crystal derived reference frequency, at least for small values of N/M.

second-order loop and I will examine that in a moment, but first a word as to why this type is preferred. Figure 9.18a compares the close-in spectrum of a crystal oscillator with that of a mechanically tuned *LC* oscillator and a VCO. Whereas the output of an ideal oscillator would consist of energy solely at the wanted output frequency f_o, that of a practical oscillator is accompanied by undesired noise sidebands, representing minute variations in the oscillator's amplitude and frequency. In a crystal oscillator these are very low, so the noise sidebands, at 100 Hz either side, are typically -120 dB relative to the wanted output, falling to a noise floor further out of about -150 dB. The Q of an LC tuned circuit is only about one-hundredth of the Q of a crystal, so the noise of a well-designed LC oscillator reaches -120 dB at more like 10 kHz off tune. In principle a VCO using a varicap should not be much worse than a conventional LC oscillator, provided the varicap diode has a high Q over the reverse bias voltage range; however, with the high value of K_o commonly employed (maybe as much as 10 MHz/volt), noise on the control voltage line is a potential source of degradation. Like any NFB loop, a phase lock loop will reduce distortion in proportion to the loop gain. 'Distortion' in this context includes any phase deviation of ω' from the phase of the reference f_c. Thus over the range of offset from the carrier for which there is a high loop gain, the loop can clean up the VCO output substantially, as illustrated in Figure 9.18b. In fact, the high loop gain will force θ_o, the instantaneous phase of ω', the output of the divider N, to mirror almost exactly ω_i, the instantaneous phase of the output of the reference divider M. Thus the purity of ω' will equal that of ω_c, the reference input to the phase detector. It follows (see Figure 8.1e and associated text) that the VCO output phase noise should be reduced to only N times that of the reference input to the phase detector, i.e. N/M times that of the crystal reference oscillator itself, as indicated in Figure 9.18b. This assumes that the phase noise sidebands of f_c are in fact 20 log M dB below the phase noise of the crystal reference f_{ref}. Unfortunately this is often not so, since dividers, whether ripple dividers or synchronous (clocked) dividers, are not themselves noise free, owing to inevitable jitter on the timing edges. Likewise, the variable ratio divider will not in practice be noise free, so that the close-in VCO phase noise in Figure 9.18b will be rather more than 20 log(f_o/f_{ref}) dB higher than the phase noise of the crystal reference.

A second-order loop enables one to maintain a high loop bandwidth up to a higher frequency, by rolling off the loop gain faster. Consider the case where the loop filter is an integrator as in Figure 9.19c; this is an example of a high-gain second-

order loop. With the 90° phase lag of the active loop filter added to that of the phase detector, there is no phase margin whatever at the unity gain frequency; as Figure 9.19b shows, disaster (or at least instability) looms at ω_1. By reducing the slope of the roll-off in Figure 9.19a to 6 dB per octave before the frequency reaches ω_1 (dashed line), a phase margin is restored, as shown dashed in Figure 9.19b, and the loop is stable. This is achieved simply by inserting a resistor R_2 in series with the integrator capacitor C at X–Y in Figure 9.19c. This is the active counterpart of a passive transitional lag. If $R_1 = (\sqrt{2})R_2$, then at the corner frequency $\omega_c = 1/CR_2$ the gain of the active filter is unity and its phase shift is 45°, whilst at higher frequencies it tends to -3 dB and zero phase shift. If ω_c equals K_oK_d/N, then ω_1 (the loop unity-gain frequency) is unaffected but there is now a 45° phase margin. As Figure 9.19b shows, at frequencies well below ω_1, the loop gain climbs at 12 dB/octave accompanied by a 180° phase shift, until the opamp runs out of open loop gain. This occurs at the frequency ω, where $1/\omega C$ equals AR_1, where A is the open loop gain of the opamp; an opamp integrator only approximates a perfect integrator. Below that frequency the loop gain continues to rise for evermore, but at just 6 dB/octave with an associated 90° lag due to the phase detector which, as noted, is a perfect integrator. This change occurs at a frequency too low to be shown in Figure 9.19a; it is off the page to the top left. I have only managed to show it in Figure 9.19b by omitting chunks of the open loop locus of the tip of the vector.

For a high-gain second-order loop, analysis by the root locus method[12] shows that the damping (phase margin) increases with increasing loop gain; so provided the loop is stable at that output frequency (usually the top end of the tuning range) where K_o is smallest, then stability is assured. This is also clear from Figure 9.19. For if K_o or K_d increases, then so will ω_1, the unity-gain frequency of the corresponding first-order loop. Thus ω_c (the corner frequency of the loop filter) is now lower than ω_1, so the phase margin will now be greater than 45°.

Having found a generally suitable filter, let's return for another look at phase detectors and

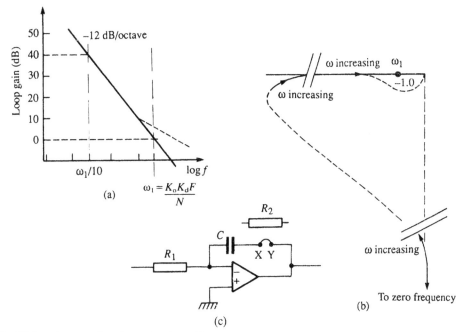

Figure 9.19 PLL with second-order active loop filter.

VCOs. Figure 9.20 shows several types of phase detector and indicates how they work. The logic types are fine for an application such as a synthesizer, but not so useful when trying to lock onto a noisy signal, e.g. from a distant tumbling spacecraft; here the ex-OR type is more suitable. Both pump-up/pump-down and sample-and-hold types exhibit very little ripple when the standing phase error is very small, as is the case in a high-gain second-order loop. However, the pump-up/pump-down types can cause problems. Ideally, pump-up pulses – albeit very narrow – are produced however small the phase lead of the reference with respect to the variable ratio divider output; likewise, pump-down pulses are produced for the reverse phase condition. In practice, there may be a very narrow band of relative phase shift around the exactly in-phase point, where neither pump-up nor pump-down pulses are produced. The synthesizer is thus entirely open loop until the phase drifts to one end or other of the 'dead space', when a correcting output is produced. Thus the loop acts as a bang-bang servo, bouncing the phase back and forth from one end of the dead space to the other – evidenced by unwanted noise

sidebands. Conversely, if both pump-up and pump-down pulses are produced at the in-phase condition, the phase detector is no longer ripple free when in lock and, moreover, the loop gain rises at this point. Ideally the phase detector gain K_d should, like the VCO gain K_o, be constant. Constant gain, and absence of ripple when in lock, are the main attractions of the sample-and-hold phase detector.

In the quest for low noise sidebands in the output of a synthesizer, many ploys have been adopted. One very powerful aid is to minimize the VCO noise due to noise on the tuning voltage, by substantially minimizing K_o, to the point where the error voltage can only tune the VCO over a fraction of the required frequency range. The VCO is pretuned by other means to approximately the right frequency, leaving the phase lock loop with only a fine tuning role. Figure 9.21 shows an example of this arrangement.[13]

Noise generators

Noise generators are widely used for measurements on receivers, frequency division multiplex

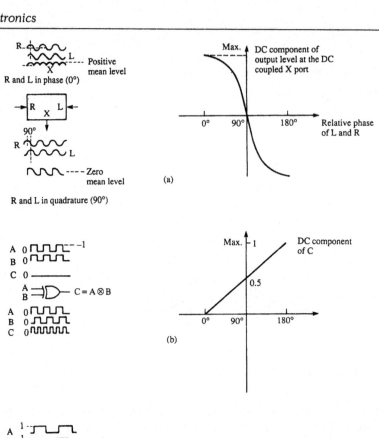

(a)

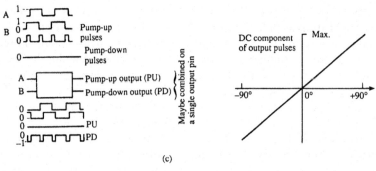

(b)

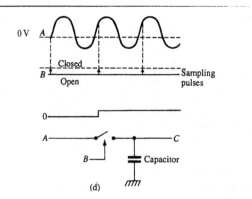

(c)

(d)

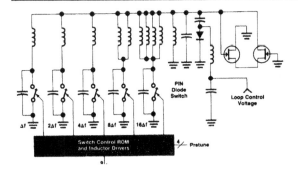

Figure 9.21 This VCO used in the HP8662A synthesized signal generator is pretuned to approximately the required frequency by the microcontroller. The PLL error voltage therefore only has to tune over a small range, resulting in spectral purity only previously attainable with a cavity tuned generator, and an RF settling time of less than 500 μs.
(Reproduced with permission of Hewlett-Packard Co.)

(FDM) lines and radio circuits, and so on. Figure 9.22 shows a variety of these. In Figure 9.22a a diode with a plain tungsten filament provides the source of noise. The tungsten filament provides a less copious supply of electrons than the usual rare earth activated cathode, so that there is no cloud of electrons surrounding it forming a virtual cathode, as in a conventional valve. This temperature limited mode of operation also avoids some sources of additional noise associated with coated cathodes. The number of electrons emitted per second (i.e. the anode current) is controlled by adjusting the filament temperature by means of the rheostat shown. All the electrons are accelerated by the HT voltage and fall directly onto the anode. This is the source of the noise, and its intensity is proportional to the rate of arrival of the individual electrons, like lead shot falling on a corrugated iron roof. The advantage of the thermionic noise diode compared with most other sources is that its noise (current) output can be predicted on theoretical grounds from the formula for shot noise

derived by Schottky. This equation, valid over the frequency range from 10 kHz to the frequency at which transit time effects become important, is

$$i_\mathrm{n}(\mathrm{RMS}) = \sqrt{(2eI_aBW)}$$

where e is the charge on an electron (1.6×10^{-19} coulomb), I_a is the DC anode current in amperes, and BW is the noise bandwidth in hertz. If the anode current is passed through a 50 Ω resistor as in Figure 9.22a, a wide band noise source with a matched output impedance is obtained. A typical example is the Rohde and Schwarz noise generator type SKTU, which provides a noise output at 50 Ω of up to 15 dB above thermal, substantially flat up to 1000 MHz.

In contrast, the once popular neon noise source of Figure 9.22b is flat only up to about 20 kHz, limiting its use to audio frequencies. The Zener noise source of Figure 9.22c provides a wider bandwidth, providing a flat frequency distribution of noise power up to 100 kHz or more. If the base/emitter junction of a 2N918 transistor in reverse breakdown is substituted for the Zener, a generally flat noise spectrum up to 800 MHz is obtained. However, this is spoilt for any serious use as a wide band noise source by the odd noise peak of 5 or 10 dB above the general level, at unpredictable discrete points in the band. This problem does not arise with certain more modern transistors such as the BFR90A, and Reference 14 describes an RF noise generator providing a level output more than 30 dB above thermal up to 1000 MHz. Figure 9.22d shows a noise source which is flat over the 20 Hz to 20 kHz audio-frequency band. It provides a low-impedance high-level noise output due to the 60 dB of gain applied to its own input noise. Many opamps exhibit a noise level which is flat across the whole of the audio band down to a particular frequency, below which the noise becomes inversely proportional to frequency: this is called the $1/f$ noise corner. The gain determining

Figure 9.20 Phase detectors used in phase lock loops (PLLs).
  (a) The ring DBM used as a phase detector is only approximately linear over say ±45° relative to quadrature.
  (b) The exclusive-OR gate used as a phase detector.
  (c) One type of logic phase detector.
  (d) The sample-and-hold phase detector. In the steady state following a phase change, this detector produces no comparison frequency ripple.

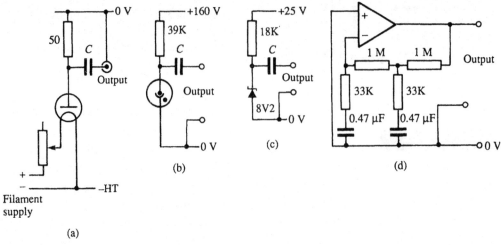

Figure 9.22 Noise generators.
(a) Thermionic.
(b) Neon tube.
(c) Zener diode.
(d) Opamp.

feedback network shown in Figure 9.22d rolls off the gain at low frequency, being 6 dB down at 10 Hz. It is therefore essential to choose an opamp with a suitably low $1/f$ noise corner, say 20 Hz or less.

Figure 9.23a shows a digital noise generator based upon a maximal length pseudo-random bit stream (PRBS) generator. If some of the outputs of a clocked shift register are exclusive-ORed together and applied at the input of the register, then a repeating bit pattern will be produced at the output. If the correct choice of outputs to be gated has been made, the pattern will last for $2^N - 1$ clock periods before repeating, this being the longest unique pattern obtainable from an N-stage register. For most lengths of register N it is possible to find a feedback connection requiring the ex-OR function of just two outputs, and a number of noise generators based upon this principle have appeared over the years.[15,16] That shown in Figure 9.23a incorporates a 20-tap finite impulse response (FIR) filter to provide a 'Gaussian' amplitude distribution.[17]

A PRBS provides only an approximation to *white noise* (noise with a flat amplitude characteristic over some specified frequency range). With the 100 kHz clock frequency shown, the 23-stage

register repeats after 84 seconds, so the noise spectrum consists of discrete spectral lines spaced 1/84th of 1 Hz apart – for all practical purposes a continuous spectrum. At the other end of the spectrum, a PRBS is 'white' up to $f_{clock}/2\pi$. However, the output is simply a stream of digital 0s and 1s, i.e. it is always at one of two discrete levels. The filter converts this to an approximation to the bell-shaped normal or Gaussian amplitude distribution. Longer runs of successive 0s or 1s will result in a larger voltage swing at the filter output than alternate 0s and 1s, since the latter look like a 50 kHz square wave, which is reduced in amplitude by the roll-off of the filter. Ideally, the filter should start to roll off at f_{clock}/N to approximate the normal curve out to the largest possible peak/mean ratio.

When reproduced through a loudspeaker system, white noise, with its contant power per unit bandwidth, sounds like the hiss of escaping steam, since the ear, like a typical tunable audio-frequency filter, reacts to the power per percentage bandwidth, e.g. power per octave. Where, for measurement or other purposes, it is necessary to convert white noise to noise having a constant power per octave (called *pink noise*), a filter with a roll-off rate of 3 dB per octave is required. This can be approximated by a

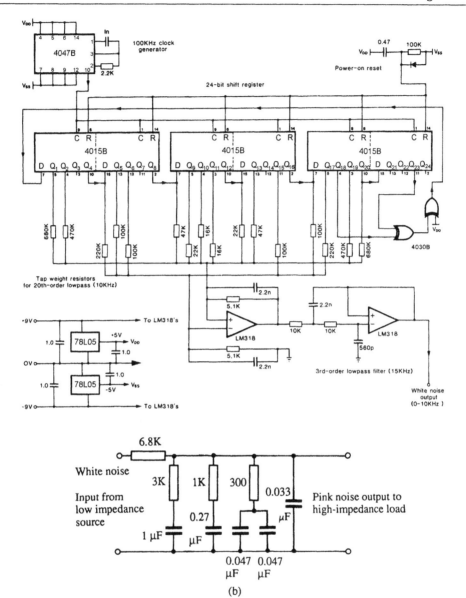

(b)

Figure 9.23 More noise circuits.
(a) An audio-frequency white noise generator with built-in digital filter (reproduced by courtesy of *Electronic Product Design*).
(b) Pink noise filter, flat within ±0.5 dB of true −3 dB/octave response from 20 Hz to 20 kHz.

filter with alternating flat and −6 dB/octave sections, as shown in Figure 9.23b.

This review of analog circuitry used in signal sources has included single-frequency generators, both sinusoidal and otherwise, and generators of signals covering a wide band of frequencies, i.e noise generators. Many other interesting techniques are used in sources of much higher frequencies, such as microwaves, but are outside the scope of this book.

References

1. Voltage Reference Circuit, I. March, p. 22, *Electronic Product Design*, April 1981.
2. Triangle Generator Reads Frequency Directly, I. March, p. 24, *New Electronics*, 13 August 1985.
3. New Sine Shaper, p. 37, *Hewlett Packard Journal*, Vol. 34, no. 5, June 1983.
4. RC Oscillators, Single-element Frequency Control, P. Williams, p. 82, *Wireless World*, December 1980.
5. Easily Tuned Sinewave Oscillators, R. C. Dobkin, p. 116, *Electronic Engineering*, December 1971.
6. FET Controlled Oscillator, I. Hickman, p. 19, *New Electronics*, 25 November 1986.
7. Good-tempered LC Oscillator, G. W. Short, p. 84, *Wireless World*, February 1973.
8. SC-Cut Quartz Oscillator Offers Improved Performance, J. R. Burgoon, R. L. Wilson, p. 20, *Hewlett Packard Journal*, Vol. 32, no. 3, March 1981.
9. A Wide Range VCO, I. March, p. 19, *New Electronics*, 14 December 1982.
10. A Linear Voltage Controlled Oscillator, J. Dearden, p. 26, *Electronic Engineering*, June 1983.
11. *Phaselock Techniques*, F.M. Gardner, John Wiley and Sons Inc., 1966.
12. *Automatic Feedback Control System Synthesis*, J. G. Truxal, McGraw Hill, New York, 1955.
13. Low-Noise RF Signal Generator Design, p. 12, *Hewlett Packard Journal*, Vol. 32, no. 2, February 1981.
14. White noise–white knight, Ian Hickman, *Electronics World*, November 1993, pp. 956–959, reproduced in Hickman's *Analog and RF Circuits*, Newnes Butterworth-Heinemann 1998, ISBN 0 7506 3742 0.
15. White Noise Generator, H. R. Beastall, p. 127, *Wireless World*, March 1972.
16. Wide-Range Noise Generator, I. Hickman, p. 38, *Wireless World*, July 1982.
17. White Noise Generator With Built-In Digital Filter, J. R. Trinder, p. 30, *Electronic Product Design*, December 1983.

Questions

1. What two separate factors must be taken into account when calculating the change of voltage ΔV across a Zener reference diode resulting from a change of current ΔI through it?
2. Name and describe the two separate feedback paths found in most astable (relaxation) oscillators. Describe in detail the operation of the unique Bowes, White or emitter coupled relaxation, with its single feedback path.
3. Show how any arbitrary waveform can be generated with 'P and Q' DACs, plus appropriate data stored in ROM or RAM.
4. Draw the circuit diagram of a basic Wien bridge oscillator using an opamp. What modification to the two-resistor negative feedback arm is necessary to ensure linear operation at a fixed amplitude?
5. Design a state variable filter based audio oscillator to cover the frequency range 0–10 kHz.
6. Add an additional stage to your design in question 5, to provide an alternative square wave output.
7. How does the manufacture of a strain compensated crystal differ from that of crystals with other cuts, such as AT etc.? What additional problem is encountered in the application of an SC cut crystal?
8. Derive from first principles the loop equation for a first-order phase lock loop. Why is a second order loop usually the preferred design?
9. Sketch the output voltage versus input phase characteristics of the following types of phase detector, (i) double balanced diode mixer, (ii) exclusive OR gate, (iii) sample-and-hold circuit.
10. A thermionic noise generator with a 50 Ω resistive source impedance is run at 5 mA anode current. What is the RMS output voltage delivered into a 50 Ω load, measured in a 10 kHz bandwidth?

Chapter

10 Power supplies

It is undoubtedly true to say that most electronic equipment is powered during normal operation from a supply of mains electricity, provided by the local or national public supply utility. However, there is a small but significant demand for the operation of equipment from battery supplies, for a number of different reasons. The most obvious is portability, a desirable feature of items as diverse as a cheap 'tranny' portable and a sophisticated frequency hopping radio transceiver used in modern military exercises (war games) and doubtless also for real.

Batteries

These two applications also neatly illustrate the two different kinds of battery that are available. The tranny will probably use a tiny 006P (PP3, 6F22, 1604 etc.) 9 V *primary* battery (see Appendix 6) of the *dry* or zinc-carbon variety, these terms indicating a battery made up of Leclanché cells. These cells have a voltage of 1.5 V when new, falling steadily during their useful life. The military radio, on the other hand, will usually (but not always) use a *secondary* battery – i.e. one which, unlike primary batteries, is rechargeable. The best known secondary battery is the lead-acid battery, with a fully charged voltage of about 2.2 V/cell, universally employed in cars and trucks for starting and lighting. This is a *wet* battery, though versions with a jelled electrolyte are available for use in applications where the battery may not remain upright throughout its working life. The military radio is more likely to use a secondary battery composed of nickel-cadmium (Ni-Cd or nicad) cells, which have a fully charged voltage of around 1.2 V each.

As already noted, the voltage of dry batteries falls steadily during use, from the new value of 1.5 V/cell to some arbitrary end of useful life voltage which is determined by the designer of the host equipment. For example, in some professional applications such as telephone network test equipment, dry batteries are regarded as having reached the end of their useful service life when the on-load voltage has fallen to 1.1 V per cell. On the other hand, portable battery operated record players, shavers, toys and similar wonders of this hi-tech age are often designed to operate to specification down to 0.6 V/cell: this provides a substantial increase in the useful life, at the expense of reliability of operation.

In professional test equipment powered by dry batteries it is common to use a *stabilizer* to provide the circuitry with a constant operating voltage over the service life of the batteries, regardless of the ambient temperature. Clearly such a stabilizer should operate with the minimum of 'headroom', so as to provide the designed output voltage down to the lowest possible battery voltage, and should draw the minimum amount of 'housekeeping' current, so that virtually all the current drawn from the battery is supplied to the load. The circuit of Figure 10.1 was designed to meet these requirements, providing a 12 V output at up to 30 mA load current (with constant current limiting for overload protection) from two PP9 (6F100, M-603) batteries down to an end point of 12.4 V or 1.03 V-cell.[1] The circuit is a conventional series stabilizer with feedback. The pass transistor Tr_3 is used in the common emitter mode so that its bottoming voltage rather than its base/emitter voltage sets the minimum voltage drop lost across it; in a circuit such as this, one does not have the luxury of auxiliary supplies as one could in mains operated equipment. R_3 is capable of supplying enough base current to the pass transistor to meet the maximum rated load. The differ-

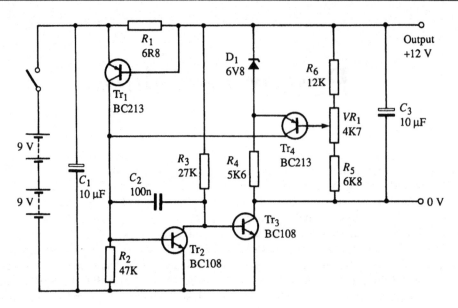

Figure 10.1 Voltage stabilizer for battery operated equipment (reproduced by courtesy of *Electronic Engineering*).

ence amplifier Tr_4 compares a sample of the output voltage with the Zener reference voltage and turns on Tr_2 as required to reduce the base current of the pass transistor to the appropriate level. Note that the reference voltage across D_1 is entirely independent of the state of the battery voltage, since it is supplied via R_4 from the stabilized voltage. Supplying the reference voltage from the stabilized output is a convenient ploy but can sometimes give rise to start-up problems on switch-on, depending on the particular circuit arrangement. For example it may be bistable, with first no output, no reference, and then no reference, no output. That problem cannot arise here due to the direct supply of pass transistor base current via R_3, but it is used to good effect in the circuit of Figure 10.3, of which more in a moment. If the output is short-circuited or overloaded, the voltage drop across R_1 turns on Tr_1. This turns on Tr_2, which reduces the base drive to the pass transistor sufficiently to limit the output current to about 80 mA.

The 6F100 style has the largest capacity of any of the readily available dry batteries and should always be first choice for dry battery operated equipment, if it is at all possible to accommodate

its not inconsiderable size. For, in terms of cost per milliampere hour of useful service life, it is by far the cheapest solution, except in a few specialized applications where the load current is so small (microamperes) that a 6F100 battery would become unserviceable due to expiry of its shelf life rather than exhaustion. Under favourable conditions, dry batteries can remain serviceable for two or three years, but routine replacement after one year is recommended. Note also that dry cells become unserviceable due to the rise in internal resistance. So if the design load current is doubled, the useful service life of a given style of dry battery in that application is considerably less than half, as can be seen from Figure 10.2, which has been drawn up from the published performance data for the premium quality 6F100 style batteries produced by a well-known UK manufacturer. It also shows the surprising longevity of this battery, when operated in an intermittent use regime: if, when you come to use a piece of battery operated equipment, you invariably seem to find that the batteries are flat, this is probably due not so much to Murphy's law as to the failure of the previous user to switch the kit off after use.

The circuit of Figure 10.3 was designed specific-

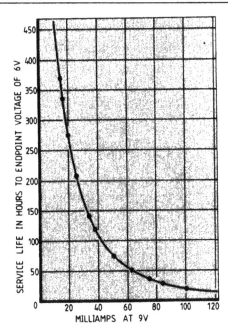

Figure 10.2 Battery service life versus (constant) load current (reproduced by courtesy of *Electronics World*).

ally to extend battery life by turning the supply off after a fixed period of use, whilst at the same time producing a stabilized output voltage.[2] The 5.6 V reference is derived from the stabilized output voltage, so that if the BC109 pass transistor is cut off, then the BC214 error amplifier cannot supply any base current to it: the circuit is bistable.

The push-button S_1 does duty instead of a conventional on/off switch. When it is pressed, base current is supplied to the pass transistor via R_1 and a diode, switching the circuit on. Via the other diode, the CD4060 counter (with built-in clock oscillator) is reset. When a count of 2^{13} is reached, the output at pin 3 of the CD4060 rises to the positive rail, switching off the PNP transistor via the diode. The 10 kΩ resistor at pin 3 of the IC is necessary to guarantee the switch-off of the PNP, since the P channel output device in the CD4060 cannot achieve this unaided when the voltage between pins 8 and 16 falls to a low value. In use, the on period can be updated to the full value at any time simply by pressing the button again. With no load connected, the output voltage will equal the battery voltage whilst the on push-button is closed. This applies equally at switch-on and when updating the on period. However, given the load current drawn by the instrument, a value for R_1 can be chosen which will reliably initiate the circuit without the output voltage exceeding the designed value.

One of the advantages of secondary batteries (apart from being rechargeable), and of some types of primary batteries, over the ubiquitous dry battery is a smaller degree of variation of terminal voltage over their useful life. However, even nicad batteries show a significant variation of terminal voltage over the full discharge cycle between charges, if ambient temperature variations are

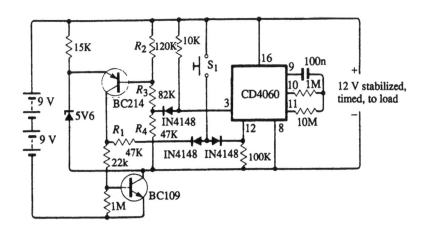

Figure 10.3 Combined stabilizer and time-out circuit (reproduced by courtesy of *Electronics World*).

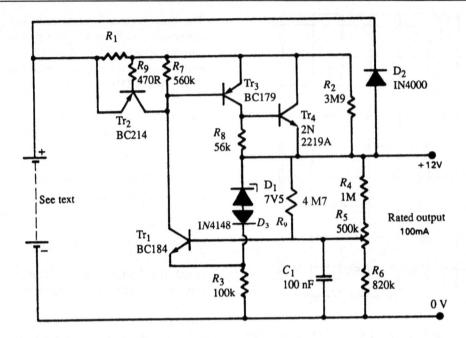

Figure 10.4   Stabilizer for use with rechargeable batteries, protected against extended short-circuit (reproduced by courtesy of *New Electronics*).

taken into account as well. Thus it is not unusual to include a stabilizer even in instruments powered by rechargeable batteries such as nicads. When easy and rapid fitting of fresh batteries is required, a clip-on battery pack is used, rather than fitting separate batteries into an internal battery compartment in the equipment. A novel idea in this case is to incorporate the stabilizer circuit in the battery pack rather than in the instrument. This has the advantage that the battery pack can be protected from damage or destruction by inadvertent short-circuit when not fitted to an instrument, e.g. during storage or handling. The circuit of Figure 10.4 does just this.[3] R_2 ensures stabilizer start-up on first fitting batteries and after removal of a short-circuit across the output, by raising the base of Tr_1 above 0.6 V. On short-circuit the current is initially limited to about 1 A by the action of Tr_2 and the 0.5 Ω resistor R_1; the circuit will reliably turn on into a load which includes a 2000 μF capacitor across the supply. On an extended short-circuit, however, C_1 discharges, Tr_1 turns off and the available short-circuit current falls to 4 μA or so. This is actually much less than

the 55 μA housekeeping current drawn when there is no load. The housekeeping current will consume about 15% of the capacity of a 3 AH battery in one year, or less than 1.5% if the battery pack is stored short-circuited. This is a smaller proportion of the charge than the battery could be expected to lose due to normal self-discharge in a year in any case. Sufficient cells must be included to ensure adequate headroom for the stabilizer down to end of life (primary battery) or end of discharge cycle (secondary battery). Where secondary cells are used, D_2 is included to permit recharging.

Rectifiers

Turning now to AC mains power, most equipment thus powered needs a supply of direct current. This is provided in the first instance by a raw supply' which turns the AC into DC. There are two basic means of obtaining a direct current supply from the alternating supply voltage, namely capacitor input and choke input rectifier circuits – of which the former is by far the commonest.

Figure 10.5 shows three *capacitor input rectifier*

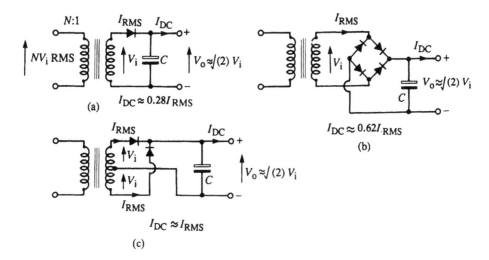

Figure 10.5 Various capacitor input rectifier circuits, showing the permissible load DC for a transformer rated at I_{RMS} amperes.
(a) Half-wave rectifier.
(b) Full-wave (bridge) rectifier.
(c) Full-wave (biphase) rectifier. Centre tapped secondary winding provides $V_i - 0 - V_i$ volts. Each half of the secondary is rated at I_{RMS}. V_o will approach $\sqrt{(2)}V_i$ off load.

circuits together with the current and voltage relationships therein. Note that these are approximate, making no allowance for the full-load voltage drop within the transformer due to winding resistance or voltage drops across the rectifiers. Figure 10.6 shows several more rectifier circuits which are used in more specialized applications. The voltage doubler in Figure 10.6a provides twice the output voltage that a half-wave or bridge rectifier would provide from a given winding. It consists in effect of two half-wave outputs in series, each topping up its own reservoir capacitor on alternate half-cycles. Thus although the ripple at the output is at twice supply frequency, the regulation is worse than for a true full-wave circuit: further, the effective size of the capacitor is only $C/2$, since there are two of them in series. The circuit of Figure 10.6b can be regarded as a full-wave voltage doubler, as is clear from comparison with Figure 10.6a, but equally it is simply a bridge circuit with centre tapped transformer and split capacitor. It is frequently used, with the centre tap regarded as $0V$, as a two-rail supply providing (say) $+40V$ and $-40V$ to a hi-fi amplifier.

If the junction of the two diodes in Figure 10.6a

is supplied not direct from a transformer winding, but via a capacitor as in Figure 10.6c, the junction cannot fall below $0V$ by more than the forward voltage drop of the lower diode. The input alternating voltage is said to be *DC restored* positive-going with respect to $0V$. Thus the output will be a voltage level positive with respect to $0V$ by an amount equal to the peak-to-peak value of the input voltage, regardless of the waveform of the latter. In practice, ordinary (non-reversible) electrolytic capacitors are frequently used in this circuit (in which case the polarities should be as marked), despite the fact that if the first half-cycle of input voltage at switch-on should be positive-going, the input capacitor will be temporarily reverse biased. They always seem to survive this, but in a very high-voltage circuit it might be worth while taking the precaution of connecting an auxiliary diode in parallel with the input capacitor, its cathode connected to the capacitor's positive terminal.

If the connection between points A and B in Figure 10.6c is broken, A can be connected to another supply, e.g. $+100V$, thus providing an even higher output voltage. The process can be

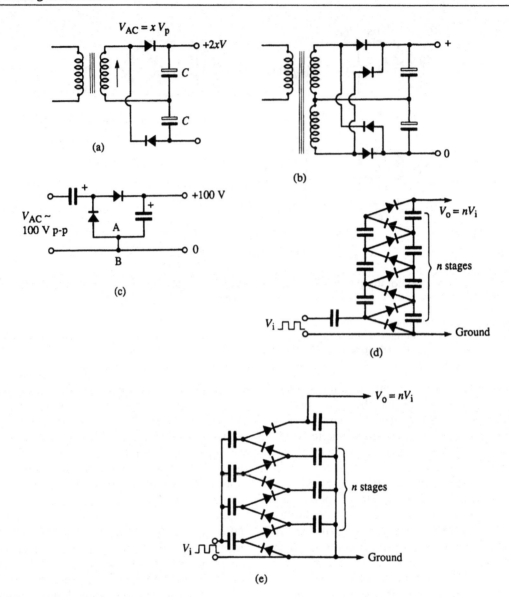

Figure 10.6   Other rectifier circuits.
   (a) Voltage doubler.
   (b) Full-wave voltage doubler.
   (c) Restoring voltage doubler.
   (d) Series (Cockcroft-Walton) voltage multiplier.
   (e) Parallel multiplier.
(Parts (d) and (e) reprodued by courtesy of *New Electronics*.)

repeated indefinitely (in theory!) to supply an output voltage greatly in excess of the peak-to-peak value of the AC input voltage. This is called a *Cockcroft-Walton multiplier* and is illustrated in

Figure 10.6d. The arrangement is widely used in power engineering to raise voltages of hundreds of kilovolts or more, for proof voltage testing of transformers, circuit breakers etc. It has the great advantage that the voltage rating of each rectifier and capacitor individually need only exceed the peak-to-peak input voltage by a modest safety margin. The disadvantage is that with so many capacitors effectively in series, the output resistance is high, leading to substantial voltage sag as soon as any load current is drawn, unless very large-value capacitors are used. In lower-voltage applications, such as are commonly encountered in electronics, capacitors rated at the full output voltage can be used as in the arrangement of Figure 10.6e.[4] This avoids the high output resistance problem. Further, as all the capacitors are charged in parallel, each diode carries the same current, unlike the circuit in Figure 10.6d, where the two input diodes carry N times the current of the two output stage diodes.

Returning to Figure 10.5, the half-wave circuit makes the worst use of the copper in the transformer and the bridge circuit makes the best. For although the DC/RMS current ratio is unity for the biphase case as against only 0.62 for the bridge, for the same total secondary voltage, the biphase circuit provides only half the output voltage obtained with the bridge connection. The rectifier current flows in pulses at the crests of the input voltage waveform, as shown in Figure 10.7a. The better the regulation (the lower the per unit winding resistance) of the transformer and the larger the reservoir capacitor, the higher the current peaks, the narrower the current pulses and the nearer the output voltage approaches $(\sqrt{2})V_i$, but the higher the ratio of I_{RMS} to I_{DC}. For small transformers such as those commonly used in electronic equipment, the regulation (percentage voltage drop from off-load to rated full resistive load) is usually in the range 7 to 10% for 100 VA rated transformers, rising to as much as 25 to 30% for very small transformers rated at 1 or 2 VA. As a rough and ready rule of thumb, the smoothing capacitance for a low-voltage full-wave (bridge or biphase) rectifier circuit with a 50 or 60 Hz input should not be less than 2200 μF per ampere of DC load current: on the other hand there is little point in going beyond 4700 μF/A except in the case of very low-voltage supplies. More detailed design data, usually based upon graphical data that first appeared in print as long

ago as 1943,[5] can be found in the appropriate application data book of any of the manufacturers of power semiconductor diodes and rectifiers.[6]

Choke input filters make use of an inductor to increase the angle of conduction of the rectifiers from 20° or so to 180°, thus greatly reducing the peak current and considerably increasing the DC current that can be supplied by a transformer secondary of a given RMS current rating. Figure 10.7b shows two full-wave rectifier circuits dating from the days of valves, but the thermionic double diode could as easily be a pair of silicon rectifier diodes or equally a bridge circuit. Nowadays, the capacitor input circuit would usually forgo the luxury of a smoothing choke and smoothing capacitor, using just the reservoir capacitor C_{R}, but the choke input circuit is entirely representative. Figure 10.7c shows the voltage that would appear across the choke (shaded) on the assumption that the voltage at the output of the choke (i.e. the voltage across the reservoir capacitor) is perfectly constant, as would be the case if the capacitor were infinitely large. The voltage at the input to the choke is simply the full-wave rectified sine wave and the value of output voltage is the average of the input voltage, or $2/\pi$ times the peak input.

Now the voltage across an inductor is proportional to the rate of change of current through it; so to support a ripple voltage equal to that indicated, the magnetizing (ripple) current through the choke must have the waveform shown in Figure 10.7d, this ripple being superimposed upon the DC component of current drawn by the load. The larger the inductance, the smaller the ripple current will be as a percentage of the full-load current. On the other hand, if the DC load current is progressively reduced, a point will be reached where it is less than the mean value of the magnetizing current. At this point – called the *critical current* – the magnetizing current must cease entirely at the troughs, and the output voltage will start to rise. Ultimately, with no load current at all, the choke might as well not be there and the output voltage will be the same as for a capacitor input circuit on no load (see Figure 10.7e, which also shows the lower effective output impedance of the choke input circuit). If

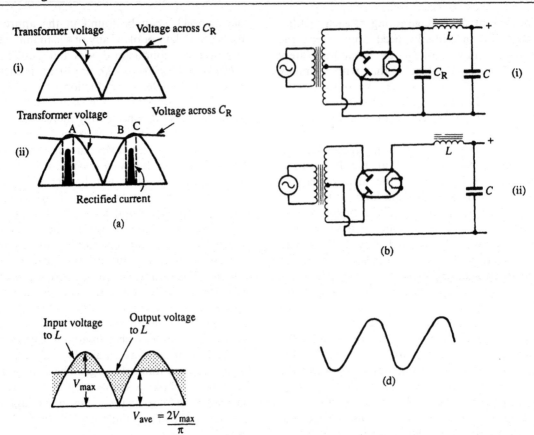

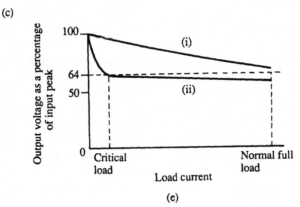

Figure 10.7   Capacitor and choke input supplies compared.
   (a) Working conditions of capacitor input circuit with (i) no current; (ii) load current.
   (b) Full-wave rectifier circuits: (i) capacitor input; (ii) choke input.
   (c) Approximate voltage relationships in a choke input circuit with very large choke inductance, when some current is flowing.
   (d) Waveform of ripple current corresponding to voltage waveform (c).
   (e) Output voltage/current curves for (i) capacitor; (ii) choke input.
(Reproduced by courtesy of *Electronics World*.)

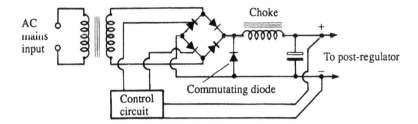

Figure 10.8 Choke input phase-controlled preregulator. Note that only two of the bridge rectifier diodes need to be replaced by SCRs.

the load current is not fixed but may vary considerably, even down to zero, it is advisable to take some precautions to avoid the large rise in output voltage below the critical current. One simple step is to fit a bleed resistor or 'bleeder' to draw a current equal to the critical current. The resultant waste of power can be minimized by using a choke with a very large inductance, to make the required magnetizing current, and hence the critical current, very small. However, a larger inductance will be more expensive; worse, it will have a higher winding resistance, increasing the output impedance of the raw supply. But the lower output impedance is one of the main attractions of the choke input filter in the first place. Fortunately, the necessary lateral thinking has already been done for us; the choke need only have a high inductance at the critical current. At full rated output current, its inductance may have fallen, due to saturation, by a factor of ten or so, but the only result will be a larger ripple current than would otherwise have been the case. A choke with this characteristic is called a *swinging choke*, indicating that its effective value varies according to the circuit conditions. In addition to its lower variation in output voltage with load current, the choke input filter improves the efficiency of utilization of the mains transformer by reducing the RMS current required for a given DC output current. This is brought about by the increased conduction angle of the rectifiers.

Far from being a relic from the past, choke input filters are used in modern high-power laboratory power supplies where high efficiency is required, but where high-frequency switching mode supplies are not acceptable owing to the electromagnetic interference and radio-frequency interference (EMI/RFI) that they generate. Figure 10.8 shows a phase-controlled choke input filter raw supply. The conduction angle of the bridge rectifier is controlled so as to increase or reduce the average input voltage to the choke in sympathy with the desired raw supply voltage and the load current. In this way, the raw supply can be preregulated so as always to supply just sufficient headroom to a following stabilizer circuit. As the choke's magnetizing current will attempt to adjust the voltage across it so that current flow is uninterrupted, a commutating diode is fitted across the bridge output. This provides a path for the magnetizing current during those parts of each half-cycle when the controlled rectifiers in the bridge are not conducting. When the complete regulated supply is required to supply maximum output voltage and current at minimum mains voltage, the conduction angle of the silicon-controlled rectifiers in the bridge will be increased right up to 180° in each half-cycle. Under all other conditions, the firing of the SCRs will be delayed, reducing the average voltage applied to the choke by the required amount.

Mains transformers

Before leaving mains derived raw supplies and moving on to the linear stabilized supplies which they so often power, a word about mains transformers is in order. In today's international electronics market, many manufacturers design their equipment as a matter of course to accept the mains voltages found in almost any country in the world. The commonest variations are 220 V as

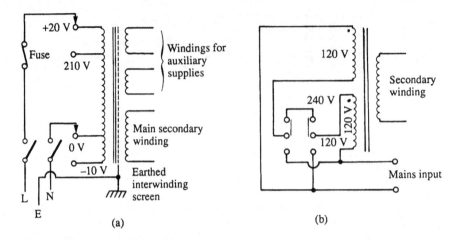

Figure 10.9   Mains transformer arrangements (typical).

in Europe and 240 V as in the UK, both at a frequency of 50 Hz, and 110 or 120 V at 60 Hz as in North America. Figure 10.9a shows a mains transformer with primary taps arranged to provide for any nominal input from 210 to 240 V AC inclusive, in 10 V steps. The tap labelled −10 V is only negative in the sense that it is a 10 V overwind on the end of the primary arbitrarily designated 0 V. The popular arrangement of Figure 10.9b is designed to cover both European and North American mains voltages, with selection by means of a double-pole double-throw (DPDT) slide switch; this is often only capable of adjustment with the point of a screwdriver, or is mounted internal to the instrument, to minimize the likelihood of inadvertent misoperation. More complicated primary switching arrangements providing for operation at both 210–240 V in 10 V steps and 105–120 V in 5 V steps are sometimes encountered, but the tendency nowadays is to use the simpler arrangement of Figure 10.9b or even to make the instrument operate from any voltage in the range 90–250 V without the need for any user adjustments whatever.

In Figure 10.9a an interwinding screen, as was almost invariably employed at one time, is shown interposed between the primary and the secondary winding(s). It consists of a single non-shorting turn of copper foil, or sometimes of a single-layer winding with only one end brought out. The older design of conventional transformer used E- and I-shaped laminations with full-width primary and secondary windings, interleaved with waxed paper. In addition to reducing the capacitive coupling of voltage spikes and other disturbances from primary to secondary, the earthed screen formed an important safety barrier, preventing a short-circuit fault developing between the primary and the secondary windings. Modern transformers of the traditional E and I core shape do not use full-width windings: the primary is wound in one half of a moulded plastic split bobbin with the secondaries alongside in the other half, there being usually no screen. A screen is often omitted also on the increasingly popular toroidal type of mains transformer, though it can be provided where capacitive shielding is required. The advantages of the toroidal transformer are many. It is generally smaller than the conventional transformer and its lower profile is particularly useful given the modern fashion for low, squat, wide instruments. It has a lower external leakage field, giving rise therefore to less by way of hum problems in hi-fi amplifiers, for example. It is also more efficient than the conventional transformer, owing to its much lower 'iron' or core loss.

The national safety regulations of many countries, relating to electronic equipment, demand a double-pole mains on/off switch as shown in Figure 10.9a. Some authorities say that the switch should protect the fuse as shown, whilst others reckon that the fuse should protect the switch, by being located on the mains input side. Either way, the fuse holder should be constructed and connected in such a way that, with the instrument plugged in and switched on, there is no possibility of the outer end of a new fuse being inserted into the holder becoming live. Another important safety point concerns the fuse rating where the transformer has more than one secondary winding, of very different VA ratings. Suppose that, in Figure 10.9a, the rating of each of the auxiliary windings is just 5% of the rating of the main winding. If all the windings have the same regulation, then a short-circuit condition on one of the auxiliaries may result in less primary current than that due to rated load on the main winding. Thus the fuse fitted in the primary circuit cannot be relied upon to clear the fault. However, the auxiliary winding would be grossly overloaded, leading to failure of the transformer and possibly even a fire hazard. The only prudent course in this case is to fit an appropriately rated secondary circuit fuse in each auxiliary circuit.

Linear stabilized supplies

Figure 10.10a shows the circuit diagram of a low-cost linear stabilized DC laboratory bench power supply of good electrical performance, designed by the author some years ago. It uses the industry standard LM324 quad opamp; this costs no more than four of the cheapest small-signal NPN transistors available, so the fact that two of the opamps are unused is of no consequence. The raw supply is an entirely conventional bridge rectifier circuit, and D_6 is a light emitting diode (LED) used as a mains on indicator: a green LED was used, since this is the standard colour for a mains indicator in Europe. Opamp IC_{1a} is used to produce a stabilized reference voltage of 12.5 V at low output impedance. The $10\,M\Omega$ resistor R_{23} guarantees start-up of the stabilizer at switch-on. IC_{1b} is the error amplifier, comparing the output voltage with the reference voltage, which is applied to its inverting input. If the output voltage is too low, then IC_{1b}'s output voltage will fall, supplying more base current to the Darlington PNP pass

transistor Tr_1 via R_{10} and R_{11}. Tr_1 therefore passes more collector current, restoring the output voltage to the set value. The LM324 quad opamp is internally compensated for gains down to unity, i.e. 100% NFB. However, the attenuation provided by R_7 and R_{12} at the input to the opamp is included to reduce the effective loop gain provided by the opamp, to allow for the extra gain introduced into the loop by Tr_1.

Overload protection is provided by constant current limiting at either of two levels, 200 mA or 2 A, as selected by front panel switch S_2. In the event of overload, the voltage drop across R_{13} turns on Tr_3. This in turn turns on Tr_2, which raises the potential at the junction of R_{10} and R_{11}, reducing the drive to the pass transistor Tr_1, and lighting the red current limit indicator LED D_8. C_4 stabilizes the constant current feedback loop by rolling off all the loop gain before the loop phase shift reaches $180°$. Its value was derived empirically by observing the smallest value of capacitance which would suppress loop oscillations under all conditions, and then trebling it. The current limiting circuit has been carefully designed to provide accurate indication of whether the supply is or is not operating in the constant voltage mode, as explained later. The circuit reduces a $\pm 10\%$ variation in mains voltage to an output voltage variation of about 1 mV, a stabilization ratio of about 20 000 : 1, whilst at 8 mΩ the output impedance is not excessively high. However, the dynamic performance of a power supply is at least as important as the static characteristics just quoted. At full output voltage, the circuit of Figure 10.10a exhibits less than 10% voltage drop or rise when the output load current is switched from 50% to 100% (of the rated 2 A) or vice versa respectively, and the recovery time is less than 20 μs in either case.

There are various types of *overload protection*, of which the current limit used in the circuit of Figure 10.10a is only one variety. However, used in conjunction with an output on/off switch, and an output bypass capacitor, it provides a very useful and beneficial action when the output is switched on. The connection of the uncharged capacitor across the output of the stabilizer presents a momentary effective short-circuit, causing current

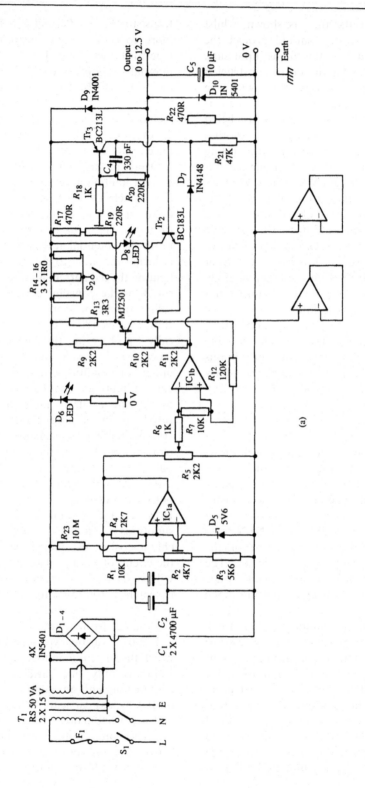

(a)

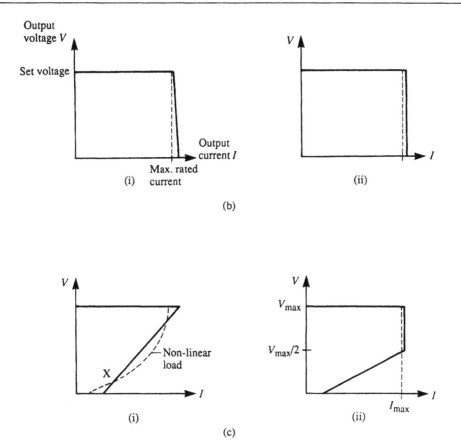

Figure 10.10 Linear laboratory bench variable voltage PSUs with overload protection by current limiting.

(a) Laboratory stabilized power supply with choice of 2 A or 200 mA current limit (S_2). Output impedance less than 8 mΩ, ripple less than 1 mV RMS, less than 10% drop or rise on full-load/half-load current change, settling in less than 20 µs.

(b) When the load resistance seen by the power supply unit falls to the point where the current drawn exceeds the maximum rating, current limiting sets in. The output voltage falls as the load resistance is reduced, until both reach zero. Normally the limiting current rises slightly as the output voltage falls, as in (i). If the current limit loop input is modified by a proportion of the output voltage, the rise can be avoided as in (ii): see R_{20} in (a).

(c) The dissipation within the PSU on full current can be reduced with 'foldback' current limiting, but now maximum rated current can only be drawn at maximum output voltage. Non-resistive loads may also present problems: see text. The foldback in (ii) is useful on a two-range switched raw voltage supply, as rated I_{max} is available down to 0 V if the lower output voltage range is selected.

limiting. The output voltage therefore promptly collapses to zero and then ramps up linearly to the set value at a rate determined by the size of the capacitor and the limiting value of current. If the circuit powered from the supply is misconnected, having for example a transistor junction connected directly across the power leads, this will be protected from the application of excessive voltage,

and provided that a suitably low current limit has been selected, no damage to the circuit under test will result.

The *thermal design* of any power supply is an important consideration. One can reckon that, in a linear stabilized supply like Figure 10.10a, the maximum power P_{max} to be dissipated by the pass transistor on short-circuit is equal to the short-circuit current times the raw supply voltage at top mains voltage. This is a rather pessimistic worst case, since it ignores the drop in the current sensing resistor, but it is a useful starting point. The thermal resistance of the power transistor from the collector junction to the mounting base or case of the transistor, θ_{jc} is quoted by the manufacturer, as often is the case to *heatsink* thermal resistance θ_{ch}. A typical figure for the latter, in the case of a T03 transistor such as used in the circuit Figure 10.l0a mounted with a silicone-based thermal grease, is $0.1\,^\circ$C/watt without a mica insulating washer or $0.3\,^\circ$C/W with. The appropriate figure plus θ_{jc} gives the junction to heatsink thermal resistance θ_{jh} and this figure times P_{max} gives the junction rise above the heatsink temperature. Working back from the top permitted junction temperature, one knows the maximum allowable heatsink temperature and hence, given the maximum ambient temperature in which the unit has to operate, the permissible heatsink temperature rise ΔT. The thermal resistance θ_{ha} of the heatsink chosen must be such that $P_{max}\theta_{ha}$ does not exceed ΔT. Be careful what figure you take as the top ambient temperature. With a transformer dissipation (due to full-load copper and iron losses) of around 10% of P_{max}, the ambient temperature in the vicinity of the heatsink could be considerably higher than room ambient, unless the case design can vent the heat from the transformer separately.

Many laboratory bench linear power supplies are designed to work either as a constant voltage supply with ideally zero output impedance, or as a constant current supply with ideally infinite output impedance – within the limits of the unit's *voltage compliance*, defined as the range (i.e. zero to maximum output voltage) over which it can adjust its output voltage in response to a change of load resistance in order to maintain the set current constant. As to which mode the supply is operating in, constant voltage or constant current, this is often indicated by a 'constant *V*' or 'constant *I*' (CV or CI) LED being illuminated as appropriate. In some stabilized supplies the two lamps are operated by the same driver, so that one or other of them will be illuminated at any time. However, there will always be a grey area at the changeover point from CV to CI operation, where the output impedance is neither virtually zero nor almost infinite. Ideally, the CV lamp should only be lit when the output impedance is below the specification limit; likewise, the CI lamp should only be lit when the output impedance is above the minimum specified limit for that mode. The circuit of Figure 10.10a is designed only as a constant voltage supply, the constant current mode being included only as the overload protection mechanism. If the current limit LED D_7 is not lit, then the output impedance is in specification. At the changeover point, the CV control loop via IC_{1b} and the CI control loop via Tr_3 and Tr_2 fight each other, resulting in changeover from CV to CI operation for a very small percentage increase in current above the selected limit of 200 mA or 2 A, but the grey area must always exist with any design. With further reduction of the load resistance, the output voltage falls but there is usually an accompanying slight rise in current, due to the finite loop gain of the CI loop. This is counteracted in the circuit of Figure 10.l0a by the inclusion of a judicious amount of positive feedback into the current loop from the output voltage, via R_{20}, which tends to turn Tr_3 on more as the output voltage falls (see Figure 10.10b).

Where, for reasons of economy or space, the heatsinking included is inadequate to permit the unit to supply the full rated output current at zero output voltage, extra feedback of this sort can be incorporated to provide a *foldback* characteristic, either from full output voltage or, more usually, from about 30% of full output voltage, as in Figure 10.10c. Supplies with foldback current limiting are designed to be *re-entrant*, that is to say that when the load resistance rises to a level at which the unit can supply the current demanded at the set output voltage, be that maximum or some lower demanded voltage, the actual output voltage will automatically rise to the set level. If, however, the load is non-linear, as can happen with semiconductor circuits, and the demanded current initially rises faster as the voltage increases, there is the possibility of latch-up where the demand and supply curves cross, indicated by the point X in Figure 10.10c.

Another form of overload protection, as exceedingly effective as it is infuriating to the user, is the

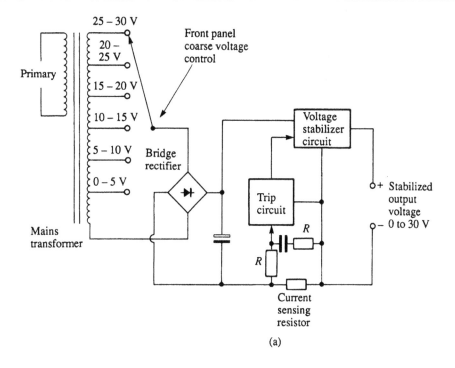

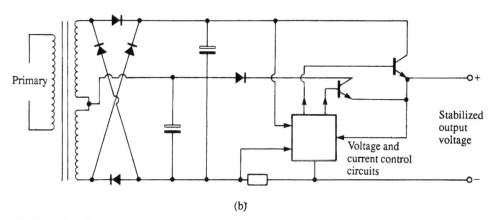

Figure 10.11 PSUs with reduced internal dissipation.
(a) Voltage subranges with overload current cut-out trip.
(b) Arrangement providing full rated current down to zero volts (no foldback) but with maximum dissipation in the PSU halved.

electronic overload cut-out. Here, should the current exceed the rated maximum, a trip instantly cuts off the pass transistor completely and lights an overload trip light. The user then has manually to reset the trip, which will only be possible if the overload that caused the supply to cut out in the first place is no longer present. The problem of

nuisance tripping due to purely transient overloads can be mitigated by partially bypassing the overload current sensing resistor at AC, so that the supply can provide current peaks up to 200% of the rated current but is still limited to the rated mean current, as indicated in Figure 10.11a. If a large uncharged capacitor is connected across the output of the supply when it is already on, tripping may still occur, but if the capacitor is connected before the supply is switched on there should be no problem, since with this type of supply the output voltage is usually ramped up linearly by the control circuit, precisely to cope with loads containing capacitors. With all these ifs and buts, it might be asked why trip protection is ever used: the answer is simply economy. The prospective internal dissipation, on short-circuit, in a stabilized supply is simply equal to the maximum raw voltage (at top mains voltage) times the short-circuit current (somewhat in excess of the rated maximum current). If this can be reduced by any means, less waste heat will be generated and the size and cost of the necessary heatsinks will therefore be less. One ploy, in the form of a foldback characteristic, has already been noted. But in practice, the worst case dissipation will still be around 40% of maximum rated current times the raw volts at top mains voltage. The scheme shown in Figure 10.11a can cut the heatsinking required dramatically, depending upon how many secondary taps are provided. The raw voltage available is set to the minimum necessary for the range of output voltage selected by the user, so that for example with six 5 V subranges providing 0 to 30 V, the heatsinks need only cater for a maximum dissipation of approximately I_{max}A times $5 + H$ V, where H is the voltage headroom needed by the pass transistor. H must of course be adequate on minimum mains volts and so will be considerably larger at top mains volts, but even so the scheme can cut the maximum dissipation by up to 75% or so – a very worthwhile improvement if one is willing to do without the convenience of re-entrant operation.

In principle, one could have the best of both worlds by automating the selection of the appropriate raw voltage setting in sympathy with the demanded output voltage or the actual output voltage (as determined by the load resistance and the current limit loop), whichever was the lower. However, for certain types of cyclically varying load, the selection mechanism would be continually hunting up and down between taps. Thus relay selection of raw voltage would be impractical on reliability grounds, whilst the additional drop

introduced by SCR selection would result in additional losses but, more seriously, would stress the reservoir capacitor with repeated switching transients. A limited number of taps can nevertheless be employed, by switching between completely separate raw supplies. Figure 10.11b shows a scheme in outline where two raw supplies are derived from a standard transformer with two secondary windings, the load current automatically being drawn from the lower voltage supply whenever possible. This occurs entirely automatically owing to the action of the voltage control loop (or indeed the current limit loop), with the higher voltage supply 'helping out' at the troughs of the raw ripple voltage, when the set voltage sits across the changeover point.

Another popular scheme is the *dual-rated* supply, shown in Figure 10.12a. Here, maximum use is made of the available heatsinking by switching the raw supply so as to provide an alternative lower output voltage range which can supply double the current available on the higher range. Note that the reservoir capacitor must have a voltage rating appropriate to the higher output voltage range, but a capacitance adequate for the higher current drawn on the lower voltage range. If the rating of the transformer is increased, the higher current can be supplied right up to the full output voltage, with no increase in the heatsinking required. This is achieved by arranging for heavy current foldback below half output voltage on the higher output voltage range (see Figure 10.10c). The full output current is still available at the lower voltages by selecting the lower output voltage range.

The size and expense of the heatsinking required in a stabilized supply is largely due to the comparatively low temperature limit on semiconductor devices. The junction temperature of a silicon power transistor must not be allowed to exceed 150°C or 175°C, according to type, limiting the junction temperature rise above a top ambient temperature of say 50°C to 125°C or less. If the unwanted power could be dissipated in a resistor, the cost of getting rid of heat could be greatly reduced, since a vitreous wirewound resistor can reject heat at a surface temperature of 350°C, a rise above ambient of 300°C. Figure 10.12b shows

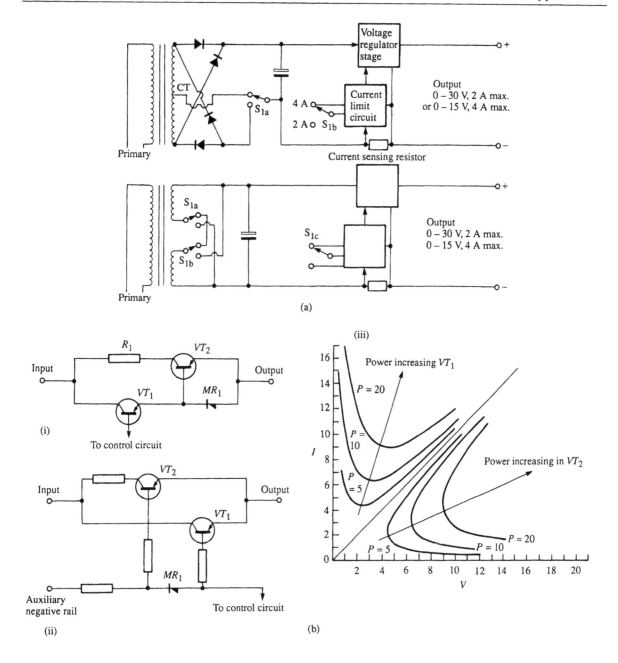

Figure 10.12   Other PSUs with reduced internal dissipation.

(a) Lower circuit provides a full doubling of rated current on the lower voltage range, but requires a three-pole switch. Upper circuit makes less efficient use of the secondary rating by operating as a biphase circuit on the lower voltage range (see Figure 10.5b and c).

(b) In the McPherson regulator, VT_1 only conducts if VT_2 is bottomed. (i) Dual parallel section element. (ii) Dual parallel section element modified to reduce minimum voltage. (iii) Constant transistor power contours for circuit.

(Reproduced by courtesy of *Electronic Engineering*.)

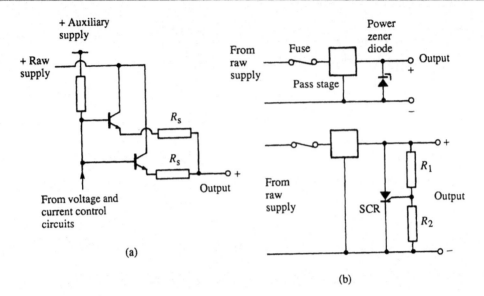

Figure 10.13   More PSU techniques.
(a) Current sharing resistors R_s used to equalize the current in paralleled pass transistors.
(b) Crowbars prevent expensive damage to a load, e.g. a computer, should a component failure result in the full raw supply voltage being applied to the output.

a scheme for doing just this.[7] The resistor R_1 is chosen such that with VT_2 bottomed, the rated short-circuit current is supplied entirely via R_1, i.e. $R_1 = V_{raw}/I_{sc}$. Thus all of the power P_{max} is dissipated in R_1 and (virtually) none in either transistor. As far as transistor dissipation is concerned, there are two worst case conditions. If the current drawn at zero output voltage is equal to $I_{sc}/2$, then the total dissipation is $P_{max}/2$ split equally between R_1 and VT_2. The other worst case condition is when the maximum current is drawn, at an output voltage of $V_{raw}/2$: VT_2 is bottomed and half of the required output current is supplied via R_1. The other half of the current must be supplied via VT_1, resulting again in a transistor dissipation of $P_{max}/4$. Thus neither transistor is ever called upon to dissipate more than one-quarter of the maximum power. Furthermore, VT_1 will never conduct (and dissipate) unless VT_2 is bottomed (and hence not dissipating significantly): consequently if the load current and output voltage are such that VT_2 is not bottomed (and hence dissipating) then VT_1 must be cut off. Thus VT_1 and VT_2 can share the same heatsink. Note that the base current to VT_1

and VT_2 is scheduled via the steering diode MR_1; the drive turns on VT_2 alone or both transistors automatically as required, under control of the combined voltage and current control loop outputs.

Figure 10.13 shows two other techniques commonly used in linear stabilized supplies. In Figure 10.13a two transistors are used in parallel, but not with a view to increasing the current rating; that is seldom the limiting factor. If a given maximum prospective dissipation is shared between two transistors, the heat flux through each is halved and so therefore is the junction temperature rise above mounting base temperature; that is, the two devices can be regarded as a single stage with an effective θ_{jc} of half that of either transistor alone. Some resistance is added in the emitter circuit of each by way of degeneration, to reduce the degree of current imbalance between the devices, which should ideally be selected for the same V_{bc} at maximum current. For should one of the two transistors hog more than half the current, its dissipation will be greater than that of the other, its junction temperature higher, and its V_{bc} consequently lower, causing further imbalance in a vicious circle. The more voltage thrown away across the emitter resistors, the lower the efficiency of the supply but the better the suppression of current hogging.

Figure 10.13b shows two forms of overvoltage protection or *crowbar*. This is commonly found in a stabilized power supply designed to produce a fixed output voltage, to supply a computer, for instance. Should the pass transistor fail short-circuit, or should a failure in the control circuit result in the pass transistor being turned on hard, a lot of expensive ICs could bite the dust. Any safety circuit designed to prevent this should, in the interests of reliability, be as simple as possible, to minimize the chance of it failing to operate when called upon to save the day. A fuse in the raw supply plus an SCR or a big fat Zener diode, of the next voltage rating above the output voltage, fill the bill simply and economically.

Stabilizer circuits in laboratory bench supplies with variable output voltage often use opamps and discrete components, in order to achieve maximum performance. But for the fixed voltage regulators in the power supply section of an electronic instrument, the economical multi-sourced three-lead regulators in the TO220 package must be the obvious first choice. They require but three connections; raw supply in, regulated output out, and ground. The 7805 and 7812 are 5 V and 12 V regulators respectively, being but two of a range of positive regulators available in a number of output voltages, all supplying 1 A maximum. Similarly, 79xx types are regulators producing a negative stabilized output from a negative raw input. Types 78/79Mxx and 78/79Lxx are medium- and low-power positive and negative regulators rated at 0.5 A and 100 mA respectively. Also available are high-voltage types accepting raw voltages up to 125 V or more; single-chip dual regulators providing ±15 V outputs from positive and negative raw inputs; high-current TO3 types supplying up to 10 A; and adjustable supplies such as the 723 which can be programmed for any output voltage and current limit with the aid of a few external resistors.

Inverters and frequency converters

Circuits designed to provide a DC output (stabil-

ized or otherwise) from an AC input (usually mains) are called, not unnaturally, power supplies. Now the term could equally well be applied to several other arrangements, but fortunately confusion is avoided by common convention as follows. A circuit which produces a DC output from a DC input of a different voltage is called a converter; a circuit which produces an AC output from a DC input is called an inverter; and a circuit which produces an AC output from an AC input of a different frequency is called a frequency converter or frequency changer.

An *inverter* could consist of a DC supply, say from a battery, powering a signal source and followed by a power amplifier of the sort featured in Chapter 4. However, even a class B amplifier would not be as efficient a solution as one could wish for, when demanding a large amount of power from a battery. Inverters therefore usually use push-pull square wave operation, where each transistor in the amplifier is either saturated or cut off, leading to efficiencies of 80 to 90% at full load. Some inverters do not use a separate signal drive source and amplifier; instead they are of the *self-oscillating* inverter variety, where positive feedback from the output is used to drive the switching transistors. Where an accurate output frequency is essential, however, a separate drive is always used.

A basic square wave output, perhaps with the sharp corners rounded off a little with a modicum of filtering, is satisfactory for a number of uncritical types of load, including many small brush and induction motors. The output voltage might for example be 240 V RMS at 50 Hz, to run a central heating pump from a car battery during mains failures. However, this type of inverter is not very suitable for loads which expect to see a sine wave. The commonest example of this is any electronic instrument with a mains transformer suppling a rectifier and capacitive filter load. For the peak voltage of 240 V AC mains, being sinusoidal, is $(\sqrt{2})240 = 339$ V, whereas the peak voltage of a 240 V RMS square wave is just 240 V, resulting in an inadequate raw supply voltage. So for the more critical sort of load, an inverter with a sine wave output is used. Many of these still use square wave switching, but with an output transformer tuned to resonance at the output frequency of 50 or 60 Hz. The 'elastic' used to accommodate the varying instantaneous voltage difference between a sine wave and a square wave is often a separate inductor, or sometimes a transformer design with a deliberately high leakage inductance between primary and secondary (output) winding. The efficiency at full load is usually not quite so high as square wave output types, but the waveform is very fair at 5% distortion or less.

Frequency converters often consist simply of a straightforward raw supply derived from the AC input, driving an inverter operating at the desired output frequency, e.g. 50 Hz in and 400 Hz out. A typical application is for the bench testing of aircraft electronic equipment.

DC converters

This brings us to the DC/DC converter, which could be regarded as an inverter driving a rectifier and smoothing circuit. Some converters operate just like that, so that the output voltage goes up and down in sympathy with the DC input voltage. Being unregulated, the output also varies somewhat with the load current drawn. The extra high-tension (EHT) supply to the picture tube in some black-and-white TVs is like this, so that the picture size varies slightly with brightness. But most converters include a feedback loop from the output circuit to the drive circuit, in order to maintain the output voltage sensibly constant. Figure 10.14 shows a number of the commoner converter arrangements, all of which use the magnetic field caused by a current in an inductor as a temporary store of energy; this energy can be supplied at one voltage and drawn off at another. The single-pole switch shown in each circuit is shorthand for an active device, a bipolar transistor or more commonly nowadays a power MOSFET. This is switched on and off at a constant frequency (usually), and at either a fixed mark/space ratio or an appropriate one as determined by a feedback control circuit designed to stabilize the output voltage.

Consider first the *buck* circuit of Figure 10.14a, with the switch operated at f switching cycles per second at a fixed 50/50 mark/space ratio. Assume

for the moment that the output voltage is half the input voltage, that some load current is being drawn, and that C is so large that there is negligible ripple voltage at the output. Each time the switch is closed for a period of $1/2f$ seconds, a voltage equal to $V_i - V_o$ is impressed across the inductor, so the current through it will increase steadily during this half-cycle at a rate given by $V_i - V_o = L \, dI/dt$. During the next half-cycle, when the switch is off, energy will be given up by the inductor, as evidenced by the current through it falling at a rate given by $V_o = L \, dI/dt$. The reversal of the rate of change of current (from rising to falling) is accompanied by a reversal of polarity of the back EMF across the inductor, so that the left-hand end is negative. In fact, for the current to continue flowing at all, the left-hand end of the inductor must fall slightly below zero volts, turning on the diode. In the steady state, the total increase in current during the on period of the switch must equal the total decrease during the off period, ready to start the next cycle all over again. Given the 50/50 mark/space ratio in this example, the rate of current increase during the on period of the switch must equal the rate of decrease during the off period. Since in general $V = L \, dI/dt$, then $V_i - V_o = V_o$, so $V_o = V_i/2$. For any other mark/space ratio, the positive dI/dt during the on period will differ from the negative dI/dt during the off period, such that the total current increase equals the total decrease during one complete cycle. Thus provided that the steady current drawn by the load exceeds half the peak-to-peak ripple current, the output voltage is determined by the mark/space ratio of the switch drive. If the steady current is less than this, the output voltage will rise, so that a lower voltage is impressed across the inductor during the on period, resulting in a smaller increase in current during the switch's conducting half-cycle. On the following half-cycle the voltage across the inductor is correspondingly greater, so the negative dI/dt is also greater. This results in the inductor running out of current and the diode ceasing to conduct, well before the switch turns on again. As with the choke input filter circuit which was examined earlier in the chapter, in the absence of any load current at all, the output voltage will rise to equal the input voltage.

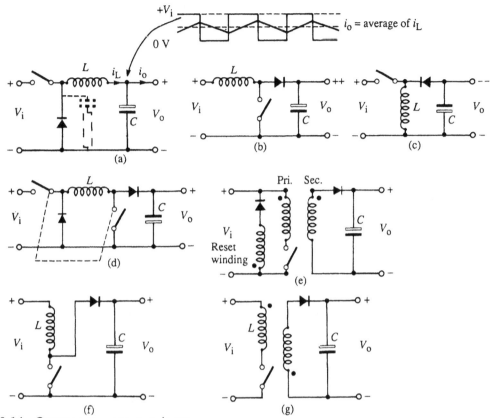

Figure 10.14 Common converter topologies.
(a) Step down (buck), non-isolating. Capacitor shown dashed may be fitted to reduce radio-frequency interference.
(b) Step up (boost), non-isolating.
(c) Step up/down (inverting, buck/boost), non-isolating.
(d) Step up/down (non-inverting, buck/boost), non-isolating.
(e) Step up/down (forward converter), isolating.
(f) Flyback converter, non-isolating (step up).
(g) Flyback converter, isolating (usually step up).

The *boost* and *buck/boost* circuits of Figure 10.14b and c also use an inductor to store energy. The buck/boost circuit is capable of operating from an input voltage which is either higher or lower than the output voltage. So for example it is capable, given a suitable control circuit to adjust the switch drive mark/space ratio, of supplying a constant 12 V output from an input of anywhere between say 9 and 15 V. However, it has two minor disadvantages. First, the polarity of the output is inverted, which could be inconvenient bearing in mind that the output voltage is not isolated from the input supply. Second, it is less

efficient than it could be owing to the large peak-to-peak voltage swing across the inductor, equal to the sum of V_i and V_o. Thus for a given magnetizing ripple current, a larger inductance is required than in Figure 10.14a or b, so the inductor will be both more expensive and more lossy. The *non-inverting buck/boost* circuit of Figure 10.14d avoids both disadvantages by using an additional switch.

Figure 10.14e shows another circuit, called the forward converter, which provides isolation between the input and output circuits and can be designed to produce an output voltage either higher or lower than V_i. Before looking at the operation of the forward converter, consider for a moment the *flyback converter* of Figure 10.14f, also known as the 'ringing choke' circuit. When the switch is closed the voltage V_i impressed across the inductor causes the current through it to rise at a constant rate, so that at the moment when the switch opens the stored energy is $LI^2/2$, where I is the current flowing at that instant. By Lenz's law, the voltage will change so as to tend to keep the current flowing; so the voltage across the now open switch will rise until the diode turns on and current flows into the capacitor C. The energy stored during the switch's on period is transferred to the capacitor and thence to the load connected to V_o. Thus for a constant operating frequency and duty cycle, the flyback inverter supplies a certain minimum amount of energy to the output circuit on each and every cycle. This is equal to the value of $LI^2/2$ at the end of the on period of the switch, when I at the start of the on period was zero. If the load current is such as just to consume this amount of energy per cycle, the ratio of input to output voltage will be the same as the ratio of the on period to the period of one complete cycle. This corresponds to the volt-second product across L during the on and off periods being the same. If the load current is reduced below this level, the output voltage will rise, reaching a dangerous level on no-load. If the load current is increased, then the mean current through the inductor increases accordingly, so that the peak-to-peak ripple (magne- tizing current) is riding on top of a larger continuous current, just as shown in Figure 10.14a.

As a moment's comparison will show, the circuit of the flyback converter is identical to that of the boost converter of Figure 10.14b, whilst the buck/boost converter of Figure 10.14c differs only in that the capacitor C and the load are connected to the other end of V_i. The *forward converter* of Figure 10.14e is similar to the flyback converter, except that the diode supplying C conducts during the on period of the switch. This has far-reaching consequences, for with this circuit the output voltage is determined by the turns ratio of the transformer. During the on period of the switch, current is supplied to the load and additionally of course a magnetizing current is built up. At switch-

off this energy must go somewhere, as otherwise there would be a dangerously large positive over-swing, as in the case of an off-load ringing choke circuit. In the forward converter, the energy associated with the magnetizing current is returned to the supply by means of the 'clamp' or reset winding, so the off-load condition poses no hazard.

The forward converter of Figure 10.14e is a crude variety, only suitable for very low-power operation, owing to the peaky nature of the current through the diode. Nevertheless it can still be very useful, as Figure 10.15a shows. This is a combined forward and flyback converter, which is self- oscillating, on the principle known as the *blocking oscillator*. The base winding is connected so as to provide positive feedback, resulting in the transistor turning on hard and feeding energy into the reservoir capacitor of the $-14\,V$ supply. However, the 1 nF base capacitor rapidly charges up, so the transistor starts to run out of base current. As soon as this happens, the rise in collector voltage is fed back negative-going to the base, resulting in a rapid and clean cut-off. This supplies energy to the $+14\,V$ circuit by flyback converter action. The capacitor is recharged via the 100 K resistor so that the negative voltage at the base rises towards turn-on and the cycle repeats. T_1 uses a 6 mm two-hole balun core and the circuit runs at about 60 kHz at around 60% efficiency. The two 10 K resistors shown represent the 1.4 mA load current available from each rail, ample for a modern micropower quad opamp. The regulation of the negative rail, being derived from a forward converter, is fairly good. That of the positive rail is less so; it can be improved by connecting a 13 V Zener diode in parallel as a shunt regulator.

Figure 10.15b shows a more sanitary forward converter configuration capable of operation up to a few hundred watts output: the diode feeds energy into an auxiliary energy storage inductor complete with freewheeling diode. Figure 10.15c shows a two-transistor converter circuit arranged so that the primary can act as its own reset winding, returning spare energy to the supply via the two diodes. It can thus be used as either a forward or a flyback converter, and even as both at once, like Figure 10.15a, but without the low-power limitation of the latter. Figure 10.15d shows four popular configurations for high-power switching supplies, for outputs in the range 100 W to 2 kW. Again transistors are shown as the switches, but high-power MOSFETs are being increasingly used.

High-frequency switching supplies

The purpose of all the converters shown is to produce a DC output from a DC input of a different voltage. Such circuits are widely used in mains powered DC supplies. The mains is rectified to produce a DC supply, an inverter chops this up into an alternating supply, and the result is rectified to arrive at the final DC output.

On the face of it, it would seem perverse – not to say bizarre – to go through three energy conversions where, in a conventional linear supply, one suffices. The reason lies in a basic law of electromagnetism noted in Chapter 1, to the effect that the EMF induced in a winding – the back EMF in the case of a primary winding – is proportional to the rate of change of flux. In the early days of the electricity supply industry, when AC was being introduced, the rotational speed chosen for alternators was 3000 RPM. With machines designed to produce one elecrical cycle per revolution, the frequency of the electricity generated was 50 Hz. This was conveniently low, in that the magnetic materials then available exhibited rather a high hysteresis loss. This loss occurs on each and every cycle and can thus be minimized by low-frequency operation, although the associated low rate of change of flux dictates fairly massive alternator and transformer cores. Later, lower-loss magnetic materials were developed, so that in the quest for lightweight generators the aviation industry was able to settle on 400 Hz for aircraft use. Later still, ferrite materials with extremely low hysteresis losses were developed, permitting efficient power transformers operating at tens or even hundreds of kilohertz to be produced. An increase in operating frequency means an increase in the induced voltage per turn for a given peak flux level, so that at the higher frequencies the peak total flux on the core can be reduced. This in turn means that (despite the somewhat lower peak flux levels

supported by ferrites before saturation sets in) a core of much smaller cross-sectional area can be used. Thus all of the linear dimensions of the transformer can be reduced when a higher operating frequency is employed, enabling a power supply operating at 100 kHz to use a transformer which is tiny by comparison with a 50 Hz transformer of the same power rating.

In a modern compact high-frequency switching power supply, the mains is rectified direct, producing a raw supply of around 160 or 320 V DC in the case of 120 or 240 V AC mains respectively. This high-voltage DC supply powers an inverter running at tens or even hundreds of kilohertz, and which can therefore use a very small transformer. The transformer serves the dual purpose of transforming the high-frequency switching waveform down to the required voltage for the output rectifier circuit (or circuits in the case of a multi-output supply), and of providing electrical isolation of the output from the primary circuit, all of which is live to the mains. The control loop, which senses the output voltage and adjusts the duty cycle (mark/space ratio) of the switch drive waveform, is also isolated from the live primary circuit by a transformer, optocoupler or other means.

Despite their very real advantages in terms of size and cost, switching supplies are notorious generators of electrical interference, both conducted and radiated. The high-frequency square waves tend to get out through the output wiring and back through the raw supply into the input power lead, polluting the mains supply, whilst the very high rate of change of current results in radiation of radio-frequency interference at harmonics of the switching frequency. Choice of configuration can help. For example, the forward converter of Figure 10.15b draws a square wave of current from the DC supply, but the current into the reservoir capacitor is continuous, with just a small triangular ripple of magnetizing current superimposed. Other configurations, derived from the boost converter of Figure 10.14b, draw a basically constant current from the DC supply, but feed a pulsating current to the output circuit. So there is likely to be unwanted high-frequency ripple, getting mostly into the mains supply in one case and mostly into the output in the other. More

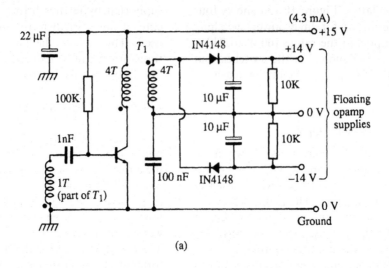

(a)

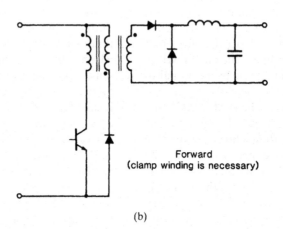

Forward
(clamp winding is necessary)

(b)

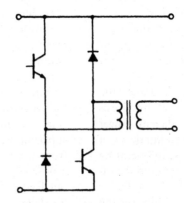

Two transistor forward or flyback
(clamp winding is not needed)

(c)

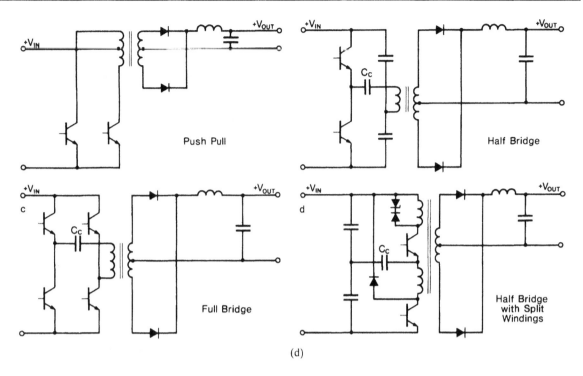

(d)

Figure 10.15 More converter circuits.

(a) Combined forward and flyback converter, providing floating supplies for low-power opamps. The 10 K resistors represent the current drawn by the opamps.

(b) Forward converter where the load current during the switch-on period is approximately constant: clamp winding is necessary.

(c) Two transistor forward or flyback: reset is provided by the primary winding itself, and clamp winding is not needed.

(d) Isolated output converters for higher output powers.

(Parts (b), (c) and (d) reproduced by courtesy of *Electronic Product Design*.)

subtle inverter topologies have been developed in which the square wave switching currents are confined entirely internally to the supply, with both the input and output currents being basically continuous.[8]

Another powerful aid in reducing conducted and radiated interference from switching mode power supplies is sine wave operation. 'Switchmode' power supplies (or, more correctly, switching power supplies: 'Switchmode' is a Motorola Inc. trademark) operating on this principle have been available for a number of years,[9] and show levels of electromagnetic interference and radiofrequency interference (EMI/RFI) of 15 to 20 dB lower than conventional supplies using square wave switching. Figure 10.16a shows how a two-

pole low-pass filter acts as a resonant tank circuit to pass the fundamental frequency component of square wave input to the load. The circuit, shown in simplified form in Figure 10.16c, maintains the output voltage constant in the face of variations both of load current and of the raw DC supply voltage derived from the mains. This is achieved by an entirely different method from the variable duty cycle control at a fixed operating frequency used in conventional switching power supplies. In the circuit of Figure 10.16c the control circuit drives the switching MOSFETs at a fixed 50/50 duty cycle, but varies the frequency to allow for both load current and supply voltage variations, as shown in Figure 10.16b. The circuit of Figure 10.16c also illustrates how a jumper connection

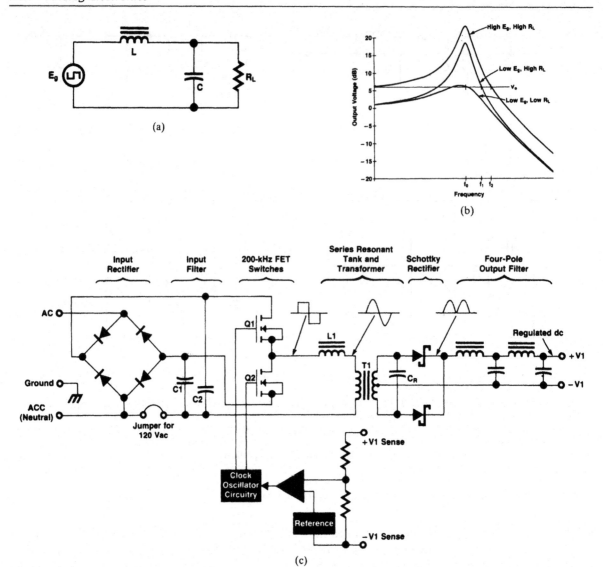

Figure 10.16   The HP65000A series power supplies use a sinusoidal operating mode, resulting in lower levels of EMI/RFI.

(a) Voltage regulation technique used. The tank circuit converts the square wave to sine waves as an integral part of the technique.

(b) Transfer function for the circuit in (a). Adjusting the frequency of the generator keeps the output voltage constant at V_o under varying line voltage and load conditions.

(c) Schematic of single-output 65000A series power supply.

(Reproduced with permission of Hewlett-Packard Co.)

can be used to derive a fixed voltage DC raw supply from either 120 V or 240 V mains. In the first case, two of the bridge diodes are unused whilst the other two operate as positive and negative half-wave rectifiers on alternate half-cycles of the input; whereas in the second case, conventional bridge operation takes place.

Many OEM switching power supplies (supplies for original equipment manufacturers to design into their products) provide several fixed output

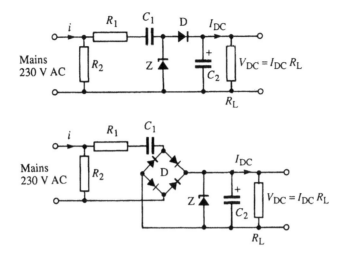

Figure 10.17   Low-voltage, low-current, non-isolated supplies, suitable for uses such as recharging a battery operated shaver. The current drawn from the mains is equal to the sum of the current through the load, represented by
R_L, and the current through the shunt regulator Zener diode Z, R_1 is a small surge limiting resistor, most of the
mains voltage appearing across the capacitor C_1: this voltage drop is reactive and hence lossless. R_2 ensures that C_1
does not remain charged to a high voltage (presenting a shock hazard at the pins of the mains plug) if the mains is disconnected at an instant of peak voltage.

voltages, e.g. $+5\,V$ at several amperes together
with $+12\,V$ and $-12\,V$ at a few hundred milliamperes each. In this case the feedback circuit
controls the main high current output: the lower
current outputs are less well regulated, and also
show *cross-regulation* effects, where a change in
load on one output causes a small change in
voltage in another output. Where higher performance is required, the auxiliary outputs are designed
for a slightly higher output, say $15\,V$, and fitted
with linear 'three-leg' (TO220) regulators. A relatively small headroom suffices for these in this
case, since the main loop takes out nearly all the
variations due to mains voltage fluctuations, leaving linear regulators with just the regulation and
cross-regulation variations to suppress.

Switching power supplies are also found in
laboratory bench linear power supplies, despite
the difficulty of adequately suppressing the EMI/
RFI. The switching supply is used as a preregulator, to produce an output voltage which is about
$4\,V$ higher than the desired output voltage. This in
turn feeds a conventional linear regulator circuit
after the style of Figure 10.10a, which provides a
very low output impedance and ripple. Expensive

heatsinking is not required, since on current–
limited short-circuit the raw supply is only some
$4\,V$ owing to the action of the preregulator. If
mains isolation is provided by a conventional
$50\,Hz$ toroidal mains transformer, a non-isolating
preregulator such as in Figure 10.14a can be used,
resulting in a very economical design for powers
up to $60\,W$ or so, e.g. $30\,V$ at $2\,A$.

Non-isolated low-voltage supplies

A final word on power supplies concerns a specialized application for low-current, low-voltage
power supplies.[10] What makes the application
unusual is that isolation of the output from the
mains is not required. Typical applications would
be supplying a microcontroller and LED display in
a domestic appliance or recharging the batteries in
a cordless electric shaver. In both cases, the
equipment is designed so that the user cannot
come into contact with the output of the supply.
Figure 10.17 shows two variations on the theme,
both using a capacitor to drop the mains voltage
down to a suitable low voltage without incurring
losses in the process. Since the circuit provides a

virtually constant current, defined by the mains voltage divided by the reactance of the capacitor at mains frequency, the output voltage can be very simply stabilized by means of a Zener diode used as a shunt regulator.

References

1. Voltage Stabiliser for Battery Operated Equipment, I. Hickman, p. 23, *Electronic Engineering*, November 1975.
2. Battery-Powered Instruments, I. Hickman, p. 57, *Wireless World*, February 1981.
3. Short-Proof Constant Voltage Batteries, I. March, p. 10, *New Electronics*, 29 September 1987.
4. Parallel Drive Voltage Multiplier, R. A. Worsley, p. 16, *New Electronics*, 16 September 1986.
5. Analysis of Rectifier Operation, O. H. Schade, p. 341, *Proc. I.R.E.*, 31.7, July 1943.
6. Linear/Switchmode Voltage Regulator Manual, Section 8, Motorola Inc., 1983.
7. Regulator Elements Using Transistors, J. W. McPherson, p. 162, *Electronic Engineering*, March 1964.
8. An Alternative dc-dc Converter Topology, D. Sheppard & B. Taylor, p. 53, *Electronic Product Design*, January 1984.
9. 200 kHz Power FET Technology in New Modular Power Supplies, R. Myers & R. D. Peck, p. 3, *Hewlett-Packard Journal*, 32.8, August 1981.
10. Designing Low-Cost PSUs with Capacitive Droppers, G. Oberg & R. Alsson, p. 65. *Electronic Product Design*, November 1983.

Questions

1. What is the new and the typical end-of-life on-load voltage of a Leclanché type primary cell? What is the usual recommended routine service replacement interval for this type of battery?
2. What are the relative advantages and disadvantages of the Cockcroft Walton multiplier, and the parallel capacitor version of the same circuit?
3. What are the advantages of a choke input filter? What is a 'swinging choke' and how does it operate?
4. Cite four advantages of the toroidal mains transformer over the conventional variety using E and I laminations.
5. In connection with Lab bench power supplies, what do the terms 'foldback' and 're-entrant' signify?
6. A 30 V linear Lab bench stabilized power supply uses a 2N3055 with a junction-to-case thermal resistance θ_{jc} of 1.5°C/W as the pass transistor. The case-to-heatsink thermal resistance θ_{ch} (including insulating mica washer) is 0.3°C/W, and the maximum ambient temperature at the heatsink is 50°C. The worst case dissipation, at maximum mains voltage tolerance, with the output short-circuited, is 40 W. Calculate the maximum permissible value of θ_{ha}, the heatsink-to-ambient thermal resistance.
7. In connection with power supplies, define the terms 'converter', 'inverter' and 'frequency converter'.
8. Sketch basic non-isolating Boost and Buck converters. What are the hazards to the input supply, and to the load circuit, of failure of the transistor or FET switch (i) as an open circuit, (ii) as a short circuit?
9. Describe in detail the operation of the forward converter. Why is a clamp (reset) winding necessary?
10. Enumerate the main means used to eliminate or at least minimize the EMI produced by switching inverters/converters.

Appendices

Colour coding

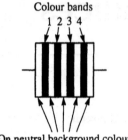

Colour bands
1 2 3 4

On neutral background colour
(varies from manufacturer to
manufacturer)

General-purpose types

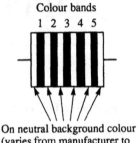

Colour bands
1 2 3 4 5

On neutral background colour
(varies from manufacturer to
manufacturer)

E96 series types (1%)

	Band						*Band*				
	1	*2*	*3*	*4*			*1*	*2*	*3*	*4*	*5*
Black	–	0	$\times 1$			BK	–	0	0	$\times 1$	
Brown	1	1	$\times 10$	1%		BR	1	1	1	$\times 10$	1%
Red	2	2	$\times 10^2$	2%		RD	2	2	2	$\times 10^2$	
Orange	3	3	$\times 10^3$			OR	3	3	3	$\times 10^3$	
Yellow	4	4	$\times 10^4$			YL	4	4	4	$\times 10^4$	
Green	5	5	$\times 10^5$			GN	5	5	5	$\times 10^5$	
Blue	6	6	$\times 10^6$			BU	6	6	6	$\times 10^6$	
Violet	7	7	$\times 10^7$			VI	7	7	7	$\times 10^7$	
Grey	8	8	$\times 10^8$			GY	8	8	8	$\times 10^8$	
White	9	9	$\times 10^9$			WH	9	9	9	$\times 10^9$	
Gold				5%		GD				$\times 0, 1$	
Silver				10%		SV					
None				20%		NO					

Examples

BN	BK	GN	SV	=	1M0, 10%		GN	BU	RD	BN	BN	=	5K62, 1%
BN	GN	BK	SV	=	15R, 10%		OR	WH	RD	BK	BN	=	392R, 1%
YL	VI	OR	RD	=	47K, 2%		BN	YL	VI	GD	BN	=	14R7, 1%

Some ranges of resistors are available in a tolerance tighter than the E range would suggest, e.g. E24 values at 1%.

E series: ranges in tolerance bands

These are selection tolerances. Additional allowance must be made for ageing: as manufacturers' data for full-load ageing or light-load ageing (shelf).

E6 ±20%	E12 ±10%	E24 ±5%	E48 ±2%	E96 ±1%
1.0	1.0	1.0	1.00	1.00
				1.02
			1.05	1.05
				1.07
		1.1	1.10	1.10
				1.13
			1.15	1.15
				1.18
	1.2	1.2	1.21	1.21
				1.24
			1.27	1.27
		1.3		1.30
			1.33	1.33
				1.37
			1.40	1.40
				1.43
			1.47	1.47
1.5	1.5	1.5		1.50
			1.54	1.54
				1.58
		1.6	1.62	1.62
				1.65
			1.69	1.69
				1.74
			1.78	1.78
	1.8	1.8		1.82
			1.87	1.87
				1.91
			1.96	1.96
		2.0		2.00
			2.05	2.05
				2.10
			2.15	2.15
2.2	2.2	2.2		2.21
			2.26	2.26
				2.32
			2.37	2.37
		2.4		2.43
			2.49	2.49
				2.55
			2.61	2.61
				2.67
	2.7	2.7	2.74	2.74
				2.80
			2.87	2.87
				2.94
		3.0	3.01	3.01
				3.09
			3.16	3.16
				3.24

E6 ±20%	E12 ±10%	E24 ±5%	E48 ±2%	E96 ±1%
3.3	3.3	3.3	3.32	3.32
				3.40
			3.48	3.48
				3.57
		3.6	3.65	3.65
				3.74
			3.83	3.83
	3.9	3.9		3.92
			4.02	4.02
				4.12
			4.22	4.22
		4.3		4.32
			4.42	4.42
				4.53
			4.64	4.64
4.7	4.7	4.7		4.75
			4.87	4.87
				4.99
		5.1	5.11	5.11
				5.23
			5.36	5.36
				5.49
		5.6	5.62	5.62
				5.76
			5.90	5.90
				6.04
		6.2	6.19	6.19
				6.34
			6.49	6.49
				6.65
6.8	6.8	6.8	6.81	6.81
				6.98
			7.15	7.15
				7.32
		7.5	7.50	7.50
				7.68
			7.87	7.87
				8.06
	8.2	8.2	8.25	8.25
				8.45
			8.66	8.66
				8.87
		9.1	9.09	9.09
				9.31
			9.53	9.53
				9.76

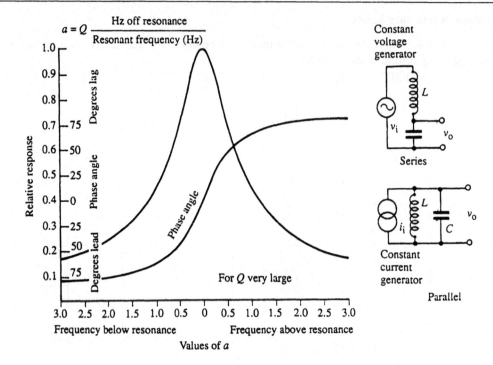

Designed for 1 ohm characteristic impedance

	T pad		π pad		Bridged T pad	
Loss D in dB	a	b	c	d	e	f
1	0.0575	8.668	0.1153	17.39	0.1220	8.197
2	0.1147	4.305	0.2323	8.722	0.2583	3.862
3	0.1708	2.838	0.3518	5.853	0.4117	2.427
4	0.2263	2.097	0.4770	4.418	0.5850	1.708
5	0.2800	1.645	0.6083	3.570	0.7783	1.285
6	0.3323	1.339	0.7468	3.010	0.9950	1.005
7	0.3823	1.117	0.8955	2.615	1.238	0.8083
8	0.4305	0.9458	1.057	2.323	1.512	0.6617
9	0.4762	0.8118	1.231	2.100	1.818	0.5500
10	0.5195	0.7032	1.422	1.925	2.162	0.4633
11	0.5605	0.6120	1.634	1.785	2.550	0.3912
12	0.5985	0.5362	1.865	1.672	2.982	0.3350
13	0.6342	0.4712	2.122	1.577	3.467	0.2883
14	0.6673	0.4155	2.407	1.499	4.012	0.2483
15	0.6980	0.3668	2.722	1.433	4.622	0.2167
16	0.7264	0.3238	3.076	1.377	5.310	0.1883
18	0.7764	0.2559	3.908	1.288	6.943	0.1440
20	0.8182	0.2020	4.950	1.222	9.000	0.1112
25	0.8935	0.1127	8.873	1.119	16.78	0.0597
30	0.9387	0.0633	15.81	1.065	30.62	0.0327
35	0.9650	0.0356	28.11	1.036	55.23	0.0182
40	0.9818	0.0200	50.00	1.020	99.00	0.0101
45	0.9888	0.0112	88.92	1.011	176.8	0.005 67
50	0.9937	0.006 33	158.1	1.0063	315.2	0.003 17

Circuit
diagram

Response
(bode plots)

Pole-zero diagrams

(i) 1st order (one pole)

(a) LP (low pass)

Plan

3D

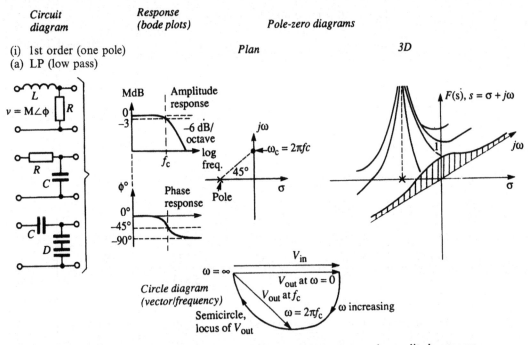

The vertical slice through the $F(s)$ surface, along the $j\omega$ axis is the same as the amplitude response, but plotted in volts against linear frequency scale instead of dB (log volts) against log frequency scale

(b) HP (high pass)

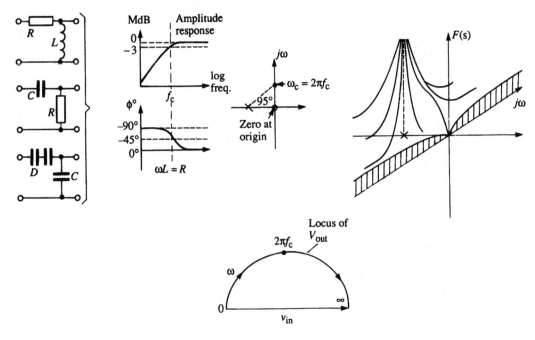

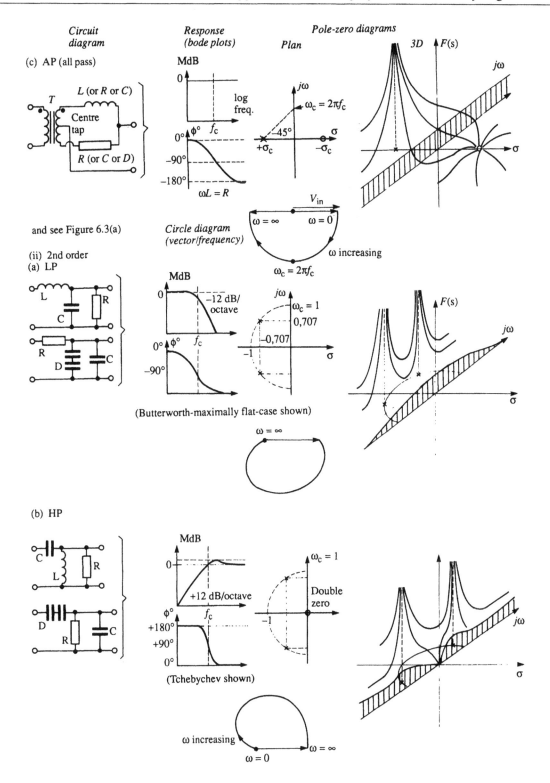

Circuit diagram

(c) AP (all pass)

L (or R or C)

T

Centre tap

R (or C or D)

and see Figure 6.3(a)

(ii) 2nd order
(a) LP

L R

C

R C

D

(b) HP

C R

L

D R C

Response (bode plots)

MdB

0

log freq.

f_c

$0°$ $\phi°$

$-90°$

$-180°$

$\omega L = R$

Circle diagram (vector/frequency)

MdB

0   -12 dB/octave

f_c

$0°$ $\phi°$

$-90°$

(Butterworth-maximally flat-case shown)

$\omega = \infty$

MdB

0

$+12$ dB/octave

f_c

$\phi°$

$+180°$

$+90°$

$0°$

(Tchebychev shown)

ω increasing

$\omega = 0$ $\omega = \infty$

Pole-zero diagrams

Plan

$j\omega$

$\omega_c = 2\pi f_c$

$-45°$

$+\sigma_c$ $-\sigma_c$ σ

V_{in}

$\omega = \infty$ $\omega = 0$

ω increasing

$\omega_c = 2\pi f_c$

$j\omega$

$\omega_c = 1$

$0,707$

$-0,707$

-1 σ

$\omega_c = 1$

Double zero

-1

3D $F(s)$

$j\omega$

σ

$F(s)$

$j\omega$

σ

$j\omega$

σ

(c) BP (band pass)

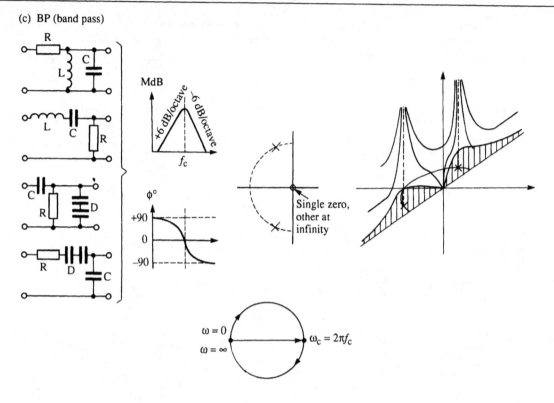

(d) AP

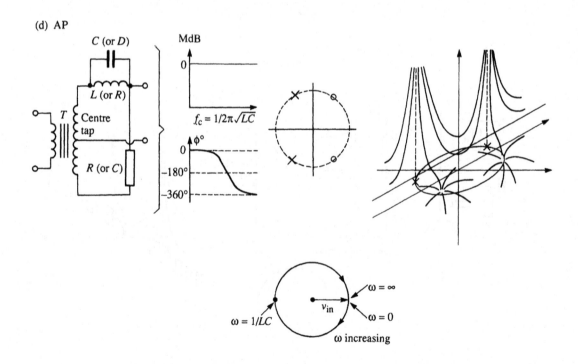

(iii) Higher order filters

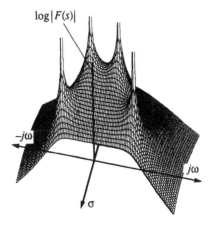

(a) Fourth order Butterworth low-pass filter

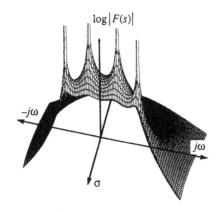

(b) Fourth order Chebychev low-pass filter

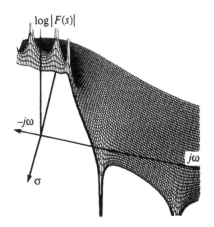

(c) Fourth order elliptic low-pass filter

In distinction to the conventional h, y and z parameters, s parameters relate to travelling wave conditions. The figure shows a two-port network with the incident and reflected waves a_1, b_1, a_2 and b_2.

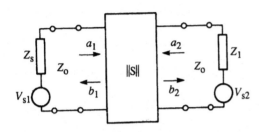

$$a_1 = \frac{V_{i1}}{\sqrt{Z_o}} \qquad a_2 = \frac{V_{i2}}{\sqrt{Z_o}}$$

$$b_1 = \frac{V_{r1}}{\sqrt{Z_o}} \qquad b_2 = \frac{V_{r2}}{\sqrt{Z_o}}$$

The squares of these quantities have the dimensions of power. Other symbols are:

Z_o characteristic impedance of the transmission line in which the two-port is connected

V_i incident voltage

V_r reflected (generated) voltage

The four-pole equations for s parameters are:

$$b_1 = s_{11}a_1 + s_{12}a_2$$

$$b_2 = s_{21}a_1 + s_{22}a_2$$

Using the subscripts i for 11, r for 12, f for 21 and o for 22, it follows that:

$$s_i = s_{11} = \left.\frac{b_1}{a_1}\right|_{a_2=0}$$

$$s_r = s_{12} = \left.\frac{b_1}{a_2}\right|_{a_1=0}$$

$$s_f = s_{21} = \left.\frac{b_2}{a_1}\right|_{a_2=0}$$

$$s_o = s_{22} = \left.\frac{b_2}{a_2}\right|_{a_1=0}$$

The s parameters can be named and expressed as follows:

$s_i = s_{11}$ input reflection coefficient: the complex ratio of the reflected wave and the incident wave at the input, under the conditions $Z_L = Z_O$ and $V_{S2} = 0$.

$s_r = s_{21}$ reverse transmission coefficient: the complex ratio of the generated wave at the input and the incident wave at the output, under the conditions $Z_S = Z_O$ and $V_{S1} = 0$.

$s_f = s_{21}$ forward transmission coefficient: the complex ratio of the generated wave at the output and the incident wave at the input, under the conditions $Z_L = Z_O$ and $V_{S2} = 0$.

$s_o = s_{22}$ output reflection coefficient: the complex ratio of the reflected wave and the incident wave at the output, under the conditions $Z_S = Z_O$ and $V_{S1} = 0$.

(Reproduced by courtesy of Philips Components Ltd)

Equivalents list

IEC	AUSTRALIA	AUSTRIA		BELGIUM	PEOPLES REPUBLIC OF CHINA	DENMARK	FINLAND	FRANCE			GERMANY		
	EVEREADY	KAPSCH	AGALUX	PLB	FIVE RAMS	HELLESENS	AIRAM	LECLANCHE	CIPEL	WONDER	VARTA PERTRIX	DAIMON	BAUM GARTEN
R20	950	R20/2	R20L			733	SP1M (Plastic)	T1	GT1L	EXPOR	211	289/ U2M6	561
R20						734					212LP	250	161
2R10	927	2R10/2	2R10L	T10		717	SP2M	T8	PT2	BATON	259	270	521
R14	935	R14/2	R14L			725	SP3M	114C	MT1	BABIX	214	287/ U11M3	540
R6	915	R6/1		E15		775	SP5M	HA6	AC1	VEBER	251	296	451
RO3	912					117			RFB	EXTAZ	239	291	
R10						729					200		
4F100	Q605												
6F22	216	6F22/2	6F22	TR1		410	LP1-9	6F22	R0603	TIBER	438	333	485
6F25	226					440	LP6-9	6F25	R0604	TONUS	29	334	488
6F50-2	246						LP2-9	6NG	R0608	TANGO	28	336/ PP6	
6F100	276					480 & 780	LP5M9	6NX	R0617	TEXAS	439	339	495
R20	1050	R20/3	R20T	E7	201	736	SP1T	R20	LF20	MARIN	232LP	251	251
R14	1035	R14/3	R14T	E9		726	SP3T	R14	LF14	ESCAL	233LP	258	
R6	1015	R6/2	R6	E18	204	728	SP5T	R6	LF6	NAVAL	244	298	447
R6		R6/1		E18				HA6	AC1			296	
3R12		3R12/2	3R12T	311		722	TP14	TLD	LF3	ORVOX	210	216	1-7
2R10		2R10/3	2R10T			757			LF10				
R20			R20S	E3		737	SP1TT	R2OL	MP20	AMIRO	222LP	253	432
R14			R14S	E9		727		R14L	P14	JUNON	235LP	259	
R6			R6	E18		738		R6L	MP6	NAVAL	280	298	448
RO3						767							
R1									PA1	RUBIS			449
3R12	703	3R12/1	3R12L	11		720	TP4	PL20	N3	BATRI	201	215	501
R40	No.6					310	PP1	1470	AD526	POSTA	2322	202	
4R25	509	4R25/2	4R25	465		555	HP16	AB200	TG4	ECOLI	430	266	412
	731									PORTO			
R1	904					114		R1-L	PC1	SAFIR	245	292	
45F40	467					436		667G	R4508	MUSIK	57	305	
10F20	411					411	LP1-15	215G	R1002		71	321	
15F20	412		15F20			409	LP1-22	222G	F1502	BETOV	72	322	600
20F20	413					413	LP1-30	230G	R2002	VERDI	73	323/ B123	
10F15	504					404	LP10-15	GB15	R1001	ADMIR	74	324	603
15F15	505					404		GB22	R1501		75	325	
						403			F0401	BOFIX	434	327	
						556							
						A505							

The information in this appendix was kindly provided by the Ever Ready Company (Great Britain) Ltd

GREAT BRITAIN					HOLLAND	HONG KONG		ITALY		
EVER READY	MALLORY	RAY-O-VAC	VIDOR	EXIDE	WITTE KAT	FIVE RAMS	FLYING BOMB	SOLE	SUPER-PILA	ZETA
		D2			153	123	360	239	60	T17
SP2	M13R	2LP	SP2		152			3239		
No. 8		8	V4	No. 8	158	789	362	235	66	T12
SP11	M14R MN1400	1LP	SP11	SP11	155		361	237	61/ U11M2	T13
U12	ZM9 MN1500	7R	SP7/SP12	T5	113		363	231	63	T22
U16	MN2400		SP16	U16		599			68	T18
				T11						
PP1			VT1	DT1/PP1						
PP3	M-1604	RR3	VT3	PP3	514	06	369	600/9d	987	R87
PP4			VT4	DT4/PP4				612/9	986	R85
PP6	M-1602	RR6	VT6	PP6				604/9d	990	R86
PP9	M-1603	RR9	VT9	PP9				592	995	A23
	MN1300				667	901			RD2	
	MN1400	RR13			668	943			RD11	
	ZM9 MN1500	RR14			666	555			433	T22R
	ZM9 MN1500								63	
					645	663		201	50 Superoro	
HP2	MN1300	HC2	HP2	HP2	615				AC2	T170
HP11	MN1400	HC1	HP11	HP11	614				AC11	
HP7	ZM9/M15R MN1500	RR15	HP7	HP7	613				AC7	T220
HP16	MN2400		HP16	HP16					72	T180
	MN9100									T250
1289	M1689	706	V5/V1289	F40/1289	101		365	203	50 ORO	Superzeta ORO
FLAG				FLAG	451		6	135	23	C1
996		941S	V18/V996	996	175	706	367	255	85	Z8
991		918	V28/V991	L18/991		766	340			
D23	MN9100			DL33	665	616		230	67	T25
B101				DM501	530				226	R7
B121				DH521	321					S10
B122	M122			DH522	511			1620/ 225	922	S11
B123				DH523					923	
B154	M154			DH554	517			1610/ 15		S30
B155				DH555					921	S31
					516					
AD35			L5040	H1184						

JAPAN					NORWAY	PORTUGAL	SINGAPORE	SOUTH AFRICA		SPAIN	
HITACHI MAXELL	NATIONAL	NOVEL	TOSHIBA	LAMINA	ANKER/ PERTRIX	TUDOR	EVEREADY	EVEREADY	VARTA	TXIMIST	TUDOR
UM1 (H) M1 (Paper)	UM1 UM1 (Paper)	N101UM-1	UM1A	UM1	411/A820	L2	950	950		R20NE	A2
UM1 (M)	UM1 (M)	N102 UM1 (M)				L2U		LP950			
		N436 PM1			259	L8		712		2R10	A8
UM2 (M)	UM2 (Paper) UM2 (M)	UM2N105 UM24N104	UM2A	UM2	214	L11	935	LP935		R14NE	A11
UM3A	UM3	UM3AN115	UM3A	UM3	AB6	L12	915	915		R6N	A12
UM4A	UM4	UM4AN116	UM4	UM4			912	HA31			
						T1					
006P	006PD(Hi-top) 006P	N556 006P	006P	006P	438	T3	216	PM3		6F22	PT3
R006	R006	N564 R006	R006	R006	29	T4	226	PM4			
4AA		N586 S106P		S106P	28		246	PM6			PT707
		N572 306			439		276	PM9	EF/EM FM9		PT673
UM1A		UM1 (M)	UM1K	UM1M	413	T2	1050	PM12		R20	A2LP
UM2A		UM2A	UM2K	UM2M	233		1035	PM13		R14	A11LP
					284	T12	1015	1015 PM14		R6	A12LP
UM3A		UM3A		UM3A				915			
					401			PM16		3R12LP	
	UM1D (Hitop)	N138 UM1 (T)			222	F2	850	HP2-1050		R20S	
	UM2D (Hitop)	N139 UM2 (T)			235		835	HP11-1035		R14S	
	UM3D (Hitop)	N142 UM3 (T)			244		815	1015		R6S	
					239			HP16			
					245		904	ES40E			
		N437 PM2			AB12	L3	703	703		3R12	A3
		N208 RM6	RM6	RM6	1423	2322	6	No.6			
		N425 4FZ			205	L996		509	EF/EM V18		A996
								991			
UM5A	UM5	N117 UM5A	UM5	UM5		S23					
BL145		N517 145		145		R101		Super No.3			
		N545 010						HB50			
BL-015	015	N504 015 (F)	015 F	015	72	S122	412	HB51			
		N505 020						HB52			
BLW10			W10F	MV10	74	S154	504	HB54			
BLMV15			W15F	MV15	75			H855			
W04					434						
								410			
								RC73			

The information in this appendix was kindly provided by the Ever Ready Company (Great Britain) Ltd

SWEDEN	SWITZERLAND			USA*						NEDA No	E.R. CELL SIZE
TUDOR	LECLANCHE	STRAHL	SANTIS	BURGESS	EVEREADY	MALLORY	RAY-O-VAC	RCA	MARATHON		
1-5S1	208	Nr.7	1013	2	950	M13F	2D	VS036		13F	2
1-5S1P	209	Nr.70	1012						121		2
3S1	222	Nr.4	1005								8
1-5S3	207	Nr.100 Super	1017	1	935	M14F	1C	VS035A	111	14F	1839
1-5S6P	201	Nr.100 (Paper)	1019	910	915	M15F	7AA	VS034A	170	15F	1915
	190			7	912	M24F	400	VS074	180	24F	12
	812										4/175
9T4	815	Nr.109	1051	2U6	216	M-1604	1604	VS323	1604	1604	6/115
9T3	814			P6	226	M-1600	1600	VS300A		1600	6/127
9T2	817			2N6	246	M-1602	1602	VS305		1602	6/147
9T1	821			D6	276	M-1603	1603	VS306		1603	6/175
1-5T9	800	Nr.77 (Plastic)	1011	230	1050	M13R	13	VS336	123	13	2
1-5T8	610	Nr.100 Transistor (plastic)	1016	130	1035	M14R	14	VS335	113	14	1839
1-5T6	602	Nr.110 (Plastic)		930	1015	M15R	15	VS334	173	15	1915
	603		1018								1915
4-5T1	810	Nr.33	1001								30
	223	Nr.44	1004								8
HD	210			220	850	M13P	210	VS736		13P	2
	605			120	835	M14P	110	VS735		14P	1839
	602			920	815	M15P	710	VS734	173	15P	1915
					812						12
	330				W468						4
4-5F1	263	Nr.1/Nr.2	1003	532	703		706	VS346		706	30
1-5E2	524			#6	No.6	M-914	61gn.C	VS006S		914	
6D3	470	Nr.57	1025	F4M	509	M-908	941	VS040C	49NL	908	60
				TW1	731	M-918	918	VS317	896	918	60
1-5S4	200	Nr.140 Super	1020	NE	904			VS073		910	4
67-5A1	712			XX45	467	M-200	200	VS016		200	
				U10	411	M-208	208	VS083		208	
22-5A3	700	Nr.122		U15	412	M-215	215	VS085		215	15/112
				U20	413	M-210	A210	VS085		210	
	760			Y10	504	M-504	220	VS704		220	10/105
				Y15	505	M-505	221	VS705		221	4/106
				F4BP	510S	M-915	942	VS040S	496S	915	
1-5G4											

Layer battery service life data

PP3 estimates service life at 20°C (M-1604, 006P, 6F22)

Milli-amperes at 9.0 V	Service life in hours to endpoint voltages of				Milli-amperes at 9.0 V	Service life in hours to endpoint voltages of			
	6.6 V	6.0 V	5.4 V	4.8 V		6.6 V	6.0 V	5.4 V	4.8 V
	Discharge period 30 min/day					Discharge period 4 hours/day			
10	26	29	32	34	1	322	355	395	412
15	17	19	21	23	1.5	215	240	260	277
25	9.2	11	12	13	2.5	132	147	162	167
50	2.3	4.1	5.1	5.8	5	63	71	77	82
					10	17	23	30	33
	Discharge period 2 hours/day				15	–	12	17	19
1.5	180	190	200	205					
2.5	112	122	132	136		Discharge period 12 hours/day			
5.0	56	62	68	72	0.5	695	750	785	815
10.0	24	28	31	34	1.0	365	390	417	427
15	14	16	20	22	1.5	240	262	277	292
25	–	7.4	9.5	11	2.5	125	152	162	170
					5.0	44	54	62	72
					7.5	18	22	29	36

Note: also available are the higher capacity PP3P for miniature dictation machines etc. and the PP3C for calculator service.

PP6 estimated service life at 20°C (M-1602, S106P, 6F50-2)

Milli-amperes at 9.0 V	Service life in hours to endpoint voltages of			
	6.6 V	6.0 V	5.4 V	4.8 V
	Discharge period 4 hours/day			
2.5	492	517	535	545
5.0	240	270	287	302
7.5	142	166	173	194
10	93	111	124	137
15	51	63	73	81
20	33	42	49	55
25	23	30	35	40
50	–	7.8	10	12
	Discharge period 12 hours/day			
0.75	2075	2200	2325	2400
1.0	1510	1650	1760	1815
1.5	965	1080	1155	1200
2.5	532	620	635	690
5.0	214	263	202	312
7.5	117	147	109	187
10	75	97	111	127
15	38	50	59	69
25	16	21	25	31
	Discharge period 30 min/day			
50	15	19	22	24
75	0.6	10	12	14
100	25	6.2	7.8	9.1
150	–	2.3	3.5	4.4
	Discharge period 2 hours/day			
7.5	169	194	205	215
10	117	140	151	161
15	67	83	93	99
25	31	38	45	48
50	8.9	12	14	16

Note: The above data is for a constant resistance load, drawing the stated current at 9 V. For a constant current load, the appropriate service life can be calculated from the constant resistance data in the figure.

(Reproduced by courtesy of *Electronics and Wireless World*)

PP9 estimated service life at 20°C (M-1603, 6F100)

Milli-amperes at 9.0 V	Service life in hours to endpoint voltages of			
	6.6 V	6.0 V	5.4 V	4.8 V
	Discharge period 30 min/day			
125	24	35	40	44
150	16	28	33	37
166.67	12	23	29	33
187.5	7.8	19	24	28
250	1.9	9.3	16	19
	Discharge period 2 hours/day			
25	193	233	269	286
33.3	150	180	209	223
37.5	122	147	168	180
50	81	99	113	124
62.5	57	71	82	92
75	41	53	62	69
83.33	33	45	53	60
100	20	32	39	44
125	9.8	19	25	29
150	6.1	13	17	20
	Discharge period 4 hours/day			
15	332	370	409	437
16.67	291	336	367	394
18.75	266	294	324	349
20	235	273	304	328
25	180	208	234	251
33.33	115	141	158	176
37.5	96	118	134	148
50	59	75	90	101
62.5	37	51	63	72
75	25	35	46	54
83.33	19	30	38	44
100	12	20	27	31
	Discharge period 12 hours/day			
15	292	340	379	407
16.67	254	294	321	352
25	127	151	178	206
33.33	73	89	105	134
37.5	58	71	86	110
50	30	40	47	65
62.5	17	24	30	42

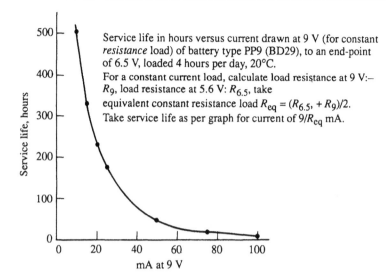

Service life in hours versus current drawn at 9 V (for constant *resistance* load) of battery type PP9 (BD29), to an end-point of 6.5 V, loaded 4 hours per day, 20°C.

For a constant current load, calculate load resistance at 9 V:– R_9, load resistance at 5.6 V: $R_{6.5}$, take

equivalent constant resistance load $R_{eq} = (R_{6.5}, + R_9)/2$.

Take service life as per graph for current of $9/R_{eq}$ mA.

Nominal diameter (mm)	Tolerance	Enamelled diameter Grade 1		Enamelled diameter Grade 2		Nominal resistance Ohms m at 20°C	Weight (kg/km)	Nominal diameter (mm)
		Min.	Max.	Min.	Max.			
0.032	±0.0015	0.035	0.040	0.035	0.043	21.44	0.0072	0.032
0.036	±0.0015	0.040	0.045	0.041	0.049	16.94	0.0091	0.036
0.040	±0.002	0.044	0.050	0.047	0.054	13.72	0.0112	0.040
0.045	±0.002	0.050	0.056	0.054	0.061	10.84	0.0142	0.045
0.050	±0.002	0.056	0.062	0.060	0.068	8.781	0.0175	0.050
0.056	±0.002	0.062	0.069	0.066	0.076	7.000	0.0219	0.056
0.063	±0.002	0.068	0.078	0.076	0.085	5.531	0.0277	0.063
0.071	±0.003	0.076	0.088	0.086	0.095	4.355	0.0352	0.071
0.080	±0.003	0.088	0.098	0.095	0.105	3.430	0.0447	0.080
0.090	±0.003	0.098	0.110	0.107	0.117	2.710	0.0566	0.090
0.100	±0.003	0.109	0.121	0.119	0.129	2.195	0.0699	0.100
0.112	±0.003	0.122	0.134	0.130	0.143	1.750	0.0877	0.112
0.125	±0.003	0.135	0.149	0.146	0.159	1.405	0.109	0.125
0.132	±0.003	0.143	0.157	0.153	0.165	1.260	0.122	0.132
0.140	±0.003	0.152	0.166	0.164	0.176	1.120	0.137	0.140
0.150	±0.003	0.163	0.177	0.174	0.187	0.9757	0.157	0.150
0.160	±0.003	0.173	0.187	0.187	0.199	0.8575	0.179	0.160
0.170	±0.003	0.184	0.198	0.197	0.210	0.7596	0.202	0.170
0.180	±0.003	0.195	0.209	0.209	0.222	0.6775	0.226	0.180
0.190	±0.003	0.204	0.220	0.219	0.233	0.6081	0.252	0.190
0.200	±0.003	0.216	0.230	0.232	0.245	0.5488	0.280	0.200
0.212	±0.003	0.229	0.243	0.247	0.260	0.4884	0.314	0.212
0.224	±0.003	0.240	0.256	0.258	0.272	0.4375	0.351	0.224
0.236	±0.003	0.252	0.268	0.268	0.285	0.3941	0.389	0.236
0.250	±0.004	0.267	0.284	0.284	1.301	0.3512	0.437	0.250
0.265	±0.004	0.282	0.299	0.299	0.317	0.3126	0.491	0.265
0.280	±0.004	0.298	0.315	0.315	0.334	0.2800	0.548	0.280
0.300	±0.004	0.319	0.336	0.336	0.355	0.2439	0.629	0.300
0.315	±0.004	0.334	0.352	0.353	0.371	0.2212	0.694	0.315
0.335	±0.004	0.355	0.374	0.374	0.392	0.1956	0.784	0.335
0.355	±0.004	0.375	0.395	0.395	0.414	0.1742	0.881	0.355
0.375	±0.004	0.395	0.416	0.416	0.436	0.1561	0.983	0.375
0.400	±0.005	0.421	0.442	0.442	0.462	0.1372	1.12	0.400
0.425	±0.005	0.447	0.468	0.468	0.489	0.1215	1.26	0.425
0.450	±0.005	0.472	0.495	0.495	0.516	0.1084	1.42	0.450
0.475	±0.005	0.498	0.522	0.521	0.544	0.09730	1.58	0.475
0.500	±0.005	0.524	0.547	0.547	0.569	0.08781	1.75	0.500
0.530	±0.006	0.555	0.580	0.579	0.602	0.07814	1.96	0.530
0.560	±0.006	0.585	0.610	0.610	0.632	0.07000	2.19	0.560
0.600	±0.006	0.625	0.652	0.650	0.674	0.06098	2.52	0.600
0.630	±0.006	0.657	0.684	0.683	0.706	0.05531	2.77	0.630
0.670	±0.007	0.698	0.726	0.726	0.748	0.04890	3.14	0.670
0.710	±0.007	0.738	0.767	0.766	0.790	0.04355	3.52	0.710
0.750	±0.008	0.779	0.809	0.808	0.832	0.03903	3.93	0.750
0.800	±0.008	0.830	0.861	0.860	0.885	0.03430	4.47	0.800
0.850	±0.009	0.881	0.913	0.912	0.937	0.03038	5.05	0.850
0.900	±0.009	0.932	0.965	0.964	0.990	0.02710	5.66	0.900
0.950	±0.010	0.983	1.017	1.015	1.041	0.02432	6.31	0.950
1.00	±0.010	1.034	1.067	1.067	1.093	0.02195	6.99	1.00
1.06	±0.011	1.090	1.130	1.123	1.155	0.01954	7.85	1.06
1.12	±0.011	1.150	1.192	1.181	1.217	0.01750	8.77	1.12
1.18	±0.012	1.210	1.254	1.241	1.279	0.01577	9.73	1.18
1.25	±0.013	1.281	1.325	1.313	1.351	0.01405	10.9	1.25
1.32	±0.013	1.351	1.397	1.385	1.423	0.01260	12.2	1.32
1.40	±0.014	1.433	1.479	1.466	1.506	0.01120	13.7	1.40
1.50	±0.015	1.533	1.581	1.568	1.608	0.009757	15.7	1.50
1.60	±0.016	1.633	1.683	1.669	1.711	0.008575	17.9	1.60
1.70	±0.017	1.733	1.785	1.771	1.813	0.007596	20.2	1.70
1.80	±0.018	1.832	1.888	1.870	1.916	0.006775	22.7	1.80
1.90	±0.019	1.932	1.990	1.972	2.018	0.006081	25.2	1.90
2.00	±0.020	2.032	2.092	2.074	2.120	0.005488	28.0	2.00

Manufacturers offer several grades of insulation material and thickness. The thicker coatings are recommended for high-voltage transformer applications. The most popular coating materials are 'self-fluxing', i.e. do not require a separate end stripping operation before soldering.

No.	SWG		BWG		AWG or B & S		No.	SWG		BWG		AWG or B & S	
	in	mm	in	mm	in	mm		in	mm	in	mm	in	mm
4/0	0.400	10.160	0.454	11.532	0.4600	11.684	24	0.022	0.559	0.022	0.559	0.0201	0.511
3/0	0.372	9.449	0.425	10.795	0.4096	10.404	25	0.020	0.508	0.020	0.508	1.0179	0.455
2/0	0.348	8.839	0.380	9.652	0.3648	9.266	26	0.018	0.457	0.018	0.457	1.0159	0.404
0	0.324	8.230	0.340	8.636	0.3249	8.252	27	0.0164	0.417	0.016	0.406	1.0142	0.361
1	0.300	7.620	0.300	7.620	0.2893	7.348	28	0.0148	0.376	0.014	0.356	0.0126	0.320
2	0.276	7.010	0.284	7.214	0.2576	6.543	29	0.0136	0.345	0.013	0.330	1.0113	0.287
3	0.252	6.401	0.259	6.579	0.2294	5.827	30	0.0124	0.315	0.012	0.305	0.0100	0.254
4	0.232	5.893	0.238	6.045	0.2043	5.189	31	0.0116	0.295	0.010	0.254	0.0089	0.226
5	0.212	5.385	0.220	5.588	0.1819	4.620	32	0.0108	0.274	0.009	0.229	0.0080	0.203
6	0.192	4.877	0.203	5.156	0.1620	4.115	33	0.0100	0.254	0.008	0.203	0.0071	0.180
7	0.176	4.470	0.180	4.572	0.1443	3.665	34	0.0092	0.234	0.007	0.178	0.0063	0.160
8	0.160	4.064	0.165	4.191	0.1285	3.264	35	0.0084	0.213	0.005	0.127	0.0056	0.142
9	0.144	3.658	0.148	3.759	0.1144	2.906	36	0.0076	0.193	0.004	0.102	0.0050	0.127
10	0.128	3.251	0.134	3.404	0.1019	2.588	37	0.0068	0.173			0.0045	0.114
11	0.116	2.946	0.120	3.048	0.0907	2.304	38	0.0060	0.152			0.0040	0.102
12	0.104	2.642	0.109	2.769	0.0808	2.052	39	0.0052	0.132			0.0035	0.090
13	0.092	2.337	0.095	2.413	0.0720	1.829	40	0.0048	0.122			0.0031	0.079
14	0.080	2.032	0.083	2.108	0.0641	1.628	41	0.0044	0.112			0.0028	0.071
15	0.072	1.829	0.072	1.829	0.0571	1.450	42	0.0040	0.102			0.0025	0.063
16	0.064	1.626	0.065	1.651	0.0508	1.290	43	0.0036	0.091			0.0022	0.056
17	0.056	1.422	0.058	1.473	0.0453	1.151	44	0.0032	0.081			0.0020	0.051
18	0.048	1.219	0.049	1.245	0.0403	1.024	45	0.0028	0.071			0.001 76	0.045
19	0.040	1.016	0.042	1.067	0.0359	0.912	46	0.0024	0.061			0.001 57	0.040
20	0.036	0.914	0.035	0.889	0.0320	0.813	47	0.0020	0.051			0.001 40	0.036
21	0.032	0.813	0.032	0.813	0.0285	0.724	48	0.0016	0.041			0.001 24	0.031
22	0.028	0.711	0.028	0.711	0.0253	0.643	49	0.0012	0.030			0.001 11	0.028
23	0.024	0.610	0.025	0.635	0.0226	0.574	50	0.0010	0.025			0.000 99	0.025

Elliptic filters

The following small subset of tables with their schematics are reprinted with permission from 'On the Design of Filters by Synthesis' by R. Saal and E. Ulbricht, *IRE Transactions on Circuit Theory*, December 1958, pp. 284–328. (© 1958 IRE (now IEEE)). The tables are normalized to $f = 1 \, \text{rad/s} = 1/(2\pi) \, \text{Hz}$, $Z_0 = 1 \, \Omega$, L in henrys, C in farads.

(Note: In using the following tables, the 'a' schematics, for example, schematic a on page 287, correspond with the top line of column headings of Tables A8.1–3. Similarly, schematics b correspond with the bottom line of column headings of the tables.)

The original gives designs for filters up to the eleventh order. Designs are presented here for third and fifth order filters with 1 dB, 0.5 dB and 0.1 dB pass-band ripples. A more extensive subset of tables from the original may be found in *The Practical RF Handbook*, 2nd edition, by the same author, ISBN 07506 34472, Butterworth-Heinemann, 1977.

3 pole

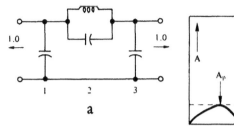
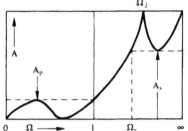
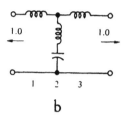

a
b

Table A8.1 $A_p = 1$ dB

Ω_s	A_s (dB)	C_1	C_2	L_2	Ω_2	C_3
1.295	20	1.570	0.805	0.613	1.424	1.570
1.484	25	1.688	0.497	0.729	1.660	1.688
1.732	30	1.783	0.322	0.812	1.954	1.783
2.048	35	1.852	0.214	0.865	2.324	1.852
2.418	40	1.910	0.145	0.905	2.762	1.910
2.856	45	1.965	0.101	0.929	3.279	1.965
Ω_s	A_s (dB)	L_1	L_2	C_2	Ω_2	L_3

(© 1958 IRE (now IEEE))

Table A8.2 $A_p = 0.5$ dB

Ω_s	A_s (dB)	C_1	C_2	L_2	Ω_2	C_3
1.416	20	1.267	0.536	0.748	1.578	1.267
1.636	25	1.361	0.344	0.853	1.846	1.361
1.935	30	1.425	0.226	0.924	2.189	1.425
2.283	35	1.479	0.152	0.976	2.600	1.479
2.713	40	1.514	0.102	1.015	3.108	1.514
Ω_s	A_s (dB)	L_1	L_2	C_2	Ω_2	L_3

(© 1958 IRE (now IEEE))

Table A8.3 $A_p = 0.1$ dB

Ω_s	A_s (dB)	C_1	C_2	L_2	Ω_2	C_3
1.756	20	0.850	0.290	0.871	1.986	0.850
2.082	25	0.902	0.188	0.951	2.362	0.902
2.465	30	0.941	0.125	1.012	2.813	0.941
2.921	35	0.958	0.0837	1.057	3.362	0.958
3.542	40	0.988	0.0570	1.081	4.027	0.988
Ω_s	A_s (dB)	L_1	L_2	C_2	Ω_2	L_3

(© 1958 IRE (now IEEE))

5 pole

Table A8.4 $A_p = 1$ dB

Ω_s	A_s (dB)	C_1	C_2	L_2	Ω_2	C_3	C_4	L_4	Ω_4	C_5
1.145	35	1.783	0.474	0.827	1.597	1.978	1.487	0.488	1.174	1.276
1.217	40	1.861	0.372	0.873	1.755	2.142	1.107	0.578	1.250	1.427
1.245	45	1.923	0.293	0.947	1.898	2.296	0.848	0.684	1.313	1.553
1.407	50	1.933	0.223	0.963	2.158	2.392	0.626	0.750	1.459	1.635
1.528	55	1.976	0.178	0.986	2.387	2.519	0.487	0.811	1.591	1.732
1.674	60	2.007	0.141	1.003	2.660	2.620	0.380	0.862	1.747	1.807
1.841	65	2.036	0.113	1.016	2.952	2.703	0.301	0.901	1.920	1.873
2.036	70	2.056	0.0890	1.028	3.306	2.732	0.239	0.934	2.117	1.928
Ω_s	A_s (dB)	L_1	L_2	C_2	Ω_2	L_3	L_4	C_4	Ω_4	L_5

(© 1958 IRE (now IEEE))

Table A8.5 $A_p = 0.5$ dB

Ω_s	A_s (dB)	C_1	C_2	L_2	Ω_2	C_3	C_4	L_4	Ω_4	C_5
1.186	35	1.439	0.358	0.967	1.700	1.762	1.116	0.600	1.222	1.026
1.270	40	1.495	0.279	1.016	1.878	1.880	0.840	0.696	1.308	1.114
1.369	45	1.530	0.218	1.063	2.077	1.997	0.627	0.795	1.416	1.241
1.481	50	1.563	0.172	1.099	2.300	2.113	0.482	0.875	1.540	1.320
1.618	55	1.559	0.134	1.140	2.558	2.188	0.369	0.949	1.690	1.342
1.782	60	1.603	0.108	1.143	2.847	2.248	0.291	0.995	1.858	1.449
1.963	65	1.626	0.0860	1.158	3.169	2.306	0.230	1.037	2.048	1.501
2.164	70	1.624	0.0679	1.178	3.536	2.319	0.182	1.078	2.258	1.521
Ω_s	A_s (dB)	L_1	L_2	C_2	Ω_2	L_3	L_4	C_4	Ω_4	L_5

(© 1958 IRE (now IEEE))

Table A8.6 $A_p = 0.1$ dB

Ω_s	A_s (dB)	C_1	C_2	L_2	Ω_2	C_3	C_4	L_4	Ω_4	C_5
1.309	35	0.977	0.230	1.139	1.954	1.488	0.742	0.740	1.350	0.701
1.414	40	1.010	0.177	1.193	2.176	1.586	0.530	0.875	1.468	0.766
1.540	45	1.032	0.140	1.228	2.412	1.657	0.401	0.968	1.605	0.836
1.690	50	1.044	0.1178	1.180	2.682	1.726	0.283	1.134	1.765	0.885
1.860	55	1.072	0.0880	1.275	2.985	1.761	0.241	1.100	1.942	0.943
2.048	60	1.095	0.0699	1.292	3.328	1.801	0.192	1.148	2.130	0.988
2.262	65	1.108	0.0555	1.308	3.712	1.834	0.151	1.191	2.358	1.022
2.512	70	1.112	0.0440	1.319	4.151	1.858	0.119	1.225	2.619	1.044
Ω_s	A_s (dB)	L_1	L_2	C_2	Ω_2	L_3	L_4	C_4	Ω_4	L_5

(© 1958 IRE (now IEEE))

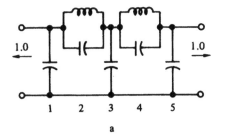

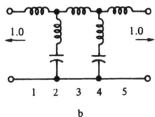

a   b

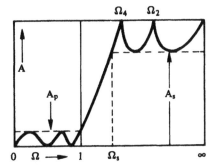

Index

Lightning Source UK Ltd.
Milton Keynes UK
UKOW05f1558010817
306465UK00002B/26/P